R. Snider

RAY S. SNIDER, Ph.D.
Department of Anatomy

THE BRAIN OF
THE TIGER SALAMANDER

The tiger salamander, Ambystoma tigrinum, natural size. The figure at the top is a midwestern adult form, after G. K. Noble ('31, by permission of the McGraw-Hill Book Co.). Below this is a photograph of a midlarval stage by Professor S. H. Bishop. The two lower figures show late larval and adult stages of the eastern form, the adult with brilliant yellow "tiger" stripes. (Courtesy of the American Museum of Natural History.)

THE BRAIN
OF THE
TIGER SALAMANDER
Ambystoma tigrinum

By

C. JUDSON HERRICK

Professor Emeritus of Neurology
The University of Chicago

THE UNIVERSITY OF CHICAGO PRESS
CHICAGO · ILLINOIS

THE UNIVERSITY OF CHICAGO COMMITTEE
ON PUBLICATIONS IN BIOLOGY AND MEDICINE

LESTER R. DRAGSTEDT · R. WENDELL HARRISON
FRANKLIN C. McLEAN · C. PHILLIP MILLER
THOMAS PARK · WILLIAM H. TALIAFERRO

The University of Chicago Press, Chicago 37
Cambridge University Press, London, N.W. 1, England
W. J. Gage & Co., Limited, Toronto 2B, Canada

*Copyright 1948 by The University of Chicago. All rights reserved
Published 1948. Composed and printed by* The University
of Chicago Press, *Chicago, Illinois, U.S.A.*

PREFACE

THIS work reports the results of a search, extending over fifty years, for the fundamental plan of the vertebrate nervous system as revealed in generalized form in the amphibians. In these small brains we find a simplified arrangement of nerve cells and fibers with a pattern of structural organization, the main features of which are common to all vertebrates. From this primitive and relatively unspecialized web of tissue it is possible to follow the successive steps in progressive elaboration as the series of animals from salamanders to men is passed in review.

This is a record of personal observation, not a compilation of the literature. It is presented in two parts, which differ in content and method of treatment of the data.

The first part gives a general over-all view of the structure without details, followed by physiological interpretation and discussion of some general principles of embryologic and phylogenetic morphogenesis. This part, with the accompanying illustrations, can be read independently of the histological details recorded in the second part.

The second part presents the evidence upon which conclusions are based, drawn from my own previously published descriptions, to which references are given, together with considerable new material. This part is designed for specialists in comparative neurology and as a guide for physiological experiments. The second part supplements the first, to which the reader must make frequent reference.

Grateful acknowledgment is made to many colleagues for generous assistance and criticism, and particularly to Doctors Elizabeth C. Crosby, Davenport Hooker, Olof Larsell, Gerhardt von Bonin, Ernst Scharrer, and W. T. Dempster. In the preparation of the manuscript invaluable help was given by Miss Anna Seaburg.

I am indebted to Dr. Paul G. Roofe for permission to copy one of his pictures, shown here as figure 86A. The American Museum of Natural History, New York, generously furnished the two photographs, taken from life, shown at the bottom of the Frontispiece. These are copyrighted by the Museum. One of them has been previously published by the Macmillan Company in Hegner's *Parade of the Animal Kingdom* (p. 289). The photograph of the midlarval stage was kindly supplied by Professor Sherman C. Bishop of the University of Rochester. The upper figure is from G. K. Noble's *Biology of the Amphibia*, courtesy of the McGraw-Hill Book Company (copyrighted, 1931). Figures 1, 2C, and 86–113 are reproductions of figures previously published by the author in the *Journal of Comparative Neurology* and used here by courtesy of the Wistar Institute of Anatomy and Biology, publishers of that *Journal*. The other figures are originals prepared for this work.

Money for the prosecution of the work and for financing its publication was liberally supplied by the Dr. Wallace C. and Clara A. Abbott Memorial Fund of the University of Chicago.

CONTENTS

PART I. GENERAL DESCRIPTION AND INTERPRETATION

I. SALAMANDERS AND THEIR BRAINS 3
   The salamanders, 3.—The scope of this inquiry, 4.—The plan of this book, 6.—Sources and material, 10.—Development of the brain, 11.—The evolution of brains, 13

II. THE FORM AND SUBDIVISIONS OF THE BRAIN 18
   Gross structure, 18.—Ventricles, 24.—Meninges, chorioid plexuses, and blood vessels, 26

III. HISTOLOGICAL STRUCTURE 28
   General histology, 28.—The neuropil, 29.—The ventrolateral peduncular neuropil, 35

IV. REGIONAL ANALYSIS 40
   The subdivisions, spinal cord to pallium, 41.—The commissures, 55.—Conclusion, 56

V. FUNCTIONAL ANALYSIS, CENTRAL AND PERIPHERAL 57
   The longitudinal zones, 57.—The sensory zone, 58.—The motor zone, 60.—The intermediate zone, 64.—The functional systems, 65

VI. PHYSIOLOGICAL INTERPRETATIONS 70
   Apparatus of analysis and synthesis, 70.—The stimulus-response formula, 72.—Reflex and inhibition, 73.—Principles of localization of function, 82

VII. THE ORIGIN AND SIGNIFICANCE OF CEREBRAL CORTEX . . . 91
   The problem, 91.—Morphogenesis of the cerebral hemispheres, 92.—The cortex, 98.—Physiology and psychology, 106

VIII. GENERAL PRINCIPLES OF MORPHOGENESIS 109
   Morphogenic agencies, 109.—Morphological landmarks, 116.—The future of morphology, 120

PART II. SURVEY OF INTERNAL STRUCTURE

IX. SPINAL CORD AND BULBO-SPINAL JUNCTION 125
   The spinal cord and its nerves, 125.—The bulbo-spinal junction, 129

X. CRANIAL NERVES 131
   Development, 131.—Survey of the functional systems, 132

XI. MEDULLA OBLONGATA 153
   Sensory zone, 153.—Intermediate zone, 156.—Motor zone, 157.—Fiber tracts of the medulla oblongata, 158.—The lemniscus systems, 162

CONTENTS

XII. CEREBELLUM 172
    Brachium conjunctivum, 176.—The cerebellar commissures, 177.—
    Proprioceptive functions of the cerebellum, 178

XIII. ISTHMUS 179
    Development, 179.—Sensory zone, 181.—Intermediate zone, 182.
    —Motor zone, 182.—White substance, 186.—Isthmic neuropil, 187.
    —Physiological interpretation, 189

XIV. INTERPEDUNCULAR NUCLEUS 191
    Comparative anatomy, 191.—Histological structure, 193.—
    Afferent connections, 197.—Efferent connections, 201.—Interpretation, 202.—Conclusion, 210

XV. MIDBRAIN 212
    Development, 212.—Sensory zone, 214.—Intermediate zone, 215.
    —Motor zone, 216

XVI. OPTIC AND VISUAL-MOTOR SYSTEMS 219
    Optic nerve and tracts, 219.—Tectum opticum, 222.—Tecto-oculomotor connections, 226.—Visual functions, 227

XVII. DIENCEPHALON 230
    General features, 230.—Development, 231.—Epithalamus, 234.—
    Dorsal thalamus, 236.—Ventral thalamus.—239.—Hypothalamus, 241

XVIII. THE HABENULA AND ITS CONNECTIONS 247
    The di-telencephalic junction, 247.—Fornix, 254.—Stria terminalis, 255.—Stria medullaris thalami, 256.—Fasciculus retroflexus, 261

XIX. THE CEREBRAL HEMISPHERES 265
    Subdivisions of the hemisphere, 265.—The olfactory system, 266

XX. THE SYSTEMS OF FIBERS 270
    The basal forebrain bundles, 271.—The tegmental fascicles, 275.—
    Fasciculus tegmentalis profundus, 286

XXI. THE COMMISSURES 289
    General considerations, 289.—The dorsal commissures, 292.—The
    ventral commissures, 294

BIBLIOGRAPHY

BIBLIOGRAPHY 307

ILLUSTRATIONS

ILLUSTRATIONS 321

ABBREVIATIONS FOR ALL FIGURES

ABBREVIATIONS FOR ALL FIGURES 391

INDEX

INDEX 399

PART I
GENERAL DESCRIPTION AND INTERPRETATION

CHAPTER I

SALAMANDERS AND THEIR BRAINS

THE SALAMANDERS

SALAMANDERS are shy little animals, rarely seen and still more rarely heard. If it were not so, there would be no salamanders at all, for they are defenseless creatures, depending on concealment for survival. And yet the tiger salamander, to whom this book is dedicated, is appropriately named, for within the obscurity of its contracted world it is a predaceous and voracious terror to all humbler habitants.

This salamander and closely allied species have been found to be so well adapted for a wide range of studies upon the fundamental features of growth and differentiation of animal bodies that during the last fifty years there has been more investigation of the structure, development, and general physiology of salamanders than has been devoted to any other group of animals except mankind. The reason for this is that experimental studies can be made with these amphibians that are impossible or much more difficult in the case of other animals. This is our justification for the expenditure of so much hard work and money upon the study of the nervous system of these insignificant little creatures.

The genus Ambystoma is widely distributed throughout North America and the tiger salamander, A. tigrinum, is represented by several subspecies. The individuals vary greatly in size and color, and the subspecies have different geographical distribution, with some overlap of range (Bishop, '43). The subspecies, A. tigrinum tigrinum (Green), ranges from New York southward to Florida and westward to Minnesota and Texas. It has a dark-brown body crossed by bright-yellow stripes, as shown in the lower figure of the Frontispiece. The species probably was named for these tiger-like markings, not for its tigerish ferocity. The upper figure of the Frontispiece is an adult of a western form, with less conspicuous markings. Other subspecies range as far to the northwest as Oregon and British Columbia. Several other species of Ambystoma are found in the same areas as A. tigrinum.

Zoölogical names.—The approved names of the genus and larger groups to which reference is here made, as given in a recent official list (Pearse, '36), are as follows:

> Salienta, to replace Anura
> Caudata, to replace Urodela
> Ambystoma, to replace Amblystoma

Ambystoma (or Siredon) maculatum has priority over A. punctatum. The names, Anura, Urodela, and Amblystoma, are used throughout this text because they are so commonly found in current literature that they may be regarded as vernacular terms.

THE SCOPE OF THIS INQUIRY

From the dawn of interest in the minute structure of the human brain, it was recognized that the simpler brains of lower vertebrates present the fundamental features of the human brain without the numberless complications which obscure these fundamentals in higher animals. This idea motivated much research by the pioneers in neuroanatomy, and it was pursued systematically by L. Edinger, H. Obersteiner, Ramón y Cajal, C. L. Herrick, J. B. Johnston, Ariëns Kappers, and many others. In 1895, van Gehuchten wrote that he was engaged upon a monograph on the nervous system of the trout, "impressed by the idea that complete information about the internal organization of the central nervous system of a lower vertebrate would be of great assistance as our guide through the complicated structure of the central nervous system of mammals and of man." The few chapters of this monograph which appeared before his untimely death intensify our regret that he was not permitted to complete this work. Van Gehuchten's ideal has been my own inspiration.

Our primary interest in this inquiry is in the origins of the structural features and physiological capacities of the human brain and the general principles in accordance with which these have been developed in the course of vertebrate evolution. This is a large undertaking. What, then, is the most promising approach to it? My original plan was, first, to review all that has been recorded about the anatomy and physiology of the nervous systems of backboned animals, to arrange these animals in the order of their probable diverse specialization from simple to complex in the evolutionary sequence, then to select from the series the most instructive types and subject them to intensive study, in the expectation that the principles underlying these morphological changes would emerge.

So ambitious a plan, however, is far too big to be encompassed within the span of one man's lifetime. The fact-finding research is extremely laborious and exacting; and, during the fifty years which have elapsed since my project was formulated, the descriptive literature has increased to enormous volume. This literature proves to be peculiarly refractory to analysis and interpretation. Until recently this vast accumulation of factual knowledge has contributed disappointingly little to the resolution of the fundamental problems of human neurology. Nevertheless, the method is sound, and this slow growth is now coming to fruition, thanks to the conjoint labors of specialists in many fields of science. What no individual can hope to do alone can be done and has been done in co-operative federation, as illustrated, for instance, by the Kappers, Huber, and Crosby team and their many collaborators.

Traditionally, comparative neurology has been regarded as a subdivision of comparative anatomy, and so it is. But it is more than this. The most refined methods of anatomical analysis cannot reveal the things that are of greatest significance for an understanding of the nervous system. Our primary interest is in the behavior of the living body, and we study brains because these organs are the chief instruments which regulate behavior. As long as anatomy was cultivated as a segregated discipline, its findings were colorless and too often meaningless.

Now that this isolationism has given way to genuine collaboration among specialists in all related fields—physiology, biochemistry, biophysics, clinical practice, neuropathology, psychology, among others—we witness today a renaissance of the science of neurology. The results of the exacting analytic investigations of the specialists can now be synthesized and given meaning. The task of comparative anatomy in this integrated program of research is fundamental and essential. The experimentalist must know exactly what he has done to the living fabric before he can interpret his experiment. In the past it too often happened that a physiologist would stab into a living frog, take his kymograph records, and then throw the carcass into the waste-jar. This is no longer regarded as good physiology. Without the guidance of accurate anatomical knowledge, sound physiology is impossible; and, without skilful physiological experimentation, the anatomical facts are just facts and nothing more.

Early in my program the amphibians were selected as the most favorable animals with which to begin a survey of the comparative anatomy of the nervous system. Time has proved the wisdom of this choice, and the study of these animals has been so fruitful that by far the larger part of my research has been devoted to them.

In this work it was my good fortune to be associated with the late G. E. Coghill, whose distinguished career pointed the way to an original approach to the problems of the origins and growth of the nervous organs and their functions. The record of phylogenetic history spans millions of years and is much defaced by time; but the record of the embryonic development of the individual is measured in days and hours, and every detail of it can be watched from moment to moment. The internal operations of the growing body are not open to casual inspection, but Coghill showed us that the sequence of these changes can be followed.

He selected the salamanders for his studies for very good reasons, the same reasons that led me to take these animals as my own point of departure in a program of comparative neurology. My intimate association with Coghill lasted as long as he lived, and the profound influence which his work has had upon the course of biological and psychological events in our generation has motivated the preparation of a book devoted to his career ('48). This influence, though perhaps unrecognized at the time, was doubtless largely responsible for my persistent efforts to analyze the texture of the amphibian nervous system, for his studies of the growth of patterns of behavior and their instrumentation in young stages of salamanders brought to light some principles which evidently are applicable in phylogenetic development also.

While Coghill's studies on the development of salamanders were in process, we were impressed by the importance of learning just how these processes of growth eventuate in the adult body. This was my job, and so we worked hand in hand, decade after decade, for forty years. Progress was slow, but our two programs fitted together so helpfully that my original plan for a comprehensive study of the comparative anatomy of the nervous system was abandoned in favor of more intensive study of salamanders.

THE PLAN OF THIS BOOK

The preceding details of personal biography are relevant here because they explain the motivation and plan of this book. The significant facts now known about the internal structure of the brain of the tiger salamander in late larval and adult stages are here assembled. The observations on this and allied species previously recorded by the writer and many others are widely scattered, often in fragmentary form, and with confusing diversity in nomenclature and interpretation. As observations have accumulated, gaps in knowledge have been filled, early errors have been corrected, the nomenclature has been systematized, and now, with the addition of considerable new observation here reported, the structure may be viewed as a whole and interpreted in relation with the action system of the living animal. Many of my observations during the last fifty years confirm those of others; and, since references to these are given in the papers cited, this account is not encumbered with them except where they supplement my own experience or deal with questions still in controversy. I here describe what I myself have seen, with exceptions explicitly noted. This explains the disproportionate number of references in the text to my own papers.

Two genera of urodeles have been studied intensively to find out what is the instrumentation of their simple patterns of behavior. All observations on the brain of the more generalized mudpuppy, Necturus, were assembled in a monograph ('33b) and several following papers. The present work is a similar report upon the brain of the somewhat more specialized tiger salamander. The original plan was to follow this with an examination of the brain of the frog, for which abundant material was assembled and preliminary surveys were made; but this research, which is urgently needed, must be done by others.

In this book the anatomical descriptions are arranged in such a way as to facilitate interpretation in terms of probable physiological operation. Though the amount of experimental evidence about the functions of the internal parts of the amphibian brain is scanty, there is, fortunately, a wealth of such observation about the brains of other animals; and where a particular structural pattern is known to be colligated with a characteristic pattern of action, the structure may be taken as an indicator of the function. The reliability of this method depends upon the adequacy of our knowledge of both the function and the structure. The present task, then, is an assembly of the anatomical evidence upon which the interpretations are based.

The first part of this work is written to give biologists and psychologists an outline of the plan of organization of a generalized vertebrate brain and some insight into the physiological principles exemplified in its action. These eight chapters, with the accompanying illustrations, can be read independently of the rest of the book.

Part II is written for specialists in comparative neurology. It covers the same ground as the first part, reviewing each of the conventional subdivisions of the brain, giving details of the evidence upon which conclusions are based, with references to sources and much new material. This involves some repetition, which is unavoidable because all these structures are interconnected and in action they co-operate in various ways. Many structures must be described in several contexts and, accordingly, the Index has been prepared with care so as to enable the reader to assemble all references to every topic.

The most important new observations reported in Part II relate to the structure and connections of the isthmus (chap. xiii), interpeduncular nucleus (chap. xiv), and habenula, including analysis of the stria medullaris thalami and fasciculus retroflexus (chap. xviii). In chapter xx a few of the more important systems of fibers are described, including further analysis of the tegmental fascicles as enumerated in the paper of 1936 and references to other lists in the literature. The composition of the commissures of the brain is summarized in chapter xxi. The lemniscus systems are assembled in chapter xi, and other tracts are described in connection with the structures with which they are related.

Since most neurologists are not expert in the comparative field, where the nomenclature is technical and frequently unintelligible except to specialists, the attempt is made in Part I to present the salient features with a minimum of confusing detail and, so far as practicable, in terms of familiar mammalian structure. This is not an easy thing to do, and no clear and simple picture can be drawn, for the texture of even so lowly organized a brain is bafflingly complicated and many of these structures have no counterparts in the human body. Homologies implied by similar names are rarely exact, and in many of these cases the amphibian structure is regarded as the undifferentiated primordium from which the mammalian has been derived. This is emphasized here because homologies are usually defined in structural terms and because organs which are phylogenetically related are regarded as more or less exactly homologous, regardless of radical changes in their functions. Thus the "dorsal island" in the acousticolateral area of the medulla oblongata of Necturus is regarded as the primordium of the dorsal cochlear nucleus of mammals, despite the fact that Necturus has no recognizable rudiment of a cochlea or cochlear nerve. This is because, when the cochlear rudiment and its nerve appear in the frog, the tissue of the "dorsal island" receives the cochlear nerve with radical change in the functions performed (p. 138).

The histological texture of these brains is so different from that of mammalian brains that the development of an intelligible nomenclature presents almost insuperable difficulty—a difficulty exacerbated by the fact that in the early stages of the inquiry it was necessary to apply descriptive terms to visible structures before their relationships were known. With increase of knowledge, errors were corrected, and unsuitable names were discarded; but terms already in use are still employed so far as possible, even though they are in some cases cumbersome and now known to be inappropriate. In all these descriptions I have consistently used the word "fissure" to designate visible furrows on the external surface of the brain and "sulcus" for those on ventricular surfaces. Attention is called to the list of abbreviations (p. 391) and to previous lists there cited where synonyms are given. In all my published figures of brains of urodeles the intent has been to use the same abbreviations for comparable structures. This intention has been approximately realized, but there are some inconsistencies, in most of which the differences express a change in emphasis rather than a correction of errors of observation.

Many well-defined tracts of fibers seen in fishes and higher animals are here represented in mixed collections of fibers of diverse sorts, here termed "fasciculi," or they may be dispersed within a mixed neuropil. The practice here is to define as a "tract" all fibers of like origin or termination, whether or not they are segregated in separate bundles. The customary self-explanatory binomial terminology is used wherever practicable—a compound word with origin and termination separated by a hyphen. But, since a single fiber of a tract may have collateral connections along its entire length, the fully descriptive name may become unduly cumbersome ('41a, p. 491). Thus, in accordance with strict application of the binomial terminology, tractus strio-tegmentalis would become tractus strio-thalamicus et peduncularis et tegmentalis dorsalis, isthmi et trigemini. The chemists seem to be able to manipulate similar enormities even without benefit of hyphens or spaces, but not many anatomists are so hardy. Few of the named tracts are sharply delimited, and all of them are mixtures of fibers with different connections. Any analysis is necessarily somewhat arbitrary.

Simple action systems of total-pattern type, wherever found

(cyclostomes, primitive ganoid fishes, urodeles), are correlated with a histological texture of the brain which is characteristic and probably primitive (chap. iii).

The external configuration of the urodele brain also is generalized, much as in a human embryo of about 6 weeks. In the next chapter special attention is directed to this comparison to assist the reader in identifying familiar parts of the human brain as they are seen in the simplified amphibian arrangement.

In our comparison of the amphibian brain with the human, two features are given especial emphasis, both of which are correlated with differences in the mode of life of the animals in question, that is, with the contrast between the amphibian simplicity of behavior with stereotyped total patterns of action predominating and the human complexity of movement in unpredictable patterns. The correlated structural differences are, first, in Amblystoma the more generalized histological texture to which reference has just been made, and notably the apparent paucity of provision for well-defined localization of function in the brain; and, second, the preponderant influence of motor patterns rather than sensory patterns in shaping the course of differentiation from fishlike to quadrupedal methods of locomotion and somatic behavior in general.

SOURCES AND MATERIAL

The material studied comprises gross dissections and serial sections of about five hundred specimens of Amblystoma from early embryonic to adult stages. About half these brains were prepared by the Golgi method and the remainder by various other histological procedures. Most of these are A. tigrinum, some are A. maculatum (punctatum), and a few are A. jeffersonianum. In early developmental stages some specific differences have been noted in the embryological papers of 1937–41, but no systematic comparative study has been made. The late larval and adult brains under consideration in this book are of A. tigrinum. In former papers there are comments on this material and the methods of preparation ('25, p. 436; '35a, p. 240; '42, p. 193).

In the study of these sections the cytological methods of Nissl and others are less revealing than in more highly differentiated brains because of the unspecialized structure of the nervous elements. Some modifications of the method of Weigert which decolorize the tissue sufficiently to show the myelinated fibers and also the arrangement of cell bodies prove to be most useful for general orientation. Other details can then be filled in by study of reduced silver preparations and especially of Golgi sections. A favorable series of transverse Weigert sections (no. IIC; see p. 321) has been chosen as a type or standard of reference, and the median section as reconstructed from this specimen (fig. 2C) has been used as the basis for many diagrams of internal structure. For reference to published figures of this brain and other details concerning it see page 321. Figures 2A and B are similar diagrams of the median section of the specimen from which figures 25–36 were drawn. The topography shown in these median sections is the basis for the descriptive terms used throughout this text.

Except for scattered references to details, the only systematic descriptions of the brain of Amblystoma are in my papers, Bindewald's ('14) on the forebrain, and Larsell's ('20, '32) on the cerebellum. Mention should also be made of Roofe's account ('35) of the endocranial blood vessels and Dempster's paper ('30) on the endolymphatic organ.

Kingsbury's admirable paper on Necturus in 1895 may be taken as a point of departure for all further investigation of the brains of urodeles, including my monograph of 1933 and several preceding and following papers. Some of the more important descriptions of the brains of other urodeles are cited in the appended bibliography, notably the following: Salamandra (Kuhlenbeck, '21; Kreht, '30), Proteus (Kreht, '31; Benedetti, '33), Cryptobranchus (Benzon, '26), Gymnophiona (Kuhlenbeck, '22), Siren (Röthig, '11, '24, '27), and several other urodeles in Röthig's later papers, Hynobius, Spelerpes, Diemyctylus (Triturus), Cryptobranchus, Necturus.

For the Anura the excellent description of the frog by E. Gaupp in 1899 laid a secure foundation for all subsequent work, and the time is now ripe for a systematic restudy of this brain with the better methods now available and the correlation of the histological structure with physiological experiments specifically designed to reveal the action of this structure. Aronson and Noble ('45) have published an excellent contribution in this field.

DEVELOPMENT OF THE BRAIN

No comprehensive description of the development of the brain of Amblystoma has been published. The difficulties met in staging specimens by criteria defined by Harrison, Coghill, and others I have discussed elsewhere ('48, chap. x). Griggs ('10) described with excellent illustrations the early stages of the neural plate and neural tube. Baker ('27) illustrated dorsal and ventral views of the open neural plate, and Baker and Graves ('32) described six models of the brain of A. jeffersonianum from 3 to 17 mm. in length. Burr ('22) described briefly the early development of the cerebral hemispheres.

Successive stages of the brain of A. punctatum have been illustrated by Coghill and others (some of which I have cited, '37, p. 391, and '38, p. 208), and at the Wistar Institute there are other models of the brains of physiologically tested specimens. Coghill's papers include a wealth of observation on the development of the mechanisms of the action system, and these were summarized in his London lectures, published in 1929. His reports were supplemented by a series of papers which I published from 1937 to 1941, but these fragmentary observations (of the younger stages particularly) were based on inadequate material and are useful only as preliminary orientation for a more systematic investigation. In Coghill's papers there are accurate projections of all mitotic figures and neuroblasts of the central nervous system in nonmotile, early flexure, coil, and early swimming stages and the arrangement of developing nerve fibers of the brain in the last-mentioned stage (Coghill, '30, Paper IX, fig. 4). On the basis of these data he divided the embryonic brain in front of the isthmus into sixteen regions, each of which is a center of active and characteristic differentiation. These regions are readily identified in our reduced silver preparations of these and later stages. Using a modification of this analysis, I have distinguished and numbered twenty-two such regions in the cerebrum and cerebellum ('37, p. 392), and the development of each of these can be followed through to the adult stage. In my papers of 1937–39 some salient features of these changes are recorded; but this account is incomplete, and more thorough study is urgently needed. In the present work some details only of this development are given in various contexts as listed in the Index under "Embryology."

The most detailed description of the development of the urodele brain is the paper by Sumi ('26) on Hynobius. Söderberg ('22) gave a brief description of the development of the forebrain of Triturus (Triton) and a more detailed account of that of the frog, and Rudebeck ('45) has added important observations.

The successive changes in the superficial form of the brain can be interpreted only in the light of the internal processes of growth and differentiation. The need for a comprehensive study of the development of the histological structure of the brain of Amblystoma, including the differentiation of the nervous elements and their fibrous connections, is especially urgent in view of the very large number of experimental studies on developmental mechanics which have been in progress for many years and will probably continue for years to come. Amblystoma has proved to be an especially favorable subject for these studies, and in many of them a satisfactory interpretation of the findings cannot be achieved without more complete knowledge than we now possess of the development of both the nervous tissues and other bodily organs.

THE EVOLUTION OF BRAINS

The nervous systems of all vertebrates have a common structural plan, which is seen most clearly in early embryonic stages and in the adults of some primitive species. But when the vertebrate phylum is viewed as a whole, the nervous apparatus shows a wider range of adaptive structural modifications of this common plan than is exhibited by any other system of organs of the body. In order to understand the significance of this remarkable plasticity and the processes by which these diverse patterns of nervous organization have been elaborated during the evolutionary history of the vertebrates, it is necessary to find out what were the outstanding features of the nervous system of the primitive ancestral form from which all higher species have been derived.

Since the immediate ancestors of the vertebrate phylum have been extinct for millions of years and have left no fossil remains, our only recourse in this search is to examine the most generalized living species, compare them one with another and with embryonic stages, and so discover their common characteristics. This has been done, and we are now able to determine with a high degree of probability the primitive pattern of the vertebrate nervous system.

The most generalized living vertebrates (lampreys and hagfish) have brains which most closely resemble that of the hypothetical primordial vertebrate ancestor. The brains of the various groups of fishes show an amazing variety of deviations from the generalized pattern. The paleontological record shows that the first amphibians were derived from one of the less specialized groups of fishes; and there is evidence that the existing salamanders and their allies have preserved until now a type of brain structure which closely resembles that of the most primitive amphibians and of the generalized fishes ancestral to them.

The internal texture of the brains of the generalized amphibians which are described in this work closely resembles that of the most primitive extant fishes; but the brain as a whole is organized on a higher plane, so that it can more readily be compared with those of reptiles, lower mammals, and man. For this reason the salamanders occupy a strategic position in the phylogenetic series. This examination has brought to light incipient stages of many complicated human structures and some guiding principles of both morphogenesis and physiological action that are instructive.

When the first amphibians emerged from the water, they had all the land to themselves; there were no living enemies there except one another. During aeons of this internecine warfare they carried protective armor; but in later times, during the Age of Reptiles, these more efficient fighters exterminated the clumsy armored amphibians. The more active frogs and toads survived, and so also did the sluggish salamanders and their allies, but only by retiring to concealment in sheltered places.

In Devonian times, probably about three hundred million years ago, various species of fishes made excursions to the land and acquired structures adapted for temporary sojourn out of water. Some of the primitive crossopterygian fishes went further and, after a fishlike larval period, experienced a metamorphosis into air-breathing tetrapods. They became amphibians. These were fresh-water species, and the immediate cause of this evolutionary change was extensive continental desiccation during the Devonian period. While their streams and pools were drying up, those fishes which had accessory organs of respiration in addition to the gills of typical fishes, were able to survive and, through further transformations, become air-breathing land animals. An excellent summary of the paleontological evidence upon which the history of the evolution of fishes has been reconstructed has been published by Romer ('46).

Two prominent features of this revolutionary change involved the organs of respiration and locomotion, with corresponding changes in the nervous apparatus of control. These systems of organs are typical representatives of the two major subdivisions of all vertebrate bodies and their functions—the visceral and the somatic. The visceral functions and the visceral nervous system will receive scant consideration in this work, for the material at our disposal is not favorable for the study of these tissues. Here we are concerned primarily with the nervous apparatus of overt behavior, that is, of the somatic adjustments.

The most important change in these somatic adjustments during the critical evolutionary period under consideration is the transition from swimming to walking. The fossil record of the transformation of fins into legs is incomplete, but it is adequate to show the salient features of the transformation of crossopterygian fins into amphibian legs (Romer, '46). In the individual development of every salamander and every frog the internal changes in the organization of the nervous system during the transition from swimming to walking can be clearly seen. And these changes are very significant in our present inquiry because they illustrate some general principles of morphogenesis of the brain more clearly than do any other available data.

In fishes, swimming is a mass movement requiring the co-ordinated action of most of their muscles in unison, notably the musculature of the trunk and tail. The paired fins are rudders, not organs of propulsion. The young salamander larva has no paired limbs but swims vigorously. This is a typical total pattern of action as defined by Coghill. The adult salamander after metamorphosis may swim in the water like the larva; and he can also walk on land with radically different equipment. Some fishes can crawl out on land, but the modified fins are clumsy and inefficient makeshifts compared with the amphibian's mobile legs.

Quadrupedal locomotion is a very complicated activity compared with the simple mass movement of swimming. The action of the four appendages and of every segment of each of them must be harmoniously co-ordinated, with accurate timing of the contraction of many small muscles. These local activities are "partial patterns" of behavior, in Coghill's sense. From the physiological standpoint there is great advance, in that the primitive total pattern is supplemented, and in higher animals largely replaced, by a complicated system of co-ordinated partial patterns. This is emphasized here because it provides the key to an understanding of many of the differences between the nervous systems of fishes, salamanders, and mammals. Motility, and particularly locomotion, have played a major role in vertebrate evolution, as dramatically told by Gregory ('43). This outline has been filled in by Howell's ('45) interesting comparative survey of the mechanisms of locomotion, and I have elsewhere discussed ('48) Coghill's contributions to this theme.

In the history of vertebrate evolution there were four critical periods: (1) the emergence of the vertebrate pattern of the nervous system from invertebrate ancestry; (2) the transition from aquatic to terrestrial life; (3) the differentiation within the cerebral hemispheres of primitive cerebral cortex; (4) the culmination of cortical development in mankind, with elaboration of the apparatus requisite for language and other symbolic (semantic) instrumentation of the mental life.

1. The extinct ancestors of the vertebrates in early Silurian times were probably soft and squashy creatures, not preserved as fossils. Some of their aberrant descendants may be recognized among the Enteropneusta, Tunicata, and Amphioxi; but the first craniate vertebrates preserved as fossils were highly specialized, heavily armored ostracoderms, now all extinct.

2. The salient features of the second critical period have been mentioned, and here the surviving amphibians recapitulate in ontogeny many instructive features of the ancestral history.

3. Amphibians have no cerebral cortex, that is, superficial laminated gray matter, in the cerebral hemispheres. This first takes definitive form in the reptiles, though prodromal stages of this differentiation can be seen in fishes and amphibians, a theme to which we shall return in chapter vii.

4. The fourth critical period, like the first, does not lie within the scope of this work, though study of the second and third periods brings to light some principles of morphogenesis which may help us to understand the more recondite problems involved in human cortical functions.

It is probable that none of the existing Amphibia are primitive in the sense of survival of the original transitional forms and that the urodeles are not only aberrant but in some cases retrograde (Noble, '31; Evans, '44); yet the organization of their nervous systems is generalized along very primitive lines, and these brains seem to me to be more instructive as types ancestral to mammals than any others that might be chosen. They lack the highly divergent specializations seen in most of the fishes; and, in both external form and internal architecture, comparison with the mammalian pattern can be made with more ease and security. So far as structural differentiation has advanced, it is in directions that point clearly toward the mammalian arrangement.

Amphibian eggs and larvae are readily accessible to observation and experiment; they are easily reared; they tolerate experimental operations unusually well; and, in addition, the amphibian neuromuscular system begins to respond to stimulation at a very early age, so that successive stages in maturation of the mechanism are documented by changes in visible overt movement. The adult structure is instructive; and, when the embryological development of this structure is compared with that of higher brains and with the sequence of maturation of patterns of behavior, basic principles of nervous organization are revealed that can be secured in no other way. In the absence of differentiated cerebral cortex, the intrinsic structure of the stem is revealed. Experimental decortication of mammals yields valuable information, but study of such mutilations cannot tell us all that we need to know about the normal operations of the brain stem and the reciprocal relationships between the stem and the cortex.

In brief, the brains of urodele amphibians have advanced to a grade of organization typical for all gnathostome vertebrates, Amblystoma being intermediate between the lowest and the highest species of Amphibia. This brain may be used as a pattern or template, that is, as a standard of reference in the study of all other vertebrate brains, both lower and higher in the scale.

CHAPTER II

THE FORM AND SUBDIVISIONS OF THE BRAIN

GROSS STRUCTURE

REFERENCE to figures 1–5, 85, and 86 shows that the larger subdivisions of the human brain are readily identified in Amblystoma, though with remarkable differences in shape and relative size. When this comparison is carried to further detail, the sculpturing of the ventricular walls shown in the median section is especially instructive. It is again emphasized that the application of mammalian names to the structures here revealed rarely implies exact homology; these areas are to be regarded as primordia from which the designated mammalian structures have been differentiated. The relationships here implied have been established by several independent lines of evidence: (1) The relative positions and fibrous connections of cellular masses and the terminal connections of tracts. In so far as these arrangements conform with the mammalian pattern, they may be regarded as homologous. (2) Embryological evidence. The early neural tubes of amphibians and mammals are similar, and subsequent development of both has been recorded. On the basis of Coghill's observations of rates of proliferation and differentiation in prefunctional stages, the writer ('37) gave arbitrary numbers to recognizable sectors of the neural tube in early functional stages, and the subsequent development of each of these is, in broad lines, similar to that of corresponding mammalian parts. (3) The relationships of supposed primordia of mammalian structures may be tested by the comparative method. In an arrangement of animal types which approximates the phylogenetic sequence from the most generalized amphibians to man, there are many instances of progressive differentiation of amphibian primordia by successive increments up to the definitive human form.

Many pictures of the brains of adult and larval Amblystoma and other urodeles have been published, some of which I have cited ('35a, p. 239). The most accurate pictures of the brain of adult A. tigrinum are those of Roofe ('35), showing dorsal, ventral, and lateral aspects and the distribution of endocranial arteries and veins. The outlines of the brain were drawn from specimens dissected after preservation for 6 weeks in 10 per cent formalin. One of these is shown here (fig. 86A). Figure 1B is drawn from a dissection made by the late Dr. P. S. McKibben, showing the sculpturing of the ventricular surfaces. Figures 1A and 85 are drawn from a wax model in which there is some distortion of the natural proportions. Not all the differences seen in these pictures and in the proportions of sections figured are artifact, for the natural variability of urodele brains is surprisingly large (Neimanis, '31). Brains of larval stages have been illustrated by many authors and in my embryological papers of 1937–39.

The somewhat simpler brain of the mudpuppy, Necturus, has been described in a series of papers as completely as available material permits, and comparison with the more differentiated structure of Amblystoma is instructive. The sketches shown in figures 86B and C illustrate the differences between the form of this forebrain and that of Amblystoma. The monograph of 1933 contains a series of diagrams ('33b, figs. 6–16) of the internal connections of the brain of Necturus similar to those of Amblystoma shown here (figs. 7–24). In 1910 I described the general features of the forebrain of A. tigrinum, with a series of drawings of transverse Weigert sections, no. IIC, which has subsequently been used as the type specimen. Though this paper contains some errors, and some morphological interpretations which I now regard as outmoded ('33a), most of the factual description has stood the test of time, and additional details and reports on other parts of the brain have been published in a series of papers.

The most conspicuous external fissures of the brain of Amblystoma are: (1) the longitudinal fissure separating the cerebral hemispheres; (2) the deep stem-hemisphere fissure; (3) a wide dorsal groove separating the epithalamus from the roof (tectum) of the midbrain; (4) the ventral cerebral flexure or plica encephali ventralis, which is a sharp bend of the floor of the midbrain, where it turns downward and backward into the "free part" of the hypothalamus; and (5) the fissura isthmi, extending downward and forward from the anterior medullary velum between midbrain and isthmus. The middle part of the fissura isthmi is at the anterior border of the auricle, which is more prominent in the larva than in the adult ('14a, figs. 1–3). Here in the adult it lies near the posterior border of the internal isthmic tissue, some distance posteriorly of the ventricular sulcus isthmi; but, like the latter, it really marks the anterior border of the isthmus, as will appear in the description of the development of the isthmic sulcus (p. 179).

The obvious superficial eminences on the dorsal aspect of the brain are the small cerebellum, the dorsal convexity of the roof of the midbrain (tectum mesencephali), the habenular nuclei of the epithalamus, and the two cerebral hemispheres. Posteriorly of the habenulae in the early larvae is the membranous pineal evagination, which in the adult is a closed epithelial vesicle detached from the brain except for the few fibers of the parietal nerve. The lateral aspect of the thalamus, midbrain, and isthmus is a nearly smooth convexity, posteriorly of which is the high auricle, composed of tissue which is transitional between the body of the cerebellum and the acousticolateral area of the medulla oblongata. This auricle contains the primordia of the vestibular part of the cerebellar cortex (flocculonodular lobe of Larsell), and most of its tissue is incorporated within the cerebellum in mammals. On the ventral aspect there is a low eminence in front of the optic chiasma, which marks the position of the very large preoptic nucleus, and a similar eminence behind the chiasma formed by the ventral part of the hypothalamus. The latter is in the position of the human tuber cinereum but is not exactly comparable with it. Most of the hypothalamus is thrust backward under the ventral cerebral flexure as the pars libera hypothalami. The large pars glandularis of the hypophysis envelops the posterior end of the infundibulum and extends spinalward from it, not anteriorly as in man.

The primary subdivisions of the human brain as defined from the embryological studies of Wilhelm His are readily identified in adult Amblystoma, as shown in the median section (fig. 2A).

At the anterior end of each cerebral hemisphere is the very large olfactory bulb, the internal structure of which shows some interesting primitive features (p. 54; '24b). The bulbar formation extends backward on the lateral side for about half the length of the hemisphere, but on the medial side only as far as the anterior end of the lateral ventricle (figs. 3, 4). Bordering the bulb is an undifferentiated anterior olfactory nucleus, and posteriorly of this the walls of the lateral ventricle show early stages of the differentiation of the major subdivisions of the mammalian hemisphere—in the ventrolateral wall a strio-amygdaloid complex, ventromedially the septum, and dorsally the pars pallialis. In the pallial part no laminated cortical gray is differentiated, but there are well-defined pallial fields: dorsomedially, the primordial hippocampus; dorsolaterally, the primordial piriform lobe; and between these a primordium pallii dorsalis of uncertain relationships.

The boundaries of the diencephalon, as here defined and shown in figure 2A, are: anteriorly, the stem-hemisphere fissure and the posterior border of the anterior commissure ridge and, posteriorly, the anterior face of the posterior commissure and the underlying commissural eminence and, more ventrally, the sulcus, s, which marks the anterior border of the cerebral peduncle. The inclusion of the preoptic nucleus is in controversy; but, whether or not this inclusion is justifiable morphologically, its relationships with the hypothalamus are so intimate that it is practically convenient to consider these parts together. The four primary subdivisions of the diencephalon as I defined them in 1910 are: (1) the dorsal epithalamus, containing on each side the habenula and pars intercalaris, the latter including the pretectal nucleus; (2) pars dorsalis thalami, which is the primordium of the sensory nuclei of the mammalian thalamus; (3) pars ventralis thalami, the motor zone of the thalamus, or subthalamus; (4) hypothalamus. The mammalian homologies of these areas are clear, though their relative sizes and fibrous connections exhibit remarkable differences.

The posterior boundary of the mesencephalon is marked by the external fissura isthmi, the ventricular sulcus isthmi (fig. 2B, $s.is.$), and ventrally in the floor plate a pit, the fovea isthmi ($f.i.$). These are all more prominent in the larva than in the adult. This sector includes the posterior commissure, the tectum mesencephali (primordial corpora quadrigemina), the underlying dorsal tegmentum (subtectal area), and the area surrounding the tuberculum posterius at the ventral cerebral flexure, termed the "nucleus of the tuberculum posterius." On embryological grounds and for convenience of description, this ventral area, which is bounded by the variable ventricular sulcus s, is here called the "peduncle" in a restricted sense ('36, p. 298; '39b, p. 582). This is a primordial mesencephalic structure which is not the equivalent of the peduncle of human neurology. Amblystoma has nothing comparable with the human basis pedunculi, and its "peduncle" is incorporated within the tegmentum of the human brain. The III cranial nerve arises within the "peduncle" and emerges near the fovea isthmi. The nucleus of the IV nerve is in the isthmus. In the human brain there are no definite structures comparable to the amphibian dorsal and isthmic tegmentum.

The isthmus is much more clearly defined than in adult higher brains, it is relatively larger, and its physiological importance is correspondingly greater, as will appear later. It is bounded anteriorly by the sharp isthmic sulcus and posteriorly by the cerebellum, auricle, and trigeminal tegmentum. The so-called "pons" sector of the human brain stem is named from its most conspicuous component, but this name is meaningless in comparative anatomy. In man it is the pons and the sector of the stem embraced by it; but in no two species of mammals is the part embraced by the pons equivalent; and below the mammals the pons disappears entirely. The medulla oblongata, on the other hand, is a stable structure, extending from the isthmus to the spinal cord, and for it the shorter name "bulb" is sometimes used, especially in compounds.

I outlined the development and morphological significance of the urodele cerebellum ('14, '24), and this was followed by detailed descriptions of the development and adult structure of this region of Amblystoma by Larsell ('20, '32), whose observations I have subsequently confirmed, including his fundamental distinction between its general and its vestibular components.

Some features of the larval medulla oblongata and related nerves have been described ('14a, '39b) and, more recently ('44b), additional details of the adult, particularly the structures at the bulbospinal junction. Much remains to be done to clarify the organization of the medulla oblongata and spinal cord.

The cranial nerves and their analysis into functional components (chap. v) were described by Coghill ('02). The embryological development of these components also has been extensively studied (chap. x). The arrangement and composition of these nerves are fundamentally similar to those of man, with a few notable exceptions. The internal ear lacks the cochlea, which is represented by a very primitive rudiment; a cochlear nerve, accordingly, is not separately differentiated. There is an elaborate system of cutaneous organs of the lateral lines, whose functions are not as yet adequately known. These are supplied by very large nerves commonly assigned to the VII and X pairs, though it would be more appropriate to regard them as accessory VIII nerves, for all these nerve roots enter a wide zone at the dorsolateral margin of the medulla oblongata known as the "area acusticolateralis." There is no separate XI cranial nerve, this being

represented by an accessorius branch of the vagus. The XII nerve is represented by branches of the first and second spinal nerves. The first spinal nerve in some specimens has a small ganglion; the second nerve always has a large dorsal root and ganglion. In this connection a passage in the comprehensive work on the anatomy of Salamandra by Francis ('34, p. 134) is worthy of mention: "After making due allowance for the absence of a lateralis component in the adult salamander, the correspondence between the cranial nerves of this animal and those of Amblystoma is very close indeed."

The configuration and mutual relations of the gross structures just surveyed can be seen only in sections, of which many, cut in various planes, have been illustrated in the literature. Only a few selected examples are included in the present work, with references in subsequent chapters to many others. For general orientation the following figures may be consulted: a series of selected transverse sections from the spinal cord to the olfactory bulb (figs. 87–100); a series of horizontal sections through the middle part of the brain stem (figs. 25–36); a few sagittal sections (figs. 101–4). Figures 6–24 show the chief fibrous connections of each well-defined region of the brain stem.

The diencephalon, mesencephalon, and isthmus have the form of three irregular pyramids oppositely oriented (fig. 2A). The broad base of the diencephalon extends from the anterior commissure to the hypophysis, and the apex is at the epiphysis. The tectum forms the base of the mesencephalic pyramid, and the apex is at the ventral tip of the tuberculum posterius, which borders the ventral cerebral flexure. The base of the pyramidal isthmus is formed by the massive tegmentum isthmi of each side and the median interpeduncular nucleus in the floor plate. It narrows dorsally into the anterior medullary velum between the tectum and the cerebellum.

The middle sectors of the brain stem—diencephalon, mesencephalon, and isthmus—contain the primordial regulatory and integrating apparatus controlling the fundamental sensori-motor systems of adjustment. The most important peripheral connections are with the eyes, and these in most vertebrates play the dominant role in maintaining successful adjustment with environment. From this topographical feature it naturally followed that, during the course of phylogenetic differentiation of the brain, the chief centers of adjustment of the other exteroceptive systems were elaborated in close juxtaposition with the visual field in the midbrain and thalamus.

Here they are interpolated between the primary sensory and motor apparatus of the medulla oblongata and spinal cord below and the great olfactory field and suprasegmental apparatus of the cerebral hemispheres above.

In all lower vertebrates the roof of the midbrain, the tectum, is the supreme center of regulation of motor responses to the exteroceptive systems of sense organs. The hypothalamus is similarly elaborated for regulation of olfacto-visceral adjustments. The patterning of motor responses for both these groups of receptors is effected in the cerebral peduncle and tegmentum. In the region of the isthmus, between the tectum and the primary vestibular area of the medulla oblongata and above the tegmentum, the cerebellum was elaborated as the supreme adjustor of all proprioceptive systems.

At the rostral end of the brain, within and above the specific olfactory area of the cerebral hemisphere, there gradually emerged a synthetic apparatus of control, adapted to integrate the activities of all the other parts of the nervous system and to enlarge capacity to modify performance as a result of individual experience. In the lowest vertebrates this "suprasensory" and "supra-associational" apparatus, as Coghill termed it, is not concentrated in the cerebral hemispheres, but it is dispersed, chiefly in the form of diffuse neuropil. In the amphibian cerebral hemispheres this integrating apparatus is more highly elaborated than elsewhere, with some local differentiation of structure. The hemispheres are larger than in fishes, and the primordia of their chief mammalian subdivisions can be recognized. A dorsal pallial part is distinguishable from a basal or stem part of the hemisphere, though the distinctive characteristics of the pallium are only incipient. There is no cerebral cortex, and, accordingly, the mammalian cortical dependencies in the thalamus, midbrain, and cerebellum have not yet appeared. The primordial thalamus is concerned chiefly with adjustments within the brain stem, though precursors of the thalamic radiations to the hemispheres are present.

VENTRICLES

The lateral ventricles of the cerebral hemispheres have the typical form except at the interventricular foramen, where the amphibian arrangement is peculiar. The anterior and hippocampal commissures do not cross as usual in or above the lamina terminalis, but in a more posterior high commissural ridge; and between these structures there is a wide precommissural recess, into which the interventricular foramina open. This results in some radical differences from reptilian and mammalian arrangements of the related fiber tracts and membranous parts, as elsewhere described (p. 291; '35). The third ventricle is expanded dorsally into the complicated membranous paraphysis and dorsal sac. Ventrally, the great elongation of the preoptic nucleus gives rise to a large preoptic recess between the anterior commissure ridge and the chiasma ridge, and in front of the latter there is a lateral optic recess (fig. 96), which in early larval stages extends outward as far as the eyeball, as a patent lumen of the optic nerve ('41), an arrangement which persists in the intracranial part of the nerve of adult Necturus ('41a). In the hypothalamus the ventricle is dilated laterally ('35a, p. 253; '36, figs. 10–14), and posteriorly it is a wide infundibulum with membranous roof and thin but nervous floor and posterior wall. The latter is the pars nervosa of the hypophysis and is partly enveloped by the pars glandularis (figs. 2, 101; '35a, p. 254; '42, p. 212 and figs. 56–65; Roofe, '37). The aqueduct of the midbrain is greatly expanded dorsoventrally. Its ventral part is contracted laterally by the thick peduncles and tegmentum, and the dorsal part is dilated as an optocoele. The sulcus lateralis mesencephali marks its widest extent, and tectal structure reaches far below this sulcus. The fourth ventricle is of typical form except anteriorly, where the wide lateral recess with membranous roof extends outward and forward to cover the whole dorsolateral aspect of the auricular lobe (figs. 90, 91; '24, p. 627). The rhombencephalic chorioid plexus is elaborately developed in interesting relation with the peculiar endolymphatic organs of this animal ('35, p. 310). The ventricular systems of adult Triturus (Diemyctylus) and of larval and adult stages of Hynobius have been described and illustrated with wax models by Sumi ('26, '26a).

The ventricular surface of both larvae and adults is clothed with very long cilia. These are not preserved in ordinary preparations and in our material are seen only in Golgi sections, where their impregnation is erratic and local ('42, p. 196). They are most frequently seen in the infundibulum and optocoele under the tectum. In the vicinity of the posterior commissure the ciliated ependyma is thickened (subcommissural organ of Dendy), and to it the fiber of Reissner is attached ('42, p. 197). This thick, nonnervous fiber extends backward through the ventricle to the lower end of the spinal cord and, like the cilia, is apparently an outgrowth from the internal ependymal membrane.

MENINGES, CHORIOID PLEXUSES, AND BLOOD VESSELS

The meninges of Amblystoma were described in 1935. This account should be compared with that of Salamandra published in 1934 by Francis, whose description was based on the investigation of Miss Helen O'Neill ('98), done under the direction of Wiedersheim and Gaupp. In Amblystoma the meninges are intermediate between the meninx primitiva of the lower fishes and those of the frog. Over the spinal cord and most parts of the brain a firm and well-defined pachymeninx, or dura, closely invests the underlying undifferentiated pia-arachnoid. The meninges of the frog have been described by others, and recently Palay ('44) has investigated their histological structure in the toad. The most interesting feature of these amphibian membranes is their intimate relation with the enormous endolymphatic organ described by Dempster ('30) and the associated blood vessels.

The vascular supply of these brains is peculiar in several respects. The distribution of arteries and veins has been described by Roofe ('35, '38), and I have added some details from the adult ('35) and the larva ('34d). The endocranial veins form a double portal system of sinusoids of vast extent and unknown significance. Between the cerebral hemispheres and the epithalamus the nodus vasculosus (Gaupp) is permeated by a complicated rete of sinusoids, which receives venous blood from the entire prosencephalon—chorioid plexuses, brain wall, and meninges. The efferent discharge from this rete is by the two oblique sinuses, which pass backward across the midbrain to enter a similar rete of wide, anastomosing sinusoids spread over the chorioid plexus of the fourth ventricle and the lobules of the endolymphatic organs. This rete also receives the veins from all posterior parts of the brain, meninges, and chorioid plexus. The common discharge for all this endocranial venous blood is by a large sinus, which emerges from the cranium through the jugular foramen and joins the jugular vein. These membranous structures are readily observable in the living animal without serious disturbance of normal conditions, and they provide unique opportunities for experimental study of some fundamental problems of vascular physiology.

It was pointed out by Craigie ('38, '38a, '39, '45) that within the substance of this brain the penetrating blood vessels are arranged in two ways—a capillary net of usual type and simple loops, which enter from the meningeal arterial network. Our preparations confirm this observation and also the fact that the vascular pattern varies in different parts of the brain. Both isolated loops and the capillary net may be seen in the same field, as in the dorsal thalamus (fig. 44), or one of these patterns may prevail, with few, if any, instances of the other. In the tectum and dorsal tegmentum of the midbrain, for instance, the tissue is vascularized by simple loops with only occasional anastomosis (fig. 48), while in the underlying peduncle and isthmic tegmentum the vascular network prevails, with occasional simple loops. In the meninges and chorioid plexuses only the network has been observed.

The telencephalic and diencephalic chorioid plexuses have an abundant arterial blood supply through the medial hemispheral artery; but the elaborately ramified tubules of the paraphysis seem to have no arterial supply or capillary net, the accompanying vessels being exclusively venous sinusoids ('35, p. 342). The same seems to be true of the endolymphatic sacs ('34d, p. 543). The chorioid plexus of the fourth ventricle has abundant arterial blood supply. In all plexuses the capillaries unite into venules, which discharge into wide sinusoids, which ramify throughout the plexus and have very thin walls. All arterioles of the chorioid plexuses are richly innervated, but it has not been possible to get satisfactory evidence of the sources of these nerve fibers ('36, p. 343; '42, p. 255; Necturus, '33b, p. 15).

The enormous development of the chorioid plexuses and associated endolymphatic organ of urodeles is apparently correlated with the sluggish mode of life and relatively poor provision for aeration of the blood. In the more active anurans the plexuses are smaller; but in the sluggish mudfishes, including the lungfishes, with habits similar to those of urodeles, we again find exaggerated development of these plexuses. Existing species in the border zone between aquatic and aerial respiration are all slow-moving and relatively inactive. The enlarged plexuses and sinusoids give vastly increased surfaces for passage of blood gases into the cerebrospinal fluid; and, correlated with this, the brain wall is thin everywhere, to facilitate transfer of metabolites between brain tissue and cerebrospinal fluid. Massive thickenings of the brain wall occur in many fishes and in amniote vertebrates, but not in mudfishes and urodeles.

CHAPTER III

HISTOLOGICAL STRUCTURE

GENERAL HISTOLOGY

IN AMPHIBIAN brains the histological texture is generalized, exhibiting some embryonic features; and it is at so primitive a level of organization as to make comparison with mammals difficult. Most of the nerve cells are small, with scanty and relatively undifferentiated cytoplasm. There are some notable exceptions, such as the two giant Mauthner's cells of the medulla oblongata and related elements of the nucleus motorius tegmenti. With the exceptions just noted, Nissl bodies are absent or small and dispersed.

Almost all bodies of the neurons are crowded close to the ventricle in a dense central gray layer, with thick dendrites directed outward to arborize in the overlying white substance (figs. 9, 99). The axon usually arises from the base of the dendritic arborization, rarely from its tip, and sometimes from the cell body; it may be short and much branched or very long, with or without collateral branches. The ramifications of the short axons and of collaterals and terminals of the longer fibers interweave with dendritic arborizations to form a more or less dense neuropil, which permeates the entire substance of the brain and is a synaptic field. Some of the nerve fibers are myelinated, more in the peripheral nerves, spinal cord, and medulla oblongata than in higher levels of the brain. Both myelinated and unmyelinated fibers may be assembled in definite tracts, or they may be so dispersed in the neuropil as to make analysis difficult. The arrangement of recognizable tracts conforms with that of higher brains, so that homologies with human tracts are in most cases clear. These tracts and the gray areas with which they are connected provide the most useful landmarks in the analysis of this enigmatic tissue.

In the gray substance there are few sharply defined nuclei like those of mammals, but the precursors of many of these can be recognized as local specializations of the elements or by the connections of the related nerve fibers. In most cases the cells of these primordial nuclei have long dendrites, which arborize widely into surrounding fields (figs. 9, 24, 61, 66), so that the functional specificity of the nucleus is, at best, incomplete. This arrangement facilitates mass movements of "total-pattern" type, but local differentiations serving "partial patterns" of action (Coghill) are incipient. Localized reflex arcs are recognizable, though in most cases these are pathways of preferential discharge within a more dispersed system of conductors (chap. vi).

Tissue differentiation is more advanced in the white substance than in the gray. The most important and diversely specialized synaptic fields are in the alba, and this local specialization is correlated with differences in the physiological properties of the nervous elements represented. This means, as I see it, that functional factors must be taken into account in both ontogenetic and phylogenetic differentiation and that in the long view the problems of morphogenesis are essentially physiological, that is, they resolve into questions of adaptation of organism to environment (chap. viii). This is the reason why in this work the histological analysis is made in terms of physiological criteria, even though these criteria are, in the main, based on indirect evidence, namely, the linkage of structures in functional systems of conductors.

The nonnervous components of this tissue comprise the blood vessels, ependyma, and a small number of cells of uncertain relationships which are regarded as undifferentiated free glial cells or transitional elements ('34, p. 94; '33b, p. 17). The ependymal elements everywhere span the entire thickness of the brain wall with much free arborization. They assume various forms in different regions, and their arrangement suggests that they are not merely passive supporting structures, though if they have other specific functions these are still to be discovered. For illustrations see figures 63, 64, 70, 79, and 81.

More detailed descriptions of the histological structure of urodele brains may be found in earlier papers ('14a, p. 381; '17, pp. 232, 279 ff.; '33b, pp. 16, 268; '33c; '33d; '34; '34a; '34b; '42, p. 195; '44a). In the olfactory bulbs of Necturus ('31) and Amblystoma ('24b) we find an interesting series of transitional cells between apparently primitive nonpolarized elements and typical neurons, as described on page 54.

THE NEUROPIL

In the generalized brains here under consideration the neuropil is so abundant and so widely spread that it evidently plays a major role in all central adjustments, thus meriting detailed description.

Only the coarser features of this tissue are open to inspection with presently available histological technique. In my experience its texture is best revealed by Golgi preparations, and very many of them, for the erratic incidence of these impregnations may select in different specimens now one, now another, of the component tissues—blood vessels, ependyma, dendrites, or axons. In each area of neuropil these components are independent variables, and in most of these areas axons from many sources are so intricately interwoven that the tissue can be resolved only where fortunate elective impregnations pick out one or another of the several systems of fibers in different specimens. It is difficult to picture the neuropil either photographically or with the pen, and the crude drawings in this book and in the literature give inadequate representations of the intricacy and delicacy of its texture.

A survey of the neuropil of adult Amblystoma as a whole has led me to subdivide it for descriptive purposes and somewhat arbitrarily into four layers ('42, p. 202). From within outward, these are as follows:

1. The periventricular neuropil pervades the central gray so that every cell body is enmeshed within a fabric of interwoven slender axons (figs. 106, 107). This persists in some parts of the mammalian brain as subependymal and periventricular systems of fibers.

2. The deep neuropil of the alba at the boundary between gray and white substance knits the periventricular and intermediate neuropil together, and it also contains many long fibers coursing parallel with the surface of the gray. The latter are chiefly efferent fibers directed toward lower motor fields (fig. 93, layer 5; '42, figs. 18–21, 24, 29–45, 47).

3. The intermediate neuropil in the middle depth of the alba contains the largest and most complicated fields of this tissue. It is very unevenly developed, in some places scarcely recognizable and in others of wide extent and thickness. Its characteristics are especially well seen in the corpus striatum (figs. 98, 99, 108, 109), thalamus ('39b, fig. 81; '42, figs. 71, 81), and tectum opticum (figs. 93, layer 2, 101; '42, figs. 26, 30, 32, 79–83). Many of the long tracts lie within this layer and have been differentiated from it. Most of the specific nuclei of higher animals, including the outer gray layers of the tectum, have been formed by migration of neuroblasts from the central gray outward into this layer. Here we find much of the ap-

paratus of local reflexes and their organization into the larger, innate patterns of behavior.

4. The superficial neuropil is a subpial sheet of dendritic and axonal terminals, in some places absent, in others very elaborately organized. Here are some of the most highly specialized mechanisms of correlation in the amphibian brain, from which specific nuclei of higher brains have been developed. Notable examples are seen in the interpeduncular neuropil (chap. xiv) and the ventrolateral neuropil of the cerebral peduncle described in the next section. This neuropil seems to be a more sensitive medium for strictly individual adjustments (conditioning) than the deeper neuropil, but of this there is no experimental evidence. This hypothesis is supported by the fact that in higher animals cerebral cortex develops within this layer and apparently by neurobiotactic influence emanating from it.

In the first synapses observed in embryogenesis numerous axonic terminals converge to activate a single final common path (Coghill, '29, p. 13). This is the first step in the elaboration of neuropil. As differentiation advances, neurons are segregated to serve the several modalities of sense and the several systems of synergic muscles, and these systems are interconnected by central correlating elements. In no case are these connections made by an isolated chain of neurons in one-to-one contact between receptor and effector. The central terminals of afferent fibers from different sense organs are widely spread and intermingled. Dendrites of the correlating cells branch widely in this common receptive field, and the axons of some of them again branch widely in a motor field, thus activating neurons of the several motor systems. This arrangement is perfectly adapted to evoke mass movement of the entire musculature from any kind of sensory stimulation, and this is, indeed, the only activity observed in early embryonic stages.

It is the rare exception rather than the rule for a peripheral sensory fiber to effect functional connection directly with a peripheral motor neuron. One or more correlating elements are interpolated; and, as differentiation advances, the number of these correlating neurons is enormously increased in both sensory and motor zones and also in the intervening intermediate zone. The axons of these intrinsic elements ramify widely in all directions, and, as a rule, they or collaterals from them interweave to form the closely knit fabric which pervades both gray and white substance everywhere. This is the axonic neuropil, within the meshes of which dendrites of neurons ramify widely. These axons are unmyelinated, and every contact with a dendrite or a cell body is a synaptic junction. Every axon contacts many dendrites, and every dendrite has contact with many axons, and these may come from near or from very remote parts.

All neuropil is a synaptic field, and, since in these amphibian brains it is an almost continuous fabric spread throughout the brain, its action is fundamentally integrative. But it is more than this. It is germinative tissue, the matrix from which much specialized structure of higher brains has been differentiated. It is activated locally or diffusely by every nervous impulse that passes through the substance of the brain, and these impulses may irradiate for longer or shorter distances in directions determined by the trend of the nerve fibers of which it is composed. This web of conductors is relatively undifferentiated, but it is by no means homogeneous or equipotential, for each area of neuropil has characteristic structure and its own pattern of peripheral and central connections. Many of these fibers, which take long courses, may connect particular areas of gray substance either in dispersed arrangement or assembled as recognizable tracts. In fact, we find all gradations between an almost homogeneous web of neuropil and long well-fasciculated tracts of unmyelinated or myelinated fibers. Many of these long fibers have collateral connections throughout their length; others are well-insulated conductors between origin and termination.

The web of neuropil shows remarkable diversity in different regions. In some places it is greatly reduced, as, for instance, in the ventral funiculi, where the alba is densely filled with myelinated fibers; in other places there are local concentrations of dendrites and finest axons so dense that in ordinary haematoxylin or carmine preparations they appear as darker fields, the *Punksubstanz* of the early histologists and the glomeruli of the olfactory bulb and interpeduncular nucleus (chap. xiv). When some of these fields are analyzed, it is found that their fibrous connections conform with those of specific "nuclei" of higher brains. In Amblystoma such a field is not a "nucleus," for it contains no cell bodies; but examination of the corresponding area in anurans and reptiles may show all stages in the differentiation of a true nucleus by migration of cell bodies from the central gray outward into the alba ('27, p. 232). Such a series of phylogenetic changes can be readily followed in the corpus striatum, the geniculate bodies, the interpeduncular nucleus, and many other places and particularly in the tectum opticum and the pallial fields of the cerebral hemispheres.

In phylogeny the long, well-organized tracts seem to have been formed by a concentration of the fibers of the neuropil. The diffuse neuropil is probably the primordial form, going back to the earliest evolutionary stages of nervous differentiation (coelenterates). Local reflex arcs and specific associational tracts have been gradually differentiated within it, that is, integration precedes local specialization, the total pattern antedates the partial patterns. In ontogeny, especially of higher animals, this history may not be recapitulated, and tracts serving local reflexes may appear very early; but in Amblystoma, even in the adult stage, there are few tracts which are compactly fasciculated and free from functional connection with the surrounding neuropil. Most of the long, well-fasciculated tracts have some myelinated fibers that seem to have functional connections only at their ends; but these are accompanied by others, which are without myelin and are provided with numberless collaterals tied into the enveloping neuropil. There is, accordingly, a seepage of nervous influence along the entire length of these tracts.

From the primordial diffuse neuropil, differentiation advanced in two divergent directions. One of these, as just pointed out, led to the elaboration of the stable architectural framework of nuclei and tracts, the description of which comprises the larger part of current neuroanatomy. This is the heritable structure, which determines the basic patterns of those components of behavior which are common to all members of the species. The second derivative of the primordial neuropil is the apparatus of individually modifiable behavior—conditioning, learning, and ultimately the highly specialized associational tissues of the cerebral cortex. In both phylogeny and ontogeny, differentiation of the first type precedes that of the second. Primitive animals and younger developmental stages exhibit more stable and predictable patterns of behavior; and the more labile patterns are acquired later. In Amblystoma both these types of differentiation are at low levels, but they are sufficiently advanced to be clearly recognizable. Tissue differentiation is further advanced in the white substance than in the underlying gray.

During the course of this progressive specialization of tissue, the primordial integrative function of the neuropil is preserved and elaborated. The mechanism employed is seen in its most generalized form in the deep periventricular neuropil, layer 1 of the preceding analysis. This is a close-meshed web of finest axons, within which all cell bodies are imbedded. It is everywhere present, providing continuous activation (or potential activation) of every neuron, summation, reinforcement, or inhibition of whatever activities may be going on in the more superficial layers and affecting the general excitatory state of the whole central nervous system. It receives fibers from all sensory and motor fields and seems to be the basic apparatus of integration. In man this type of tissue survives in the periventricular gray of the diencephalon, and, as suggested by Wallenberg ('31), it probably plays an important part in determining the disposition and temperament of the individual ('34a, pp. 241, 245). This unspecialized tissue may serve the most general totalizing function. From it there have been derived the complicated mechanisms for synthesizing the separate experiences and organizing them in adaptive patterns—a process of differentiation which culminates in the human cerebral cortex. The relations of neuropil to reflex arcs are discussed in chapter vi.

The relative abundance of myelinated fibers is a rough indicator of the relation between stable and labile types of performance. Thus the myelinated white substance is relatively greater in the spinal cord than in the brain, and the ratio of myelinated to unmyelinated tissue diminishes as we pass forward in the brain, as is well illustrated by a published series of Weigert sections drawn from a single specimen (no. IIC) from the olfactory bulb to the spinal cord ('10, figs. 8–21; '25, figs. 2–9; '44b, figs. 1–6). This is because the myelinated fibers, most of which are long conductors of through traffic, tend to be compactly arranged, with relatively scanty collateral connections with the neuropil.

In the gray substance there is a similar increment in relative amount of neuropil as we pass forward from spinal cord to hemispheres, the higher levels being specialized for correlation, association, conditioning, and integration and the lower levels for stabilized total activities and reflexes. In the phylogenetic series this principle takes a form which can be quantitatively expressed as Economo's coefficient of the ratio between total gray substance and the mass of the nerve cells contained within it. The lower the animal species in the scale, the greater is the mass of the cells compared with the gray substance. This law may be expressed in the converse form: Higher animals have a larger proportion of neuropil in the gray substance, thus giving them capacity for conditioning and other individually acquired patterns of behavior (Economo, '26, '29). No mathematical precision should be claimed for these laws, but they do express the general trend of our experience. For additional data and critical comment see von Bonin on page 64 of the monograph by Bucy and others ('44).

The properties of amphibian neuropil have been discussed in several places, in addition to the summary already cited ('42, p. 199). In one of these papers ('33d) the geniculate neuropil of Necturus was described, and later ('42, p. 280) the quite different arrangement of Amblystoma (compare the corresponding structure of the frog, '25), together with a full account of the neuropil of the optic tectum and its connections (here, again, comparison with the more differentiated structure of the frog is instructive). In connection with a general survey of the neuropil of Necturus, its peculiar relations in the pallial part of the hemisphere were discussed in four papers ('33b, p. 176; '33e, '34, '34a). Amblystoma is similar ('27), though the details have not been fully explored. In the pallial field the four layers of neuropil tend to merge into a single apparatus of association. This is corticogenetic tissue within which the earliest phases of incipient differentiation of laminated cortex can be recognized in the hippocampal area, as described in chapter vii.

Some samples of the appearance of elective Golgi impregnations of amphibian neuropil are shown in the accompanying illustrations, and many others are in the literature. The related morphological and physiological problems can best be presented in the form of illustrative examples, one of which is the striatal neuropil (chap. vii), another the interpeduncular neuropil described in chapter xiv, and still another in the superficial ventrolateral neuropil of the peduncle, the area ventrolateralis pedunculi, which will next be described.

THE VENTROLATERAL PEDUNCULAR NEUROPIL

The "peduncle" in the restricted sense as here defined, together with the adjoining tegmentum, is the chief central motor pool of the skeletal muscles. Efferent fibers from it are among the first to appear in the upper brain stem in embryogenesis, descending in the primary motor path, which is the precursor of the fasciculus longitudinalis medialis, and activating the musculature of the trunk. A second efferent path, which appears very early, goes out directly to the periphery through the oculomotor nerve. The first sensory influence to act upon this pool comes from the optic tectum through the posterior commissure in the S-reaction stage. In subsequent stages it is entered by fibers from many other sources, including the basal optic tract (fig. 14) and the olfacto-peduncular and strio-peduncular tracts (fig. 6). In the adult animal, motor impulses issuing from this pool are probably concerned primarily with synergic activation of large masses of muscle, notably those concerned with locomotion and conjugate movements of the eyes. Control of the movements directly involved in seizing and swallowing food is believed to be chiefly in the isthmic tegmentum, and this apparatus matures later n ontogeny.

The chief synaptic connections of the great motor pool of the peduncle are in the intermediate, deep, and periventricular layers of neuropil. External to these is a ventrolateral band of dense neuropil extending from the root of the III nerve forward along the entire length of the peduncle. In former papers this has been termed "area lateralis tegmenti" and "nucleus ectomamillaris," but both these names now seem to me inappropriate ('42, p. 233). This band receives at the anterior end all the terminals of the large basal optic tract (figs. 14, 94) and, at the posterior end, terminals of the secondary and tertiary ascending visceral sensory and gustatory tracts (figs. 8, 23). The terminals of these two systems of fibers are intimately interlaced, and among them are dendritic arborizations of the underlying cells of the peduncle into which optic, olfactory, and visceral systems of nervous impulses converge and through which the combined effect is transmitted in the outgoing motor pathways. These seem to be the primary components of this neuropil, and to them are added axonic terminals from a wide variety of sources, the most notable of which are sketched in figure 23 (compare Necturus, '34, p. 103 and fig. 4).

The peduncular dendrites, which arborize within this neuropil, include some from the nucleus of the oculomotor nerve (fig. 24; '42, p. 275), so that here fibers of the basal optic tract coming directly from the retina may synapse with peripheral motor neurons—one of the rare examples of a two-neuron arc with only one synapse between the peripheral receptor and the effector, though even here there are at least two additional synapses in this arc within the retina. Other peduncular neurons may transmit retinal excitations downward through the ventral tegmental fascicles to all lower motor fields.

At its anterior end this neuropil is connected by fibers conducting in both directions with the dorsal (mamillary) part of the hypothalamus, as in Necturus ('34b, fig. 3), and also with its ventral (infundibular) part (fig. 23; '42, pp. 226, 227). This is an extension of the visceral-gustatory tract into the hypothalamus, and it may be that some of the visceral fibers pass without interruption through the peduncular neuropil to the hypothalamus, though this has not been satisfactorily demonstrated. In some fishes such a direct connection is evident and large; in mammals the course of the ascending visceral-gustatory path is still uncertain.

Other connections of this neuropil, as shown in figure 23, include terminals of the olfacto-peduncular tract from the anterior olfactory nucleus, probably the nervus terminalis (observed in Necturus) fibers from the tectum, pretectal nucleus, dorsal and ventral thalamus, and terminals of the fasciculus retroflexus (p. 262 and fig. 20). These terminals of fibers from surprisingly diverse sources are all closely interwoven with one another and with the terminal dendrites of peduncular neurons, a unique arrangement occurring, so far as known, only in Amphibia. I have indulged in the following speculations upon its possible physiological significance.

In the first place, it is clear that this curious tissue is either an undifferentiated primordium of a number of structures which are separately differentiated in more specialized brains, or else it is a retrograde fusion of several such structures. The former supposition seems more probable, for the phylogenetic history of one of its components is easily read ('25, pp. 443–49). In Necturus there are no cell bodies directly associated with this neuropil; in Amblystoma a few cells have migrated out of the gray layer to its border (fig. 24); and in the frog the optic component of the neuropil is separate and surrounded by a spherical shell of cell bodies, which is a true basal optic nucleus (Gaupp, '99, p. 54, fig. 19). This nucleus attains large size and complexity in some reptiles, as described by Shanklin ('33) and others. In the mammals it retains its individuality as the chief terminus of basal optic fibers, though other fibers of this system are rather widely spread in the surrounding tegmentum (Gillilan, '41). The history of the visceral and other components of the amphibian ventrolateral neuropil has not yet been written.

In my recent description ('42) of the optic system of Amblystoma, special attention was given to the central distribution of thick and thin fibers from the retina, and this was followed ('42, p. 295) by some speculations about its physiological significance. In development the thick fibers appear first, and the number of thin fibers is enormously increased in later stages, particularly at the time of metamorphosis. Thick fibers conduct more rapidly than thin fibers, and this time factor may play an important part in the central analysis of mixed retinal excitation.

Both thick and thin fibers end in the thalamus, pretectal nucleus, and tectum; and in each of these fields their terminals are mingled, not segregated. The optic fibers to the basal peduncular neuropil are moderately thick; upon retinal stimulation this field, accordingly, is the first to be activated, and its excitation of the underlying peduncular neurons may precede any influence upon these cells from the longer paths by way of the tectum and thalamus. Figure 22 shows in continuous lines the major optic tracts and the motor paths from the peduncle, and in broken lines some internuncial connections. The thick fibers which descend from the tectum and thalamus are myelinated; some of them are crossed in the posterior commissure and the commissure of the tuberculum posterius, and some are uncrossed. They connect primarily with the descending pathway in the ventral tegmental fascicles ($f.v.t.$). The thinner correlating fibers take other pathways, and they may make synaptic connection with peduncular neurons in any one or all of the four layers of the neuropil. The efferent path may be to lower motor centers through the ventral tegmental fascicles or to the muscles of the eyeball through the III nerve.

Figures 24 and 93 are diagrammatic transverse sections at the level of the III nuclei and the middle of the tectum, illustrating some of the tecto-peduncular connections. These fibers are all of medium or thin caliber and, for the most part, unmyelinated. The dendrites of the peduncular neurons ramify widely throughout the entire thickness of the brain wall, and few, if any, of them are in physiological relation with any single one of the afferent systems of fibers. Each of them may be activated by any or all of these systems. The only significant localized specialized tissue here seems to be in the neuropil, where the texture is different in the four layers and where all peduncular neurons spread their dendrites in all these layers. Since the movements which are activated from this motor pool are not disorderly convulsions, it is evident that the discriminative and well co-ordinated responses which follow its excitation are not ordered primarily by the arrangement in space of the motor elements. There are differences in the structural arrangements of the synaptic junctions, though these are not so pronounced as in most other animals, and there may be chemical and other factors yet to be determined. As I have elsewhere pointed out, a time factor can be recognized by physiological experiment, and its structural indicator can be identified histologically because large fibers have a faster rate of conduction than small fibers and the interpolation of synaptic junctions in a nervous pathway retards transmission.

In the structural setup before us it may be inferred that the first result of a retinal excitation is the activation of the entire tectum, pretectal nucleus, and dorsal thalamus through the thick myelinated fibers of the optic nerve and tract and also of the ventrolateral neuropil of the peduncle through the basal optic tract. This is presumably a generalized nonspecific effect, and it will come to motor expression, first, through the basal tract, for this is the shortest path. The resting state is changed to a state of excitation in both the peduncular gray and the peripheral musculature with which it is connected. This is immediately followed by volleys from the tectum, pretectal nucleus, and thalamus through the myelinated tecto- and thalamo-peduncular tracts; and this may contribute a spatial factor determined by the position of the exciting object in the visual field and the sector of the tectum upon which this local stimulus is projected. The first overt movement, accordingly, is an orientation of the body and the eyeballs with reference to the source of stimulus. After an appreciable time the smaller fibers from the retina deliver their volleys, and the small fibers of the correlating tracts are activated. These deliver to the peduncle, not unmixed or purely visual impulses but discharges, fired or inhibited, as the case may be, by the existing excitatory state of the correlating apparatus; and this, in its turn, is determined by numberless nonvisual features of the total situation, present and past. If, for instance, one of the visual or nonvisual components is fatigued, this will affect the pattern of tectal discharge.

In salamanders the delay between the preliminary orientation of the body and the consummation of the reaction may be, and commonly is, very long—a period of tension, which in a man we would call "attention" and which Coghill called the "regarding" reaction (p. 78; Coghill, '33, Paper XI, p. 334; '36). During this period an inconceivably complicated resolution of forces is in process within the central apparatus of adjustment. Inhibition plays an important role here, and it may be that the chief function of the ventrolateral superficial neuropil of the peduncle is control of the inhibitory phase of these activities comparable with that suggested in chapter xiv for the specific interpeduncular neuropil. Of the details of these adjustments our knowledge is scanty, but some hints may be gathered from urther examination of the structure involved; and here, to simplify the problem, we shall confine the discussion to the superficial peduncular neuropil.

This neuropil is interpolated "in series" in the visual-motor path of the basal optic tract, and this short circuit is connected "in parallel" with the longer visual-motor circuits by way of the tectum and thalamus. The latter are very much larger and more complicated, and they evidently comprise the major part of the apparatus by which behavior is regulated by visual experience. The basal optic system seems to be ancillary to the major system and to be related with it in two quite different ways, first, as a nonspecific primary activator, as already explained, and, second (and subsequently), as a device for modifying or "inflecting" (to borrow Arnold Gesell's expression) the standard patterns of behavior by intercurrent influence of present activity in other fields, including, perhaps, conditioning and the individuation of hitherto unaccustomed types of response. This second feature involves further consideration of the nonvisual components. These have already been listed, and some of them are shown in figure 23.

The largest components of this peculiar neuropil are the optic and the visceral-gustatory systems, and their fibers end exclusively here. This is why these are regarded as the primary components; all other systems of entering fibers are spread more or less widely also in the deeper layers of peduncular neuropil. In the visceral-gustatory system this neuropil is inserted in series in the ascending pathway toward the hypothalamus, and there is probably a parallel system of conduction with no synapse in the peduncle (fig. 8). As seen in this figure, peduncular neurons activated from this neuropil send axons around the tuberculum posterius into the hypothalamus; and there is a return path from both dorsal (mamillary) and ventral (infundibular) parts of the hypothalamus to this neuropil and contiguous parts of the peduncle (figs. 18, 23). All visceral-gustatory influences which reach the hypothalamus are, accordingly, previously modified or "inflected" in the superior visceral nucleus or in the peduncle. The excitatory state of the hypothalamus, in turn, affects all activity in the peduncle.

This sort of circular activity is everywhere present in these urodele brains, working through the diffuse periventricular and deep neuropil and in many places also through specially differentiated tracts (p. 76). This is the primordial apparatus of integration, and in its more differentiated form it is part of the apparatus by which individuated partial patterns of local reflexes are kept in appropriate relationship with the larger total patterns of behavior.

In a recent survey ('44a) of the optic and visceral nervous circuits here under consideration, several different patterns of linkage of the component units were listed. These include direct activation of eye muscles through the basal optic tract, similar activation of the skeletal musculature, indirect activation of either or both of these systems of muscles by visual stimuli through the tectum and thalamus, direct activation of either or both systems of muscles through the visceral-gustatory path, indirect visceral-gustatory action by way of the superior visceral nucleus, with or without intercurrent influence of the tectum upon this nucleus by way of the tecto-bulbar tract, and the variable effects of concurrent discharge of many other systems of fibers into each of the centers of adjustment involved. Since each of these patterns of linkage is complex and the structural units themselves are not simple, it is evident that in actual performance the number of ways in which the known units may be combined and recombined is practically unlimited. The simplest possible activity of stimulus-response type involves a central resolution of forces in an equilibrated dynamic system of inconceivable complexity. Oversimplification of the problem will not hasten its solution.

The preceding description illustrates the relations of a specific field of specialized neuropil to other parts of the brain with known functions. These visible connections exhibit a structural arrangement which can be interpreted as putative pathways of transmission of the several components of a complex action system, some concerned with stable reflex patterns and some with more labile individually acquired components. The relations of these two classes of components of the action system to each other and of local units to the integrated whole present the major problems of neurology.

CHAPTER IV

REGIONAL ANALYSIS

SINCE the brain of Amblystoma presents a generalized structure which is probably close to the ancestral type from which all more highly specialized vertebrate brains have been derived, the salient features of internal organization are here summarized in schematic outline. The accompanying diagrammatic figures 1–24 give the necessary topographic orientation, and the details may be filled in by reference to the corresponding sections of Part II. What is here described may be regarded as the basic organization of the brains of all vertebrates above fishes, that is, the point of departure from which various specialized derivatives have been differentiated. Amblystoma possesses the equipment of sensory and motor organs typical for vertebrates at a rather low level of specialization and in evenly balanced relations. All the usual systems are present, and none shows unusual size or aberrant features. The great lateral-line system of sense organs so characteristic of fishes is preserved, though somewhat reduced after metamorphosis. On the motor side the organs of locomotion and respiration have advanced from the fishlike to the quadrupedal form, but in very simple patterns. In early phylogeny the specialization of the motor systems seems to lag behind that of the sensory systems because the aquatic environment of primitive forms is more homogeneous than that of terrestrial animals, and, accordingly, fewer and simpler patterns of behavior are needed.

Our search in this inquiry is for origins of human structures and for an outline of the history of their evolution. From this standpoint it is evident that in the central nervous systems of all vertebrates there is a fundamental and primary difference between the cerebrum above and the rhombencephalon (and spinal cord) below a transverse plane at the posterior border of the midbrain (for further discussion of this see chaps. viii and xiii).

The spinal cord and rhombic brain contain the central adjustors of the basic vital functions—respiration, nutrition, circulation, reproduction, locomotion, among others. This apparatus is elaborately organized in the most primitive living vertebrates, as also no doubt it must have been in their extinct ancestors. The cerebrum, on the other hand, except for the olfactory component, is a later acquisition. This is suggested by what is seen in Amphioxus and by the retarded development of the cerebrum in all vertebrate embryos, as illustrated especially clearly in the early fetal development of the opossum.

At an early (and unknown) period of vertebrate ancestry a pair of eyes was differentiated. These and the olfactory organs are the leading distance receptors, and as such they gave to the vertebrate ancestors more information about their surroundings and hence greater safety in moving about freely. The nose and eyes, with the associated oculomotor apparatus, early assumed the dominant role in the recognition of food, mates, and enemies, and their cerebral adjustors were enlarged accordingly. The contact receptors are adequate for sedentary, crawling, or burrowing ancestors, and here the response to stimulation follows immediately. But, as Sherrington long ago pointed out, in a free-swimming animal there is a time lag between reception of the stimulus from a distant object and the consummation of the response. The pregnant interval between the anticipatory and consummatory phases of the reaction gives the clue to an understanding of the entire history of forebrain evolution. During this interval there is a central resolution of forces, which eventuates in appropriate behavior; and, with increasing complication of patterns of behavior, this central apparatus of adjustment assumes more and more structural complexity and physiological dominance over the entire bodily economy (chap. vii).

The details of these internal connections are not relevant here. It suffices to present two summaries, one in this chapter in topographic arrangement by regions from spinal cord to olfactory bulbs as conventionally described and one in the next chapter on a different plan, i.e., an arrangement in longitudinal zones which are functionally defined. For the present purpose it is convenient to recognize seventeen subdivisions of the central nervous system, each of which is characterized by special physiological activities, though these activities are not localized here exclusively. This subdivision might be carried further into detail indefinitely. Numbers 2–6 in the following paragraphs are in the rhombic brain; the others are in the cerebrum.

THE SUBDIVISIONS, SPINAL CORD TO PALLIUM

1. THE SPINAL CORD

The spinal cord is not described in this report except for some features closely related to the brain, to which reference is made in the next paragraph. The cord segments are organized for the regulation of local reflexes of the limbs and the integration of these reflexes with one another and with the action of the trunk musculature, as in ordinary locomotion.

2. THE BULBO-SPINAL JUNCTION

The sector of the bulbo-spinal junction includes the upper segments of the spinal cord and the lower part of the medulla oblongata. It is the first center of correlation to become functional in embryonic development (Coghill, '14, Paper I). Its dorsal part around the calamus scriptorius receives fibers from the entire sensory zone of the bulb and cord, so that this gray of the funicular and commissural nuclei is a general clearing-house for all exteroceptive, proprioceptive, and visceral functions of the body except vision and olfaction. Here these functions are integrated in the interest of control of posture, locomotion, visceral activity, and other basic components of mass-movement type. Some of these connections are shown diagrammatically in figures 3, 7, 8, 87; for details and discussion see chapter ix and a recent paper ('44b).

Efferent fibers from the dorsal nuclei are directed spinalward and forward. Most of the latter connect with motor nuclei of the medulla oblongata; some go farther forward to the cerebellum, tectum, and thalamus; and there is a strong, ascending visceral-gustatory tract to the isthmus and peduncle. The motor zone of this sector is occupied chiefly by fibers of passage. The moderately developed gray substance includes motor neurons for muscles of the neck region, for the tongue muscles, and for special visceral motor elements of the accessorius component of the vagus.

3. MEDULLA OBLONGATA

The medulla oblongata, or bulb, includes all the stem between the isthmus and the calamus scriptorius except the cerebellum, there being no pons. Its dorsal field receives all sensory fibers from the head except the optic and olfactory, fibers from lateral-line organs widely distributed over the body, and general visceral sensory fibers chiefly by way of the vagus. The general visceral sensory and gustatory root fibers are segregated from the others in the fasciculus solitarius; and this group has its own system of secondary fibers, which converge into the visceral motor nuclei of the medulla oblongata and spinal cord. There is also a strong ascending secondary visceral tract ($tr.v.a.$) to the isthmus, peduncle, and hypothalamus, through which all cerebral activities may be influenced by visceral and gustatory functions (fig. 8 and chaps. x, xi).

REGIONAL ANALYSIS

The other afferent fibers of the V to X cranial nerves, upon entrance into the brain, are fasciculated according to the functional systems represented, as outlined in the next chapter and shown in figures 7, 9, 88, 89, 90. The general cutaneous, lateral-line, and vestibular fibers are arranged in a series of fascicles bordering the external surface; the visceral sensory and gustatory fibers are assembled in a single deeper bundle, the fasciculus solitarius. The marginal fascicles of root fibers are arranged from ventral to dorsal in the following order: general somatic sensory (chiefly cutaneous), vestibular, and, dorsally of these, five or six fascicles of fibers of the lateral-line roots of the VII and X nerves. The details of this arrangement are variable within the species and from species to species of urodeles. The fascicles of vestibular and lateral-line roots, together with the underlying gray and the intervening neuropil, comprise the area acusticolateralis. The dorsal part of this, which receives only roots of the lateral-line nerves, is lobulated on the ventricular side.

Most of these root fibers divide into ascending and descending branches, and each fascicle spans the entire length of the medulla oblongata. Some of the general cutaneous and vestibular fibers extend far down into the spinal cord and upward into the cerebellum (figs. 3, 7). Terminals and collaterals of all these fibers end in a common pool of neuropil, from which secondary fibers go out to effect local connections in the medulla oblongata, to enter the cerebellum, to descend to the spinal cord, and to ascend in the lemniscus systems to the tectum and thalamus. The physiological specificity of the root fibers is largely, though not entirely, obliterated at the first synapse in the neuropil of the sensory field, in sharp contrast with the mammalian arrangement (chap. xi).

From this arrangement of the sensory systems of fibers and their central secondary connections it is clear that the bulbar structure is so organized as to facilitate mass movements of total-pattern type, which may be activated by any one of the exteroceptive or proprioceptive systems or by any combination of these. There is some provision here for local reflexes of the muscles of the head, but the structure indicates that most of these are patterned from higher centers.

The central receptive field of the visceral-gustatory system is well segregated from that of the general cutaneous and acousticolateral systems; and this is the structural expression of the fundamental distinction between the visceral and the somatic sensory physiological systems, to which further reference is made on pages 67 and 83.

Otherwise, there is little histological evidence of precise localization of function in the medulla oblongata. The visceral and somatic sensory fields are cross-connected within the sensory zone, and they converge into a common sensory field at the bulbo-spinal junction. Proprioceptive adjustments are made throughout the spinal cord, medulla oblongata, cerebellum, and tectum; and each of these regions evidently plays a different role in the adjustments. Arcuate fibers connect all parts of the sensory zone with the motor zone of the same and the opposite side, and many of these divide into long descending and ascending branches, thus activating extensive areas of the intermediate and motor zones.

The motor field of the medulla oblongata and the intimately related reticular formation contain the complicated apparatus by which the nuclei of the motor nerves are so interconnected as to act in groups, each of which may execute a series of co-ordinated actions in patterns determined by these connections. The tissues of the motor tegmentum, which effect this analysis of motor performance, are so intricately interwoven that it has not been possible to recognize the components of the several functional systems, and further analysis of this field is desirable.

4. CEREBELLUM

The cerebellum is small and very simply organized, but the chief structural features of the mammalian cerebellum are present except the pontile system, which is totally lacking. The urodele cerebellum consists of three major parts: (1) the median body, activated from the spinal cord, trigeminal nerve, and tectum (figs. 1, 3, 10); (2) the lateral auricles, which are enlargements of the anterior ends of the sensory zones of the medulla oblongata (figs. 7, 91); and (3), ventrally of the body of the cerebellum, a nucleus cerebelli, which is the primordium of the deep cerebellar nuclei of mammals (figs. 10, 32, 91).

This analysis of cerebellar structure is based on the comprehensive studies of Larsell ('20–'47) and Dow ('42), whose descriptions of Amblystoma, published in 1920 and 1932, I have confirmed in all respects. It should be noted that my definition of the amphibian auricle includes more than Larsell's, for he includes in this structure only the vestibular and lateral-line components. I find that these terminals and the related field of neuropil are so intimately related with the terminals of the trigeminus, the visceral-gustatory system, and lemniscus fibers that their segregation is not practical anatomically. The auricle, accordingly, is here regarded as the common primordium of several structures which in higher animals are diversely specialized for different functions. The most notable of these are the superior or pontile nucleus of the trigeminus and the floccular part of the flocculonodular lobe of the cerebellum. The primordia of these structures are clearly evident, and the history of their further differentiation in higher animals has been written.

Efferent fibers of tractus cerebello-tegmentalis leave all parts of the cerebellar complex for the underlying gray; and one fascicle of these—the brachium conjunctivum—passes forward to a ventral decussation and distributes its fibers to the isthmic tegmentum of both sides (figs. 10, 71). No primordium of the nucleus ruber or of the inferior olive has been recognized.

This primitive cerebellum exhibits the typical vertebrate pattern in very instructive form, with localization of the vestibular system laterally and the other systems medially. It is an appendage added to the basic sensori-motor systems; it supplements them, not as an aid in determining the pattern of performance, but to insure efficient execution. In species in which it is greatly enlarged, it contains enormous reserves of potential nervous energy, which is released during motor activity to reinforce and stabilize the operation of the effectors. For additional details see chapters x and xii.

5. ISTHMUS

The isthmus is unusually large in urodeles and is clearly circumscribed from surrounding parts. Dorsally it is small, containing in and near the superior medullary velum a special segment of the mesencephalic V nucleus and probably other peripheral connections through the nerves of the chorioid plexuses and meninges. Below this there is the superior visceral-gustatory nucleus (figs. 2B, 8, 23, 34). The nucleus isthmi, which is large in the frog, is here undifferentiated. The ventral part of the isthmus is very large, containing the nucleus of the IV nerve and a mass of tegmental cells. This isthmic tegmentum is interpolated between the primary sensori-motor systems of the medulla oblongata and midbrain, and it serves as an intermediary between them. There is a large central nucleus of small cells which receives fibers from a wide variety of correlation centers of intermediate-zone type. These enter by all the dorsal tegmental fascicles and by several other paths (figs. 16, 17, 21). This nucleus is enveloped by a group of larger cells, which is continuous posteriorly with similar tegmental cells of the region of the trigeminus (figs. 29, 30, 91). The complex as a whole is believed to have two chief functions: (1) Here are organized the patterns of the local reflexes of the musculature of the head, particularly those concerned with feeding. (2) The small-celled central nucleus is a special differentiation of the periventricular gray, which serves, in addition to the specific functions just mentioned, a more general, nonspecific, totalizing function; that is, it is a part of that integrating apparatus which appears in mammals as the dorsal tegmental nucleus and the related fasciculus longitudinalis dorsalis of Schütz (p. 208). The details of structure are given in chapter xiii.

The isthmic tegmentum occupies a strategic position between the primitive bulbo-spinal mechanisms and the higher cerebral adjustors; it plays a major role both in the patterning of local reflexes and in the integration of all bodily activities. This mass of tissue, which in urodeles is at a low level of differentiation, in higher animals is split up and distributed so that in mammals the identity of the isthmic tegmentum as an anatomical entity is lost in the adult brain, though the isthmic sector is plainly marked in the early embryonic stages.

6. INTERPEDUNCULAR NUCLEUS

The interpeduncular nucleus also is unusually large in urodeles. It is not interpeduncular but interisthmic, extending from the fovea isthmi back to the level of the V nerve roots. The histological texture is extraordinary. A well-defined, trough-shaped column of cells borders the ventral angle of the ventricle, with dendrites extending downward through the ventral commissure, to arborize with tufted endings in a ventromedian band of neuropil (figs. 65, 66, 82, 83, 91). The axonic components of this interpeduncular neuropil come from various sources: (1) terminals of the fasciculus retroflexus, which take the form of a flattened spiral (fig. 50); (2) terminals of tr. tegmento-interpeduncularis from small cells of the overlying tegmentum with tufted endings, which join with the dendritic tufts of the interpeduncular nucleus to form small glomeruli (figs. 60–66, 84); (3) collaterals of thick fibers of tr. tegmento-bulbaris from the large cells of the tegmentum with similar tufted endings in the glomeruli (fig. 68); (4) collaterals of tr. interpedunculo-bulbaris, which also enter glomeruli (figs. 83, 84); (5) terminals of tr. mamillo-interpeduncularis with dispersed free endings (figs. 60, 61); (6) similar terminals of tr. olfacto-peduncularis (fig. 59); (7) less numerous terminals from several other sources. The slender, unmyelinated axons of the interpeduncular cells branch freely in the interpeduncular neuropil and continue from the nucleus in two strands (figs. 83, 84). The ventral interpedunculo-bulbar tract descends beyond the nucleus for an undetermined distance in the lip of the ventral fissure. The dorsal tract descends dorsally of the fasciculus longitudinalis medialis and turns laterally to end in wide arborizations in the tegmentum as far back as the IX nerve roots. Associated with these dorsal fibers are interpedunculo-tegmental fibers, which end in the neuropil of the isthmic tegmentum. The dorsal and interpedunculo-tegmental fibers are regarded as comparable with the isthmic and bulbar parts of the mammalian f. longitudinalis dorsalis of Schütz.

The physiological problems suggested by this peculiar structure are puzzling. In the light of such scanty experimental evidence as we possess, I have ventured to suggest that the interpeduncular complex provides both activating and inhibitory components of reflex and general integrative activities, the actual patterns of which are elsewhere determined.

Topographically, this nucleus lies in the motor zone, but its functions clearly are of intermediate-zone type. It is present in all vertebrates at the anteroventral border of the isthmus, that is, at the boundary between cerebrum and rhombencephalon. Most of its afferent fibers come from the cerebrum, and evidently it serves chiefly as an intermediary adjustor between the sensory and intermediate zones of higher levels and the motor zone of the rhombic brain (for details see chap. xiv).

At this point in our analysis we cross the boundary between rhombencephalon and cerebrum. The radical differences in structure and physiological properties of these two chief parts of the brain are masked and in large measure overruled, especially in higher animals, by ascending and descending connectives, of which the interpeduncular system is a typical illustration.

7. TECTUM AND PRETECTAL NUCLEUS

The tectum and the pretectal nucleus, as sectors of the sensory field, together with the dorsal thalamus, form a physiological unit within which all exteroceptive sensory systems are integrated in the interest of cerebral control of all lower sensori-motor systems involved in the operation of the skeletal musculature. This unit is intimately related with the cerebral peduncle and ventral thalamus. In the most primitive vertebrates and in early embryonic stages of all vertebrates, these structures might appropriately be united as a middle subdivision of the brain, which serves as the dominant center of cerebral control of all somatic activities. But in the adult animal the parts of this natural subdivision have so many distinctive connections and physiological properties that it seems preferable to treat them separately.

In vertebrates below the mammals the tectum opticum is the chief central end-station of the optic nerve; and, since the eyes are the chief distance receptors in most of these animals, fibers of correlation of all other sensory systems concerned with external adjustment naturally converge to this station. The tectum, accordingly, becomes the dominant adjustor of all exteroceptive systems. The tectum mesencephali of Amblystoma has a larger optic part—the superior colliculus—and a small, poorly differentiated nucleus posterior—the primordial inferior colliculus. The latter is interpolated between the tectum opticum and the cerebellum, and its connections suggest that its most primitive functions are proprioceptive. It receives a small primordial lateral lemniscus and evidently also serves such generalized auditory functions as this animal possesses (chap. xv). The development of the optic nerve and adult tectal structure and connections have been described in detail ('41, '42). Chapter xvi is devoted to the visual system; for the arrangement of the mesencephalic nucleus and root of the V nerve see page 140 and figure 13.

Optic and lemniscus tracts and smaller numbers of fibers from various other sources all terminate in a broad sheet of intermediate neuropil, which is spread through the entire tectum and is nearly homogeneous in texture (figs. 11, 93). The tectum is not definitely laminated, though separation of the layers, which are conspicuous in the frog, is incipient. Fibers diverge from it in all directions (figs. 12, 18, 21–24, 93). It is inferred from this structure that movements activated directly from the tectum are of total-pattern type. Such local visual reflexes as this animal possesses are probably patterned elsewhere—in the thalamus and dorsal and isthmic tegmentum. Conditioning of reflexes is probably effected in these areas and perhaps also in the ventrolateral peduncular neuropil (p. 38). Experiments upon Triturus and anurans (Stone and Zauer, '40; Sperry, '43, '44, '45, '45b) demonstrate anatomical projection of retinal loci upon the tectum opticum. This is true also in Amblystoma (Stone, '44; Stone and Ellison, '45), though the nervous apparatus employed has not been described.

The pretectal nucleus (figs. 2B, 35, 36, *nuc.pt.*) receives fibers directly from the retina and from the tectum, habenula, and cerebral hemisphere. Its efferent fibers go to the tectum, thalamus, hypothalamus, and cerebral peduncle (figs. 11, 12, 14, 15, 16, 22, 23). Its functions are unknown, but, by analogy with mammals, this may be part of the apparatus for regulation of the intrinsic musculature of the eyeball. Doubtless other functions are represented also. This area is the probable precursor of the mammalian pulvinar and neighboring structures.

The thalamus receives many fibers from the retina, and it is broadly connected with the tectum by uncrossed fibers passing in both directions in the brachia of the superior and inferior colliculi (figs. 11, 12). There are also systems of tecto-thalamic and hypothalamic and thalamo-hypothalamic tracts which decussate in the postoptic commissure; some of these crossed fibers take longer courses to reach the peduncle and isthmic tegmentum (figs. 12, 15). This intimate thalamo-tectal relationship is radically changed in higher animals, where the thalamo-cortical connections are highly elaborated.

8. DORSAL THALAMUS

I have separated the dorsal thalamus into three sectors: (1) anteriorly, the small nucleus of Bellonci of uncertain relationships; (2) a well-defined middle part, an undifferentiated nucleus sensitivus, which is the primordium of most of the sensory nuclei of the mammalian thalamus; and (3) a vaguely delimited posterior sector, which apparently contains the undifferentiated primordium of both lateral and medial geniculate bodies (chap. xvii).

The middle and posterior sectors receive numerous terminals and collaterals of the optic tracts, terminals of the general bulbar and spinal lemnisci, and, through the brachia of the superior and inferior colliculi, these sectors are broadly connected with the tectum by fibers running in both directions. There is a similar, but much smaller, connection with the habenula.

From the middle sector a small, well-defined tr. thalamo-frontalis goes forward to the hemisphere (figs. 15, 71, 72, 95, 101, *tr.th.f.*); this is the common primordium of all the thalamo-cortical projection systems of mammals, though here few, if any, of its fibers reach the pallial area. Other efferent fibers go to the ventral thalamus, hypothalamus, peduncle, and tegmentum. These thalamic reflex connections antedate in phylogeny the thalamo-cortical connections, and they persist in mammals as an intrinsic paleothalamic apparatus, an important part of which is the periventricular thalamic contribution to the f. longitudinalis dorsalis of Schütz. The largest pathways of efferent discharge from the dorsal thalamus go backward to the peduncle and tegmentum by both crossed and uncrossed tracts (figs. 15, 18, 21, 23). The peduncular connection puts all the primary systems of total-pattern type under some measure of thalamic control. The connections with the dorsal and isthmic tegmentum probably co-operate in the patterning of local reflexes, particularly supplying the visual component of the feeding reactions.

9. PEDUNCLE

The "peduncle" described here is not the equivalent of the human cerebral peduncle (p. 21). The intimate relations of this field with the overlying tecto-thalamic field have been commented upon in the preceding paragraphs. This ventral field is a well-defined column of cells, differentiated at the anterior end of the basal plate of the embryonic neural tube. It is the head of the primary motor column (of Coghill), which in all vertebrates, from early embryonic stages to the adult, contains the nucleus of the oculomotor nerve and a much larger mass of nervous tissue, which activates the primitive mass movements of locomotion. It maintains cerebral control of the lower bulbo-spinal segments of the latter systems, and some other motor functions also are represented here. Into it fibers converge from all other parts of the cerebrum (figs. 12, 14, 15, 17, 18, 20–24), and from it efferent fibers go out in four groups: (1) Ventromedial tracts go to the medulla oblongata and spinal cord. The longest of these fibers are in the f. longitudinalis medialis (fig. 6). (2) The oculomotor nerve supplies intrinsic and extrinsic muscles of the eyeball (figs. 22, 24). Associated with these peripheral fibers are central connections with the nuclei of the IV and VI nerves, so arranged as to execute conjugate movements of the eyes. The details of the apparatus employed are unknown. (3) Visceral sensory and gustatory fibers enter the peduncle (fig. 8), and with these are related efferent fibers to the hypothalamus and to lower levels of the motor zone. The pathways taken by the latter in the amphibian brain have not been clarified. (4) From both ventral thalamus and peduncle, fibers diverge to various surrounding parts, notably to the hypothalamus and isthmic tegmentum. These probably provide for co-ordination of various local reflex activities with the basic peduncular functions.

At the ventrolateral border of the peduncle there is an area of superficial neuropil, which is the terminus of the basal optic tract, large secondary and tertiary visceral-gustatory tracts, some fibers of the f. retroflexus, and fibers from several other sources (figs. 22, 23,

24). This is an undifferentiated primordium of the basal optic nucleus and some other structures of the mammalian brain (pp. 35, 221). It is related with the olfacto-visceral functions of the hypothalamus and probably also with conditioning of the fundamental peduncular activities.

10. VENTRAL THALAMUS

There are anterior and posterior sectors of the ventral thalamus which differ in embryological origin (p. 239) and in certain connections of intermediate-zone type. Both sectors are here included in the motor zone because their chief efferent connections resemble those of the "peduncle," of which the posterior part is physiologically an anterior extension. The ventral thalamus is the primordium of the motor field of the mammalian subthalamus. The anterior sector contains a nucleus specifically related to the stria medullaris and amygdala and, above this, the eminentia thalami, which is a bed-nucleus of tracts related to the primordium hippocampi (chap. xviii; figs. 16, 17, 19, 20, 96).

The ventral thalamus and peduncle of urodeles form a single massive column, which is anatomically well defined. The specialized structures derived from it in mammals are dispersed among large masses of tissue of more recent phylogenetic origin; but in the human brain this region still retains cerebral control of the primordial coordinated movements of the musculature of the eyeballs and of the trunk and limbs.

11. THE RETINA AND ITS CONNECTIONS

In early embryonic stages the retina is part of the brain, and, as development advances, it absorbs much of the diencephalic sector of the early neural tube. This precociously accelerated development results from the dominance of vision in exteroceptive adjustment from the time that the larva begins to feed. For further details of this development and of the organization of the visual-motor apparatus see chapter xvi.

12. HABENULA

As described in chapter xviii, this specialized part of the epithalamus receives fibers from almost all parts of the telencephalon and diencephalon and from the tectum (fig. 20). The habenular commissure connects the two habenulae, and it also contains two commissures of pallial parts of the hemispheres—commissura pallii posterior and com. superior telencephali. The chief efferent path from the habenula is the f. retroflexus (chap. xviii), which terminates in the cerebral peduncle and interpeduncular nucleus. In the brains of lower vertebrates the habenular complex is one of the most widely spread and physiologically important members of the central correlating apparatus. Its primary function seems to be to integrate the activities of all parts of the brain that are under olfactory influence with the exteroceptive functions of the tectum and thalamus in the interest of higher cerebral control of the feeding reactions of the skeletal muscles.

13. HYPOTHALAMUS

In the large preoptic nucleus and hypothalamus, olfactory connections dominate the picture, as they do in the habenular system; but here the nonolfactory functions represented are interoceptive instead of exteroceptive. All parts of the cerebral hemisphere are connected with the hypothalamus by fibers passing in both directions in the medial forebrain bundles (p. 273), stria terminalis (p. 255), and fornix (p. 254) systems. The visceral-gustatory afferent paths are shown in figure 8. Large tracts from the thalamus and tectum also end here, so that all kinds of sensory experience of which the animal is capable are represented in the hypothalamus. This experience is here organized in terms of visceral responses. The efferent tracts go to the peduncle, tegmentum, interpeduncular nucleus, and descending fibers in the deep neuropil which are precursors of the f. longitudinalis dorsalis of Schütz. There is a large tract to the hypophysis for nervous control of endocrine activity. There is also evidence that some neuro-endocrine functions are performed in the hypothalamus itself (Scharrer and Scharrer, '40). The structure of the hypothalamus has been described ('21a, '27, '35a, '36, '42, and in the embryological papers, '37–'41). It is similar to that of Necturus, of which more detailed descriptions have been published ('33b, '34b). For the composition of the postoptic commissure see chapter xxi.

14. STRIO-AMYGDALOID COMPLEX

The primordial corpus striatum occupies the thickened ventro-lateral wall of the cerebral hemisphere and, like all the rest of the hemisphere, is under olfactory influence. This is stronger at its anterior and posterior ends. Anteriorly, it is divided by a striatal sulcus into dorsal and ventral parts (fig. 99), and posteriorly it is much enlarged as the amygdala (figs. 1, 96, 97), which has the typical mammalian connections (fig. 19).

The ventral sector of the anterior olfactory nucleus is interpolated between the olfactory bulb and the corpus striatum, as in Necturus (figs. 6, 86). This is the primordium of the tuberculum olfactorium ('27, p. 290), which is enormously enlarged in the lungfishes (Rudebeck, '45). Posteriorly of this nucleus is a poorly defined field, which embraces the ventral angle of the lateral ventricle and is regarded as the probable precursor of the head of the caudate nucleus (fig. 99) It is intimately connected with the rest of the striatum and the septum. The chief efferent path is the olfacto-peduncular tract to the dorsal hypothalamus, ventral border of the peduncle, and interpeduncular nucleus.

The middle sector of this complex is the undifferentiated primordium of the mammalian lentiform nucleus, as shown by its fibrous connections. It is characterized by very dense, sharply circumscribed neuropil in the white substance (p. 96; figs. 98, 99, 108, 109) and has the typical striatal connections with the overlying pallium and the thalamus, including a small sensory projection tract from the dorsal thalamus (fig. 15, *tr.th.f.*). The chief efferent path is the lateral forebrain bundle (f. lateralis telencephali, *f.lat.t.*), which contains strio-thalamic, strio-peduncular, and strio-tegmental fibers comparable with the corresponding components of the mammalian extrapyramidal system. There is also a strio-tectal and strio-pretectal connection (figs. 11, 14, 101). The separation of the lentiform nucleus into globus pallidus and putamen is incipient (p. 97).

15. SEPTUM

The septum complex occupies the ventromedial wall of the hemisphere between the anterior olfactory nucleus and the lamina terminalis and hippocampal area (figs. 4, 98, 99). Its position and connections are similar to those of mammals. It is directly connected with the olfactory organ by the nervus terminalis, and it receives fibers from the olfactory bulb, anterior olfactory nucleus, pallium, and hypothalamus. The chief efferent paths are by the medial forebrain bundle (f. medialis telencephali, *f.med.t.*) and to the overlying pallium by the f. olfactorius septi ('27, p. 291). There is also a broad connection across the ventral aspect of the hemisphere with the amygdala and the piriform area, the diagonal band of Broca (figs. 96, 97, 98, *d.b.*), and a connection with the habenula by tr. olfacto-habenularis anterior and tr. septo-habenularis (chap. xviii).

16. OLFACTORY BULB AND ANTERIOR OLFACTORY NUCLEUS

The olfactory bulb is very large, embracing the anterior end of the lateral ventricle and extending back in the lateral wall for about half the length of the hemisphere (figs. 1, 3, 4, 85, 100, 105, 110; '24b, '27). All peripheral olfactory fibers end in the glomeruli of the bulb. Fibers of the second order pass in large numbers to the anterior olfactory nucleus, and they enter longer olfactory tracts with wide distribution (fig. 6). The olfactory tracts are mixtures of fibers from the bulb and the anterior nucleus, as in Necturus ('33b, figs. 6–16). They reach all parts of the cerebral hemisphere, the habenula, and the hypothalamus. Some of these decussate in the ventral part of the anterior commissure and some in the habenular commissure (com. superior telencephali).

The histological texture of the olfactory bulb is more differentiated than that of Necturus ('31), but more generalized than that of higher vertebrates. I have contrasted this with the mammalian pattern and added a theoretic interpretation of probable differences in physiologic properties of the tissue ('24b). In brief, this tissue is interpreted as illustrating several transitional stages in the differentiation of polarized nervous elements from an unpolarized or incompletely polarized matrix. In Necturus ('31) the transitional character of this tissue is still more clearly evident. The granule cells, in particular, give no structural evidence of physiological polarity, i.e., of differentiation of dendrites from axon, though the connections of these cells in Amblystoma suggest that they have a transient and reversible polarity. In connection with this description ('24b, pp. 385–95) there are some speculations regarding possible phylogenetic stages in the differentiation of permanently polarized neurons from an unspecialized nonsynaptic nerve net or neuropil.

In Amblystoma there is a moderately developed accessory olfactory bulb, but no other evidence of local specialization in the primary olfactory center. (There are hints of this in some mammals, e.g., the mink, Jeserich, '45, and references there cited). In 1921, I described the peripheral and central connections of the accessory bulb of Amblystoma and compared these with the more specialized structures of the frog. The anatomical connections there described are, I believe, correct, but the theoretic interpretation of the relationships in vertebrates generally between the vomeronasal organ, accessory bulb, and amygdala is less secure and awaits confirmation or correction.

The anterior olfactory nucleus is a zone of relatively undifferentiated cells interpolated between the bulbar formation and the more specialized areas posteriorly of it (figs. 6, 86B and C, 105, 109; '27, p. 288). In higher animals much of this tissue seems to be specialized and added to the adjoining fields ('24d). A very large proportion of the fibers of the olfactory tracts, arising from both the bulbar formation and the anterior nucleus, are assembled in a dense superficial sheet of fibers in the medial sector of the anterior olfactory nucleus, which I have named the "fasciculus postolfactorius" (figs. 5, 100, 105, 110, *f.po.*). Here these fibers take a vertical course and then are distributed to all the olfactory tracts. In chapters vii and xix there is further discussion of the significance of the olfactory system in the morphogenesis of the hemisphere.

17. PALLIUM

The pallial part of the hemisphere can be distinguished from the stem part, though there is no laminated cortex. There are three sectors (figs. 96–99)—the dorsomedial primordium hippocampi (*p.hip.*), the dorsolateral primordium piriforme (*p.pir.*, or nucleus olfactorius dorsolateralis, *nuc.ol.d.l.*), and between these a dorsal sector of uncertain relationships (*p.p.d.*). The gray, as elsewhere in these brains, is confined to a thick periventricular layer except in the hippocampal sector, where the cell bodies are dispersed through the entire thickness of the wall and are imbedded in dense neuropil. This is evidently a first step toward differentiation of superficial cortex. The homologies of the hippocampal and piriform sectors with those of mammals are clear, as shown by substantially similar nervous connections. Further discussion will be found in chapter vii and the references there given.

THE COMMISSURES

Throughout the length of the central nervous system all parts of the two sides are broadly connected by systems of commissural and decussating fibers. These are in two series, dorsally and ventrally of the ventricles. Their composition is summarized in chapter xxi, with references to more detailed descriptions. In the aggregate they make provision for the co-ordinated action of the motor organs on right and left sides of the body.

CONCLUSION

The preceding outline of a regional analysis is framed in very general terms. The evidence upon which it is based is assembled in Part II of this work and references there given. This evidence, though far from complete, is regarded as adequate for the anatomical arrangements described. The physiological inferences drawn from these arrangements and the general theory expressed in the following chapters rest on a less secure basis. The correctness of these conclusions can be tested experimentally, and the hope that this will be done has motivated the labor expended upon this program of histological study.

CHAPTER V

FUNCTIONAL ANALYSIS, CENTRAL AND PERIPHERAL

THE brain of Amblystoma performs three general classes of functions, with corresponding local differentiation of structure. We recognize, accordingly, three zones on each side—a dorsal receptive or sensory zone; a ventral emissive or motor zone; and, between these and infiltrating them, an intermediate zone of correlation and integration.

THE LONGITUDINAL ZONES

Figures 4 and 5 are sketches of longitudinal sections of the cerebrum of adult Amblystoma tigrinum to illustrate the areas included in the motor and sensory zones as here arbitrarily defined. The zones of the medulla oblongata as seen in transverse section are shown in figure 9. The sensory zone includes those parts of the brain which receive afferent fibers from the periphery, together with more or less closely related tissue of correlation. The motor zone includes those parts from which efferent fibers go out to the periphery, together with related apparatus of motor co-ordination. Both these zones contain some areas which, though not directly connected with the periphery, are nevertheless primarily concerned with specific types of sensory or motor adjustment. What is left over is assigned to the intermediate zone, and whether a particular area will be included here depends on one's estimate of its preponderant physiological character. The body of the cerebellum and the pallial part of the cerebral hemispheres are excluded from the zones as supra-segmental structures.

The lines drawn in this analysis are frankly arbitrary, chosen primarily for convenience of description; but, as will appear, this functional analysis contributes to an understanding of the meaning of the structure, and, moreover, it has morphological justification as well. These zones are not autonomous units when viewed either structurally or physiologically. Their interconnections are intimate and complicated. The more important of these connections are shown in a number of diagrams, some in this contribution and some in previous papers.

This analysis of the more obvious structural features of an amphibian brain in terms of physiological criteria is an artificial schematization of a complicated fabric, the several parts of which are so intimately connected that there is an over-all integration of their activities. The main highways of through traffic have been mapped, with signboards pointing the way at the crossroads. Selected examples of some of the lines of fore-and-aft through traffic are illustrated in the diagrams; but this kind of analysis does not take us to our destination. It does not tell us how mixed traffic is actually sorted out and so reorganized as to give the body as a whole efficient adjustment to the momentarily changing exigencies of common experience. These problems are discussed in subsequent chapters, but, first, the schematic outline will be summarized here.

Each zone is structurally diversified, and many of these local differentiations are directly correlated on the sensory side with the modalities of sense represented in the end-organs with which they are connected and on the motor side in a similar way with synergic systems of muscles. Each sensory and motor system of nerves has a local primary central station in direct connection with the periphery, and each of these stations has widely spread connections within its own zone and with other zones, thus insuring efficient correlation of sensory data, co-ordination of motor responses, and integration of the action system as a whole. In this summary the sensory systems are given special attention because these are the most useful guides in the analysis of the structure of this brain.

THE SENSORY ZONE

The sensory zone is defined as those parts of the brain that receive fibers from peripheral organs of sense. In some fields the number of such fibers is large, in others it is very small; and some parts of the brain, like the cerebral peduncle, have both sensory and motor peripheral connections. Within the sensory zone there is a complicated apparatus of correlation, and in lower vertebrates the receptive areas and surrounding tissues dominated by them are so large that most of the mass of the brain can be blocked into fields appropriately designated "nosebrain," "eyebrain," and so on ('31a, p. 129). This feature is due to the fact that in these animals the sense organs are well developed and highly specialized, but the motor apparatus is more simply organized, chiefly for co-ordinated mass movements. The sensory zone with its own apparatus of correlation, accordingly, bulks larger than the motor zone.

Though the peripheral sensory apparatus in Amblystoma differs from the human in many details, yet the general principles of its organization are similar; the structural organization of the central apparatus of adjustment, on the contrary, is so radically different that comparisons are difficult. Here the several functional systems of peripheral nerve fibers enter the brain in fascicles of the nerve roots, which are physiologically as specific as are those of mammals; but at the first synapse this specificity may almost completely disappear, in so far as it has visible structural expression. The root fibers of all sensory systems (except, perhaps, the olfactory) terminate by wide arborizations in a few common fields of neuropil, in each of which several of these systems are inextricably mingled. This neuropil is a common synaptic pool for all entering systems. The several pools are interconnected with one another and with similar pools in other parts of the brain stem, and there is no supreme cortical regulator. The translation of sensory experience into adaptive behavior and the integration of this behavior are somehow accomplished within this interplay of the local activities of the brain stem.

The sensory zone is continuous from the dorsal gray column of the spinal cord to the olfactory field, comprising the dorsolateral part of the medulla oblongata, the auricle in the cerebellar region, the anterior medullary velum and a small amount of contiguous tissue, the tectum of the midbrain, pretectal nucleus, dorsal thalamus, olfactory bulb with the adjoining anterior olfactory nucleus, and optionally the septum and some other parts of the hemisphere, a portion of the hypothalamus, and the ventrolateral neuropil of the peduncle. The fields optionally included receive terminals of the nervus terminalis (p. 267); the hypothalamus has a small but significant connection with the optic nerve, the basal root of which also connects with the peduncle. In some vertebrates the epithalamus receives fibers from the parietal eye, but in Amblystoma these have not been seen, and in this animal the predominant functions of the "optional" areas are of intermediate-zone type. The body of the cerebellum and the pallial part of the cerebral hemisphere might be assigned to the sensory zone as here defined anatomically; yet, as previously mentioned, they are excluded from this zone because of their distinctive physiological characteristics and their remarkable specialization in higher animals.

In the sensory zone of the medulla oblongata there are two elongated synaptic pools of neuropil, into which terminals of the sensory root fibers converge (chap. xi). One of these receives terminals of all somatic sensory systems; the other lies more ventrally and internally in the visceral lobe and receives terminals of the visceral sensory and gustatory systems (figs. 9, 89). The secondary fibers which emerge from these pools are distributed locally to the motor zone of the medulla oblongata, downward to the spinal cord, and upward to higher levels. The last take different courses, some to the cerebellum, some to the tectum and thalamus, and some to the hypothalamus. Each of these pathways discharges into a higher synaptic pool of neuropil, where its terminals are in physiological relation with terminals of other related sensory systems. The relations to which reference has just been made are in terms of the types of response to be evoked. Thus the tectum becomes the dominant regulator of somatic adjustments to exteroceptive stimulation, the hypothalamus becomes the regulator of visceral responses to olfacto-visceral stimulation, and the cerebellum provides regulatory control of the action of the skeletal muscles. The dorsal thalamus is ancillary to the tectum and shows a very early stage in the evolution of the ascending sensory projection systems to the cerebral hemispheres.

These local differentiations, each with characteristic structure and connections, are receptive fields for the several systems of peripheral sensory fibers, though some of them receive few peripheral fibers and are concerned chiefly with sensory correlation.

THE MOTOR ZONE

The motor zone as here defined includes the peripheral motor neurons and those areas of the brain stem concerned with the organization of motor impulses in patterns of synergic action. It includes the following histologically different parts: (1) corpus striatum (paleostriatum); (2) anterior part of the ventral thalamus; (3) posterior part of the ventral thalamus; (4) nucleus of the tuberculum posterius ("peduncle" in the restricted sense); (5) isthmic tegmentum; (6) trigeminal tegmentum; (7) a poorly defined tegmental field extending farther posteriorly through the length of the medulla oblongata into continuity with the ventral gray column of the spinal cord.

The floor plate of the embryonic neural tube probably ends anteriorly at the fovea isthmi (fig. 2B, *f.i.*), and the adjacent basal plate, which is the primordial motor zone, extends forward of this to include the whole of the peduncle and probably more or less of the hypothalamus and ventral thalamus. This primordial zone contains not only nervous elements with peripheral connections, like the sensory zone, but also an elaborate apparatus of central co-ordination of the neuromotor systems.

Anteriorly of the peduncle the motor zone has no peripheral connections, but the apparatus of motor co-ordination extends forward through the thalamus into the lateral wall of the hemisphere. Since the present analysis is based primarily on physiological criteria, this anterior extension of the motor field is included in the motor zone. The anterior boundaries of this zone are, of course, arbitrarily drawn; that they are artificial is emphasized by the fact that the large basal optic root terminates in the peduncle, which is in the motor zone as here defined. Efferent fibers have been described as leaving the brain in many places outside the motor zone, even as here broadly defined. Vasomotor and other visceral efferent fibers have been reported in various animals associated with the nervus terminalis and the olfactory and optic nerves and in other places for distribution to meninges and chorioid plexuses. We have nothing new to report about Amblystoma in this connection.

In the spinal cord and medulla oblongata the peripheral motor neurons are so mingled with the co-ordinating neurons of the tegmentum and reticular formation and they are so similar in form that it is often impossible to distinguish the peripheral neurons except in cases where their axons are seen to enter the nerve roots. The cells of the nuclei of the eye-muscle nerves are fairly clearly segregated, and in some reduced silver preparations they react specifically to the chemical treatment (fig. 104); but even here their dendrites are widely spread and intertwined with those of tegmental cells, so that both kinds of neurons would appear to be similarly activated by the neuropil within which they are imbedded. The cell bodies are locally segregated; but their dendrites, where most of the synaptic contacts are made, are not segregated.

In the medulla oblongata the motor tegmentum contains small and large cells in an endless variety of forms, but these elements are not segregated in accordance with either size or morphological type. It is true that the arrangement of their cell bodies may show some rather ill-defined local segregation, but their dendrites and axons are so intimately intertwined in the neuropil that nothing comparable with the localized nuclei of higher brains can be recognized. Farther forward in the isthmus and peduncle the tegmental tissue of co-ordination is much increased in amount and somewhat more differentiated. In some of my former papers (e.g., '30, p. 76) the term "nucleus motorius tegmenti" was used loosely (and inaccurately) to include a tegmental zone defined topographically. This seemed to be justified in the case of Necturus by the lack of localization of the large motor elements which characterize this region; but this justification is inadequate, both factually and morphologically—see the discussion by Ariëns Kappers, Huber, and Crosby ('36, pp. 653, 666).

It is obvious that most of the tissue of the motor zone is concerned with co-ordination of the action of the peripheral elements, so that synergic groups of muscles are activated in appropriate sequence; but, with the technic available, it has not been possible to analyze this complex so as to reveal the mechanism employed. In the medulla oblongata this organization is chiefly for local control of bulbar and spinal reflexes, the intermediate zone participating. In the isthmus and peduncle the number of peripheral elements is relatively small and the co-ordinating apparatus larger, giving these areas control over all motor fields spinalward of them. This intrinsic motor apparatus is supplemented by a segregated band of correlating tissue in the intermediate zone, the subtectal dorsal tegmentum. In mammals both these zones are further specialized into separate nuclei distributed in the tegmentum.

The primary patterns of somatic movements are predetermined by the course of central differentiation within the motor and intermediate zones in premotile stages of development. After connection with the peripheral musculature is established, each of these muscles seems to exert some sort of distinctive reciprocal influence upon that motor field of the central nervous system from which its innervation is derived. The nature of this influence is unknown, but its reality is well attested by experiments of Paul Weiss ('36, '41) and colleagues upon "myotypic response" and "modulation."

In later stages the primary motor patterns may be modified, or "inflected," by sensory experience and practice. Influence of use or some other functional factors seem to be essential for maintenance of motor efficiency, as graphically shown by Detwiler's observations ('45, p. 115; '46) on the behavior of decerebrate larvae of Amblystoma (to which further reference is made on p. 118). In young larvae of stage 37, swimming movements may be perfectly executed after transection immediately below the auditory vesicle under control of the lower medulla oblongata and spinal cord (Coghill, '26, Paper VI, p. 111; '29, p. 15); but, subsequent to Harrison's stage 40, Detwiler finds that sustained motor activities, including swimming, fail rapidly if the influence of the midbrain is blocked in prefunctional stages, though feeding reactions are preserved after complete ablation of hemispheres and visual organs. The midbrain evidently supplies a factor essential for maintenance of motor efficiency.

The motor field of this brain is smaller and more simply organized than the sensory field because most of the activities are mass movements of total-pattern type. Within this larger frame of total behavior, the partial patterns of local reflexes are individuated with more or less capacity for autonomous action. The number of these local partial patterns is smaller than in higher animals, and all of them are far more closely bound to the total patterns of which they are parts. The segments of each limb, for instance, may, upon appropriate stimulation, move independently; but in ordinary locomotion they move in a sequence related to the action of the entire limb, the other limbs, and the musculature of the trunk.

The peripheral motor nerves (omitting the general visceral components of preganglionic type not here considered) are in three groups: (1) the spinal nerves; (2) the eye-muscle nerves, III, IV, and VI pairs of cranial nerves, which are somatic motor; and (3) the special visceral motor nerves of the V, VII, IX, and X pairs, innervating the striated musculature of the visceral skeleton of the head. The primary movements of trunk and limbs are organized for locomotion in the motor zone of the spinal cord. This organization is under exteroceptive and proprioceptive control locally throughout the length of the cord and more especially at the bulbo-spinal junction; it is under additional proprioceptive control from the labyrinth and the cerebellum; there is further control from the cerebrum—optic, olfactory, and the related apparatus of higher correlation. The bulbar group of special visceral motor nerves is primarily concerned with movements of the head, notably those of respiration and feeding. The feeding reactions are under visual, olfactory, somesthetic, gustatory, and general visceral afferent control, and the pattern of performance seems to be organized in the large isthmic tegmentum. The very large and complicated interpeduncular nucleus is an isthmic structure which is physiologically of intermediate-zone type (chap. xiv). Details of the structure and connections of the various parts of the motor zone are in Part II, and the peculiar features of its forward extension in the cerebral hemisphere are discussed in chapter vii.

THE INTERMEDIATE ZONE

The characteristics of this zone are implicit in the preceding account of the sensory and motor zones. It is more elaborately developed and its boundaries are more clearly defined in parts of the cerebrum than in the rhombencephalon. These boundaries are necessarily arbitrary, for all parts of the brain are involved in correlation and integration of bodily activities; but throughout the length of the spinal cord and brain there is a band of tissue between the sensory and motor zones primarily concerned with these adjustments. At lower levels I have termed this tissue the "reticular formation," and here it infiltrates the other zones with no clear boundaries (for details see chap. xi). At higher levels it increases in amount and is more clearly segregated. It would be appropriate to include in this zone most of the diencephalon and telencephalon except the specific optic and olfactory terminals; but, for reasons mentioned above, a different subdivision has been adopted, primarily for convenience of description. The dorsal tegmentum, or subtectal area, is a typical representative of this zone in position and physiological connections. In the isthmic and bulbar tegmentum the characteristics of the intermediate and motor zones are inextricably mingled. The habenula, hypothalamus, and interpeduncular nucleus, as elsewhere described, clearly belong physiologically to the intermediate zone; and the whole cerebral hemisphere, except the olfactory bulb, might appropriately be so classified in all Ichthyopsida.

In the most primitive vertebrates the intermediate zone is scarcely recognizable as an anatomical entity. As the action system becomes more complicated in higher animals, this zone shows corresponding differentiation. This specialization is more directly dependent upon complication of the peripheral motor apparatus than upon sensory differentiation, for, so long as the action system is largely confined to mass movements, the patterning of these total activities is effected in the sensory and motor zones. In tetrapods and birds more complex central adjustors are required, and these are differentiated between the two primary zones and anteriorly of them. With the appearance of more autonomy of the local reflex systems, more efficient apparatus of integration is demanded. The final result is that in the human brain the apparatus of intermediate-zone type has increased so much that it comprises more than half the total weight of the brain, for both cerebral and cerebellar cortices are derivatives of this primordial matrix, as will appear in the ensuing discussions.

THE FUNCTIONAL SYSTEMS

The preceding physiological analysis of the brain obviously rests upon the peripheral relations of its several parts. The two primary functions of the nervous system are, first, the maintenance of the integrity of the individual, with efficient co-operation of parts among themselves and with the total organization, and, second, the analysis of experience and the translation of the sensory data into appropriate behavior. The peripheral nerves are key factors in both these domains. Our knowledge of the functional analysis of the cranial nerves has been greatly increased during the last fifty years, largely by the work of the so-called American school of comparative neurologists, which I have recently reviewed ('43).

Before I discuss the components of these nerves, a few definitions are in order. In the attempt to envisage the nervous system from the operational standpoint, distinctions have been drawn between sensory correlation, motor co-ordination, and those central processes that provide integration, and some measure of spontaneity of action which might be grouped under the name "association" ('24c, p. 235; '31a, p. 35). This classification is necessarily artificial, for all these processes are interrelated. They interpenetrate, and they are not sharply localized in the structural fabric. Nevertheless, these several types of action are recognizable components of the unitary dynamic system, and there are local differentiations of the structural organization correlated with preponderance of one or another of them, more clearly so in higher vertebrates than in lower.

Sensory correlation, as the term is here employed, refers to interaction of afferent impulses within the sensory zone, that is, within the field reached by terminals of peripheral sensory fibers. The interplay of these diverse afferent impulses takes two forms: (1) in fields of undifferentiated neuropil, the activation of which results in alterations of the central excitatory state or in mass movements of large numbers of synergic muscles; (2) in more restricted areas (nuclei), which activate the neuromotor apparatus of local reflexes. The members of both groups are interconnected by systems of internuclear fibers like the lemniscus systems, all within the sensory zone, so that all activities of this zone interact one with another. These internuncials are so arranged that functional systems of afferents, which normally co-operate to effect a particular type of motor response, are more intimately associated. Thus the tectum opticum receives most of the lemniscus fibers of all somesthetic systems and minimum numbers of olfactory and visceral systems. This basic pattern as seen in Amblystoma is changed in mammals, where higher associational centers have taken over most of the functions of correlation.

Motor co-ordination is effected primarily in the motor zone, which is so organized as to activate synergic groups of muscles in appropriate sequence with inhibition of their antagonists. This grouping may be adapted for mass movements or for local reflexes. Internuclear tracts connect the various parts of the sensory zone directly with appropriate parts of the motor zone. More refined analysis and conditioning of motor responses are effected through the intermediate zone, and the tissues of the latter group are greatly enlarged and complicated in higher brains.

The activities of stimulus-response type which have just been considered are so interconnected with internuclear tracts and the interstitial neuropil as to facilitate the integration of all local activities in the interest of the requirements of the body as a whole. Every local part of the brain is a component of the apparatus of general integration, and some of these parts have this association as their dominant function. In Amblystoma most of this suprasensory and supramotor tissue is dispersed as interstitial neuropil. In mammals, higher types of associational tissue have been differentiated locally, notably in the cerebral cortex and its dependencies, with corresponding enhancement of those synthetic functions which are manifested as conditioning, educability, and reasoning. Parallel with these changes there is an enormous increase in accumulated reserves of potential nervous energy, which come to expression as spontaneity, memory, and creative imagination.

A survey of the nerve fibers of Amblystoma as a whole in view of the principles just expressed shows that they may be classified in four groups: (1) the peripheral afferent systems and associated internuclear correlating tracts within the sensory zone (lemniscus systems, etc.); (2) the peripheral efferent systems and the related co-ordinating fibers of the motor zone; (3) the central internuclear systems intercalated between the preceding two and so interconnected as to yield appropriate responses to ordinary recurring stimuli; and (4) infiltrating these mechanisms of stimulus-response type, a different sort of adjusting apparatus which insures general integration of these systems, with provision for conditioning of reflexes and other forms of individually acquired behavior and for release of accumulated reserves of nervous potential as needed. These four groups intergrade but, in general, are recognizable.

The peripheral fibers are grouped in *functional systems*, each of which is defined as comprising all nerve fibers and related end-organs, which are so arranged as to respond to excitation in a common mode, either sensory or motor. These functional systems are convenient anatomical units also, for all fibers of each sensory system, regardless of variations in the peripheral distribution of their end-organs and regardless of the particular nerve trunks or roots through which they connect with the brain, are segregated internally and converge into local areas or zones. In higher vertebrates (but less so in lower) the secondary connections of these terminal stations tend to retain their physiological specificity. From this it follows that the peripheral systems of sensory analyzers are extended into the brain as far as related central pathways are separately localized—in the human brain it may be even up to the projection areas of the cerebral cortex. Accordingly, we include in the sensory zone as here defined not only the terminal nuclei of peripheral sensory nerves but also their related nervous connections, so far as these are with other parts of the sensory zone and not directly with the motor zone. The neuromotor apparatus can be similarly analyzed into functional systems, each of which is concerned with the synergic activation of some particular group of muscles.

This anatomical segregation of the functional systems is not carried to perfection, even in the human nervous system. The various modalities of cutaneous and deep sensibility, for instance, are not completely segregated and localized either peripherally or centrally. Yet this differentiation has gone so far that it provides our most useful guide in the analysis of the structure of the brain.

The activities of the body may be divided into two major groups; (1) those concerned with adjustment to environment, the *somatic* functions, and (2) those concerned with the maintenance and reproduction of the body itself, *visceral* functions. These, of course, are not independent of each other; nutrition, for instance, involves somatic activity in the search and capture of food and visceral activity in its digestion and assimilation. Nonetheless, these types of function are so different, especially in the responses evoked, that this strictly physiological criterion marks also the most fundamental structural analysis of the nervous system. Anatomically, the somatic systems of peripheral organs and nerves and central adjustors, including the proprioceptors, are, in general, rather sharply distinguished from the visceral. The systems are cross-connected by internuclear tracts, and some sensory systems, like the olfactory, may serve, on occasion, either somatic or visceral adjustments.

A phylogenetic survey of these systems reveals remarkable plasticity in their interrelationships. Thus, taste buds, which in most vertebrates are typical interoceptors, may in some fishes be spread over the external surface of the body, where they serve exteroceptive functions, with corresponding changes in the anatomical pattern of the central apparatus of adjustment (chap. x; '44b). On the motor side the apparatus of feeding and respiration exhibits still more remarkable transformations. In fishes this musculature is connected with the visceral skeleton—jaws, hyoid, and gill arches—and the functions performed are obviously visceral, though the larger part of this musculature is striated. The related parts of the nervous system are classified as special visceral motor. With suppression of the gills in higher animals, some of these muscles undergo remarkable transformations. Those which are elaborated to form the mimetic facial musculature of mammals become physiologically somatic ('22, '43).

The classification of peripheral end-organs and their related nerves which has proved most useful grew out of the analysis of these nerves into their functional components, to which reference has just been made, by histological methods. Serial sections through the entire bodies of small vertebrates differentially stained for nerve fibers enable the investigator to reconstruct not only the courses of the nerves but also the arrangement of the functional systems represented in each of them and to follow these components to their peripheral and central terminals, a result that cannot be achieved by ever so skilful dissection. The first successful application of this method was Strong's analysis of the nerve components of the tadpole of the frog in 1895, a fundamental research which provided the generalized pattern which prevails, with endless modifications of details, throughout the vertebrate series, as has been abundantly demonstrated by numerous subsequent studies by many workers.

This was followed in 1899 by my Doctor's dissertation on the nerve components of the highly specialized teleost, Menidia. From these and subsequent studies the peripheral nervous system of the head was analyzed into functional systems as follows:

1. *Somatic sensory fibers of two groups.*—(1) Exteroceptive systems, including (a) the specialized olfactory (in part), optic, auditory, and lateral-line nerves with differentiated end-organs, and (b) the nerves of general cutaneous and deep sensibility with simple free endings, these entering chiefly in the V nerve root with some in the VII, IX, and X roots. (2) Proprioceptive fibers from specialized end-organs of the internal ear and (probably) lateral-line organs and also fibers from muscles, tendons, and other deep tissues. Here belongs also the peculiar mesencephalic root of the V nerve. See chapter x for further comments on the proprioceptive system.

2. *Visceral sensory fibers of two types.*—(1) With specialized end-organs, viz., the olfactory organ (in part) and the taste buds, the latter entering by the VII, IX, and X nerve roots. (2) Fibers of general visceral sensibility with free endings, entering in the same roots as the preceding and mingled with them.

3. *Somatic motor fibers.*—Somatic motor fibers which supply striated muscles derived from the embryonic somites, those in Amblystoma being limited to the nerves of the extrinsic muscles of the eyeball in the III, IV, and VI nerves.

4. *Visceral efferent fibers of two types.*—(1) Special visceral motor fibers of cranial nerves supplying striated muscles, not of somitic origin, related with the visceral skeleton, jaws, hyoid, branchial arches, and their drivatives (in the V, VII, IX, and X roots). (2) Preganglionic fibers of the general visceral (autonomic) system, terminating in sympathetic and parasympathetic ganglia, where they activate postganglionic fibers distributed to unstriated and cardiac muscles and glands (in the III, VII, IX, and X roots). The last system is not further considered in this work. For application of this analysis to the human nervous system see my *Introduction to Neurology* ('31a, chaps. v and ix).

This analysis has yielded our most useful clues for resolution of the complexity of both peripheral nerves and brain. Descriptions of the peripheral end-organs and the courses of the nerves do not lie within the scope of this work. Some of these details which are significant for understanding their central connections are included in chapter x.

CHAPTER VI

PHYSIOLOGICAL INTERPRETATIONS

APPARATUS OF ANALYSIS AND SYNTHESIS

IN A primitive brain like that of Amblystoma the stable framework of localized centers and tracts performs functions that are primarily analytic. The sense organs are analyzers, each attuned to respond to some particular kind of energy. The sensory systems of peripheral nerves and the related internal sensory tracts are parts of the analytic apparatus, in so far as their functional continuity with the peripheral organs of the several modalities of sense can be traced.

On the motor side similar conditions prevail. The neuromotor apparatus is organized in functional systems, each of which is adapted to call forth the appropriate sequence of action in a particular group of synergic muscles. These units are as truly analyzers as are those of the sensory systems, though in an inverse sense. Out of the total repertoire of possible movements, those, and only those, are selected which give the appropriate action. The efferent fibers are grouped, the members of each group being so bound together by central internuclear connections that they act as a functional unit adapted for the execution of some particular component of behavior, such as locomotion, conjugate movements of the eyeballs, seizing and swallowing food, and so on.

The several sensory systems are so interconnected within the sensory zone as to react mutually with one another. They form a dynamic system so organized that all discharges from this zone are resultants of this interaction. This interplay has pattern. The various modalities of sense are not discharged into a single common pool of equipotential tissue. The sensory components of the nerves are segregated, more or less completely, so that related systems converge into dominant centers of adjustment—exteroceptors in the tectum, proprioceptors in the cerebellum, olfacto-visceral systems in the hypothalamus, olfacto-somatic systems in the habenula, and so on.

A review of the internal architecture of the adult brain of Amblystoma suggests that the specifications of the general plan are drawn in terms of current *action*. The elaboration of the analytic apparatus on the sensory side is carried only so far as is requisite to facilitate responses to any combination of sensory stimuli in patterns determined by the appropriate use of such motor equipment as the animal possesses. In species with simpler action systems the central analytic apparatus is less elaborate; in species endowed with more complicated motor organs the central architecture is more elaborate. In all species the peripheral sensory equipment determines the architectural plan of the primary centers of the sensory zone; internally of this level the details of the plan are shaped by two additional factors: first, the motor equipment available and, second, the amount and quality of the apparatus of correlation and integration required for the most efficient use of such sensory excitations as the animal experiences. The cross-connections between the sensory and motor zones are quite direct and simple in early embryological stages, so arranged as to provide uniform stereotyped responses to oft recurring situations. But as development advances these connections become more and more complicated, an intermediate apparatus of correlation is interpolated, and, correlated with this, the behavior becomes more diversified and unpredictable.

In the sequence of development of behavior patterns this change can be accurately dated. For instance, in Amblystoma between the early swimming and early feeding stages, at about Harrison's stage 40, the swimming movements, which in younger stages are perfectly co-ordinated by the bulbo-spinal central apparatus alone, lose this autonomy, and participation of the midbrain is essential for the maintenance of efficient swimming, as was mentioned on page 62 in describing experiments by Detwiler ('45, '46). It is during this period that tecto-bulbar and tecto-spinal connections of essentially adult pattern are established ('39, p. 112). In human fetal development there is a similar critical period at about 14 weeks of menstrual age (Hooker, '44, p. 29). At this time the upper levels of the cerebrum acquire functional connections with the lower brain stem, and the behavior shows a corresponding change. "The fetus is no longer marionette-like or mechanical in the character of its movements, which are now graceful and fluid, as they are in the new-born."

Synthesis and integration may be effected in various ways. The most evident nervous structures employed here are the internuclear tracts which form a complicated web of conductors, which interconnect the analytic units with one another so that the entire complex forms an integrated equilibrated system. This is the apparatus of the stable heritable components of the action system—the reflexes and instincts. A second integrating apparatus is found in the all-pervasive neuropil, and a third in specialized derivatives of the latter, the associational tissues locally differentiated in the brain stem and reaching maximal development in the cerebral cortex.

The total behavior of neuromuscular type emerged within a pre-existing bodily organization, which maintained the unity of the individual by nonnervous apparatus. The nervous system is from its first appearance a totalizing apparatus. Local differentiations of tissue for the analysis of sensory experience and of motor responses arise within this integrated structure, and local reflexes similarly emerged within a total neuromuscular pattern of action adapted to maintain the unity of the organization. As development advanced, the mechanisms of the local reflexes acquired increasing autonomy, but they are never completely emancipated from some control in the interest of the behavior of the body as a whole. The organic unity of the whole is preserved while local specificity is in process of development, and this unitary control is never lost during the normal life of the individual.

THE STIMULUS-RESPONSE FORMULA

The stimulus-response formula has wide application and great usefulness as a basic concept in physiology and psychology, but its apparent simplicity is illusory and has tended to divert attention from essential features of even the simplest patterns of behavior. This I have illustrated ('44a) by an examination of the simplest reflex connection known in Amblystoma—from retina to ocular muscles by way of the basal optic tract.

The late G. E. Coghill, during a productive period of forty years, studied the development of the action system of Amblystoma and the correlated processes of bodily growth. These researches have demonstrated beyond question that in this animal the neuromuscular system is so organized in prefunctional stages that, when first activated from the sensory zone, the resulting movement is a total response of all the musculature that is mature enough to respond to nervous excitation. These "total patterns" of activity are not disorderly, and they become progressively more complicated while the apparatus of local reflexes ("partial patterns") is slowly differentiated within the larger frame of the total pattern. The development of both the total pattern and the partial patterns is initiated centrally, and throughout life all of them are under some measure of unified central control so that the body acts as an integrated whole with diverse specialization of its parts (Coghill, '29; Herrick, '29). Coghill's contributions of factual observations and the principles derived from them have been critically reviewed by the writer in a book ('48), to which the reader is referred.

The patterning of these orderly movements is determined by the intrinsic structure of the nervous system. This structural pattern is not built up during early development under the influence of sensory excitations, for in the embryo the motor and sensory systems attain functional capacity independently of each other; and when central connection between the sensory zone and the motor zone is made, the first motor responses to excitation exhibit an orderly sequence, the pattern of which is predetermined by the inherited organization then matured (Coghill, '29, p. 87; '30, Paper IX, p. 345; '31, Paper X, pp. 158, 166). This early structural differentiation goes on independently of any stimulus-response type of activity, though the latter may modify the pattern of subsequent development. This is a principle of wide import, applicable not only in lower vertebrates but in higher forms as well (Coghill, '40), including man (Hooker, '44). The stimulus-response mechanism is not a primary factor in embryogenesis; it is a secondary acquisition.

REFLEX AND INHIBITION

It has been pointed out that the functions of the sensory and motor zones are fundamentally analytic—analysis of environmental influences and analysis of performance in adjustment to those influences. How the units of the analytic apparatus are actually related so as to insure the appropriate correlated action of the separate parts is the key problem, which must be resolved before animal (and human) behavior can be approached scientifically in other than a descriptive way. Good progress has been registered. The sensory and motor analytic apparatus has been exhaustively studied and well described; and this was the appropriate place to begin, for these organs are most accessible to observation and experiment. Because these systems of peripheral end-organs and the related pathways of conduction and centers of control are, in the human nervous system, obviously interconnected in stable and definitely localized patterns, it was natural to use this structural framework as the point of departure in the elaboration of the hypothetical superstructure of current doctrines of reflexology. But reflexes can be conditioned, and this name for a well-known physiological fact is for the neurologist scarcely more than a symbol of complete ignorance of the mechanisms actually employed.

The several reflexes have been so closely colligated with specific details of central architecture that the reflex arc came to be regarded as the primary unit of nervous organization, and it was assumed that the increasing complexity of the upper levels of the brain in higher vertebrates has been brought about by progressively more intricate interconnections among these elementary units. The integrative action of the nervous system was conceived in terms of the definition of mathematical integration—"the making up or composition of a whole by adding together or combining the separate parts or elements." This conception leaves unexplained how any additive process of this sort can result in such a unique centrally controlled unitary organization as we actually observe, capable of conditioning the reflexes in terms of individual experience (learning), of abstracting some common features of mixed experience and synthesizing these into quite original patterns of response, and of maintaining some measure of "spontaneous" or self-determined directive control.

A far more serious charge against traditional doctrines of reflexology is the observed fact that in the development of Amblystoma the early responses to external stimulation are not local reflexes but total movements of the entire available musculature. The integrated total pattern precedes in time the appearance of the partial patterns. These are individuated within the total pattern; they are integral parts of it, and for an appreciable time they are subordinate to it. Even in the adult animal the local partial patterns are not completely emancipated from control by the body as a whole. It is, indeed, impossible to find in this brain any sharply defined, well-insulated reflex arcs.

What happens during the emergence of specific reflexes from the total reactions is, first, the development of an increasing number of collateral branches of the primary axons and the central linkage of sensory and motor pathways in ever more complicated patterns. Then, second, in the adjusting centers additional neurons are differentiated, the axons of which take longer or shorter courses, branching freely and participating in the formation of a nervous feltwork of extraordinary complexity. These neurons are not concerned primarily with specific reflexes but with the co-ordination and integration of all movements. Some parts of this intricate fabric, generally with thicker fibers, more or less well fasciculated, activate mass movements of primitive type, and other parts control local reflexes as these are individuated. But these systems of fibers are not segregated in completely insulated reflex arcs. They are interconnected by collateral branches with one another and with the interstitial neuropil. There are lines of preferential discharge, but whether any one of them is actually fired depends on numberless factors of peripheral stimulation and central excitatory state.

The phylogenetic history is parallel. The further down we go toward the primitive ancestral vertebrates, the less clear evidence do we find of definitely localized reflex arcs, and the overt behavior tends more toward mass movements of total-pattern type.

It must be borne in mind that the development of the individual does not exactly recapitulate the phylogenetic development (Hooker, '44, pp. 15, 33). The pattern of the sequence of structural changes which take place during prefunctional stages of growth is determined by the organization of the germ plasm and the interaction of the genetic factors with one another. This organization, in turn, has been determined during preceding evolutionary history in adaptation to the environment and habitus of the species in question. In broad lines the history of ancestral development is repeated in the growth of the embryo, but cenogenetic modifications of it may appear in adaptation to changing conditions, as illustrated, for instance, by the appearance of some local reflexes earlier in mammals than in amphibians.

The structural organization of the brain sets off in sharp relief a few important general physiological principles. First, it is to be noted that the "resting" nervous system is not inert. The body acts before it reacts. There is always some spontaneous—that is, centrally excited—activity, and the importance of this factor increases as we ascend the phylogenetic scale. There is always intrinsic activity, as demonstrated, for instance, by the Berger rhythms, and it is always acted upon by numberless extrinsic agencies. When an excitation is received from the periphery, there results a change in the central excitatory state both locally and diffusely, which involves both activation and inhibition.

Another general principle may be mentioned here. The flow of nervous impulses from receptor to effector is not one-way traffic.

The excitation of a peripheral sense organ may be followed by an efferent discharge back to the receptor. An instructive illustration of this is seen in the auditory apparatus of mammals. Excitation of the cochlea is followed by an efferent return to the tensor tympani and stapedius muscles and also to the cochlea itself (the latter pathway recently demonstrated by Rasmussen, '46). Almost all contracting muscles report back to the center by a system of proprioceptive fibers. The central nervous system is full of similar reciprocating systems. Many of the fasciculated tracts of Amblystoma are two-way conductors, transmitting in both directions, and there are numberless illustrations of a circular type of connection, efferent fibers of one center activating another, which has a return path, perhaps by a devious route, back to the first center. A neuropil may be interpolated in any of these types of circuit. The thalamo-cortical connections of the human brain are of this sort, exhibiting what Campion and Elliot Smith ('34) have aptly named a "thalamo-cortical circulation," a circulation not of blood but of nervous transmission. All parts of the cerebral hemispheres are in similar reciprocal interconnection, as has recently been emphasized and illustrated by Papez ('44).

Here reference may be made to Dewey's ('96) stimulating analysis of the reflex-arc concept or, as he prefers to say, the "organic circuit" concept. This he elaborated in terms of psychology, and nearly twenty years later I made this comment about it ('13a):

"Let us see how it may be applied to biological behavior. The simple reflex is commonly regarded as a causal sequence: given the gun (a physiologically adaptive structure), load the gun (the constructive metabolic process), aim, pull the trigger (application of the stimulus), discharge the projectile (physiological response), hit the mark (satisfaction of the organic need). All of the factors may be related as members of a simple mechanical causal sequence except the aim. For this in our illustration a glance backward is necessary. An adaptive simple reflex is adaptive because of a pre-established series of functional sequences which have been biologically determined by natural selection or some other evolutionary process. This gives the reaction a definite aim or objective purpose. In short, the aim, like the gun, is provided by biological evolution and the whole process is implicit in the structure-function organization which is characteristic of the species and whose nature and origin we need not here further inquire into. . . . . The *aim* (biological purpose) is so inwrought into the course of the process that it cannot be dissociated. Each step is an integral part of a unitary adaptive process to serve a definite biological *end*, and the animal's motor acts are not satisfying to him unless they follow this predetermined sequence, though he himself may have no clear idea of the aim. These reactions are typically organic circuits. . . . . Always the process is not a simple sequence of distinct elements, but rather a series of reactions, each of which is shaped by the interactions of external stimuli and a preformed or innate structure which has been adapted by biological factors to modify the response to the stimuli in accordance with a purpose, which from the standpoint of an outside observer is teleological, i.e., adapted to conserve the welfare of the species."

This apparent teleology is commented upon in chapter viii. Since the passage just quoted was written, control of gunfire by radar has been perfected, thus reinforcing our analogy at one weak spot. In the reflex the "aim" does not precede the stimulus that pulls the trigger; it is automatically adjusted to the stimulus as in radar. But this automatism in both cases is dependent upon the presence of a preformed structure adapted to provide it.

Our analysis of the adult structure of the brain of Amblystoma confirms and supplements the conclusions reached by Coghill from his study of the development of the same species. His major contribution, as I see it, was the demonstration of the primacy of the integrative factors in the development of behavior patterns and of some of the features of structural growth during the individuation of local partial patterns within the larger total pattern. The adult structure of the brain of Amblystoma is in perfect conformity with the conclusions to which he was led. One of these conclusions should receive special emphasis here, for it clarifies our conception of what the reflex is in general, and in particular it helps us over some hard places in our attempt to discover the actual mechanisms involved in the individuation of local reflex patterns within the frame of the total pattern.

In the central resolution of forces which eventuates in some particular pattern of overt movement there is always an inhibitory factor (Coghill, '36, '43). In discussing the individuation of partial patterns (local reflexes) from the total pattern, he wrote ('40, p. 45): "Individuation is obviously the result of organized inhibition. . . . . The major division of the total pattern must be under inhibition when a part acquires independence of action, and the same part can be inhibited while the major segment of the total pattern acts. So that the whole individual probably acts in every response, either in an excitatory or inhibitory way." This he generalized in the following statement ('30, p. 639):

"For an appreciable period before a particular receptor field acquires specificity in relation to an appropriate local reflex its stimulation inhibits the total reaction. Inhibition, accordingly, through stimulation of the exteroceptive field, begins as a total pattern. It is then in a field of total inhibition that the local reflex emerges. The reflex may, therefore, be regarded as a total behavior pattern which consists of two components, one overt or excitatory, the other covert or inhibitory. The essential anatomical basis for this is (1) in the mechanism of the total pattern of action, or primary motor system, and (2) in the mechanism of the local reflex, or secondary motor system; the mechanism of the total pattern being inhibited and that of the reflex excited. But since inhibition is not a static condition but a mode of action, the mechanism of the total pattern must be regarded as participating in every local reflex."

This conception of the reflex as involving a factor of inhibition of the total pattern linked with excitation of the partial pattern is fruitful. Total inhibition plays a more obvious role in the overt behavior of amphibians than in most other animals, not only in embryogenesis of behavior but also in the adult. This was emphasized by Whitman ('99) in his classic description of the behavior of Necturus. In my manuscript notes of a conference with Dr. Coghill on January 1, 1929, I find a record of his remarks which is here transcribed.

"The first neurons to differentiate in Amblystoma are in the floor-plate. These and others adjacent form the primary motor column, the dominant function of which is activation of muscles of the same side for mass movement of the trunk and limbs and inhibition of the musculature of the opposite side which is in the same phase of locomotor movement. In later stages, when mechanisms of specific local reflexes emerge, residual neurons in the region of the floorplate maintain their functional importance for mass movements as activators of the whole somatic motor apparatus. They may prime this neuromotor system, putting it into a subliminal excitatory state in advance of its patterned activation.

"At an age which immediately precedes the first feeding reactions and before it is possible to open the mouth and swallow, the larva will react to a moving object in front of the eyes by a total reaction, a leap forward. It cannot seize the object. The general activator mechanism here comes to overt expression before the specific local reflex patterns are mature enough to function. After feeding activities have matured there is a similar general activation, accompanied by inhibition, as illustrated by the 'regarding' reaction [p. 38]. A larva which had been feeding for several days was stimulated by moving a hair slowly across the field of vision. The animal responded by moving the head slowly following the hair. The head is bent to the side, with rotation of the eyes, movement of the fore limbs, and adduction of both hind limbs. When the hair was not too far distant, the animal finally, at the end of this 'regarding' reaction, jumped after it. Here there is a clear distinction between what Sherrington calls the anticipatory phase and the consummatory phase of the reaction, and evidently in the anticipatory phase inhibition plays the major role. This is obvious also in almost all adult behavior of these animals."

The mechanism of central inhibition is still obscure. There is some evidence that a nervous impulse impinging upon a central neuron may, on occasion, activate the element, or under other conditions of central excitatory state, strength, or timing of the afferent flow it may inhibit activity in process. Whether or not this is true, it is well known that a central neuron may exhibit a large variety of types of synaptic junctions, differing in histological structure, electrical properties, and perhaps also in chemical reactivity. These afferent fibers may come from widely separated regions with diverse functions, and the impulses delivered may differ in intensity and temporal rhythm. Bodian's description ('37, '42) of axon endings on Mauthner's cell of the medulla oblongata shows four main types of synaptic contact which vary from 0.5 to 7μ in extent, with a wide variety of arrangements. There are between four and five hundred of these endings on a single cell, and the presumption is legitimate that these diverse structures are correlated with significant differences in electrical and chemical properties, including the timing of the pulses of transmission. It has been suggested that some of the influences transmitted across the synaptic junctions are excitatory and that others are inhibitory. Synaptic junctions on dendrites are in some cases structurally different from those on the axon hillock or axon, and they may be activated from different sources. Some observers believe that excitation of dendrites is excitatory and of axons is inhibitory, a supposition supported with physiological evidence by Gesell and Hansen ('45, p. 156). In their theory of the electronic mechanism of activation and inhibition, these functions are viewed as basically similar, activation being associated with an increasing, and inhibition with a decreasing, intensity of the electronic current. The connections of horizontal cells of the retina as described by Polyak ('41, p. 385) suggest to him a different inhibitory apparatus. The horizontal cells may exert an inhibitory influence upon the synapses between the rods and cones and the bipolar cells, that is, the synapses of the horizontal cells may function as "countersynapses" to the photoreceptor-bipolar synapses.

Whatever may be the mechanism employed in central inhibition, it is clear that in some parts of the brain excitatory functions predominate, in other parts inhibitory functions. Noteworthy examples of the latter are (1) the head of the caudate nucleus (Fulton, '43, p. 456); (2) a region in the reticular formation of the medulla oblongata explored by Magoun ('44); and (3) certain specific zones of the cerebral cortex (areas $4s$, $2s$, $19s$, and some others) which are known as "suppressor bands." In all these cases, excitatory and inhibitory fields are intimately related physiologically in such a way as to secure appropriate balance of activation and inhibition of the members of synergic systems of muscles in proper sequence.

The role of general inhibition in the patterning of behavior has been under investigation for several years by Beritoff and his colleagues. The first half of the fifth volume of the *Transactions* of his institute is devoted to studies on the nature of general inhibition and its role in the co-ordination of cortical activity and reflex reactions of the spinal cord. Beritoff believes that the neuropil possesses an inhibitory function—slow changes in voltage, expressing the active state of the neuropil, show an anelectrotonic effect on the cellular bodies, lowering excitability in them and weakening the excitation. The evidence is drawn from both somatic and visceral stimulation. He writes ('43, p. 142): "Thus, during each reflex reaction in the visceral organs, taking place in response to a stimulation of the interoceptors and of visceral afferent fibers, just exactly as during somatic reflexes, the spinal cord acts as a whole, making the given reflex local and every spinal reflex reaction entire by means of general inhibition." This is essentially the same as Coghill's position as stated in the preceding quotations. In other articles in the same volume the role of the neuropil in a great variety of spontaneous and stimulated activities of the brain is emphasized by Beritoff.

The neuropil as a whole is not, in my view, a specific inhibitor. It may partipciate on occasion in either excitation or inhibition, and in the inhibitory phase it acts as part of the covert component of the reflex or of mass action, as the case may be, in Coghill's analysis as quoted above. In my discussion of the habenular system (chap. xviii) and the interpeduncular nucleus (chap. xiv) I have ventured to suggest a possible mechanism through which general inhibition effected in the interpeduncular neuropil may operate in the facilitation of both mass movement and local reflexes. On this hypothesis this local band of neuropil must be able to act as a specific inhibitor in Beritoff's sense.

The amphibian neuropil in its various forms is structurally adapted for a considerable variety of functions of different grades of specialization. There is generally a diffuse spread of terminals, so that a single incoming fiber may activate many neurons of the second order. If the receptive tissue is homogeneous, this provides for simple central summation. If the receptive tissue is heterogeneous, as in most sensory fields, this arrangement facilitates mass movement of the musculature or total patterns of action. If many fibers converge upon a single neuron, the threshold of central excitation is lowered, as in the mitral cells of the olfactory bulb and in the "motor pool," as this concept has been developed by Sherrington. If the activated motor pool is large, with wide distribution of the efferent fibers, complicated integrated mass movement may result. If the pool is small, with a single final common path, a local reflex may follow. If the outlet comprises a number of open channels with different connections and physiological properties, there is provision for discriminative response, the selection being made (presumably) in terms of the central excitatory state of the components of the system ('42, p. 295).

It has been objected that the preceding comments on the limitations of current doctrines of reflexology are based upon the amphibian organization, which is aberrant and degenerate and therefore not typical or significant in the interpretation of the behavior of higher animals. But, even so, the Amphibia live well-ordered lives, and their behavior conforms in basic patterns with that of other vertebrates. We want to know how they behave as they do. Accepting the current view that Amblystoma is a retrograde descendant of some more highly specialized ancestor now extinct and that some of its characters are aberrant, yet the evidence seems to me adequate that such retrogression as may have occurred has been toward a generalized form ancestral to modern amphibians and mammals.

Conclusion.—I have assembled in these pages some factual description of observed structure, together with speculative interpretations of its probable physiological significance. The organic structure here under consideration is not something vague and ill-defined. Its anatomical distribution, histological organization, and fibrous connections can be described with precision. Not until this has been done can our imperfect knowledge of its functions be advanced by experiments designed to reveal its physiological properties.

In a discussion of "localized functions and integrating functions" more than a decade ago ('34a), the significance of neuropil in the evolution of cerebral structure was summarized in these words:

"The neuropil is the mother tissue from which have been derived both the specialized centers and tracts which execute the refined movements of the local reflexes and the more general web which binds these local activities together and integrates the behavior. It retains something of embryonic plasticity and so is available as a source of raw material for two very different lines of specialization—first, toward the structural heterogeneity requisite for the execution of localized reflex and associational functions, and, second, toward the more generalized and dispersed apparatus of total or organismic functions of tonicity, summation, reinforcement, facilitation, inhibition, 'spontaneity,' constitutional disposition and temperament, and extra-reflex activities in general."

PRINCIPLES OF LOCALIZATION OF FUNCTION

The great advances that have been made in the diagnosis and treatment of nervous diseases have been due in large measure to the more accurate mapping of the structural features of the nervous system and recognition of the specific functions of its several parts. Before a disorder can be successfully treated we must know what it is and where it is. The most notable triumphs in this medical field have been registered with those diseases whose situs can be recognized and then subjected to appropriate therapy or surgery. Even a systemic disorder like anemia has localization in blood corpuscles and blood-forming organs; and a general infection, like poliomyelitis, spreads in preferential paths determined by the histochemical structure of the tissue. The stable heritable tissues of the nervous system are most accessible to this kind of inquiry, of diagnosis, and of treatment; conquest of the unlocalized disorders has been retarded.

Some kinds of disorder, particularly those of primary concern to psychiatrists, have resisted all attempts at localization in accordance with conventional principles, and in the field of physiology the concept of local reflex arcs has limited application. The various attempts to elaborate a comprehensive account of animal and human behavior in terms of conventional reflexology have broken down. These conspicuous failures have led some competent authorities to question the over-all significance of localization in space of nervous functions and to search for other principles in the realm of pure dynamics or chemical interaction or some as yet unknown factors which operate quite independently of stable structural patterns. But no nervous tissue is structurally homogeneous or physiologically equipotential. In this connection it is interesting to note that Lashley, the leading advocate of the equipotentiality of the nervous tissues, has given us clear demonstration of point-to-point projection of retinal loci upon the lateral geniculate body and the cerebral cortex of the rat (Lashley, '34, '34a). This is the most refined anatomical localization of function known. In a later communication ('41) he demonstrated a very precise projection of the thalamic nuclei upon the cerebral cortex and added: "A functional interpretation of the spatial arrangement of the thalamo-cortical connections is not justified on anatomic grounds alone for any sensory system."

Somewhere between the extreme views of rigid localization in spatial patterns and a labile physiological equipotentiality a practicable working conception of the meaning of the structural configuration will be found. Clearly, the nervous system does not operate, even in the case of the simplest known reflex, on the mechanical principles of an automatic telephone exchange. We get only confusion by oversimplification of the problem. On the other hand, there are no disembodied functions, and the apparatus that performs functions has locus in space and time. Our problem is, first, to observe what is done and then to find out where and when it is done and how.

The observed spatial arrangements are not meaningless, and their functional interpretation is possible and fruitful, as evidenced by their practical utility in medical diagnosis and treatment. These structural patterns are stable and heritable. Their phylogenetic development can be traced, and in broad outline this has been done. But as these patterns are followed backward in the evolutionary series they become, not more simple and sharply defined, but less so, until in the most primitive and generalized vertebrates they tend to disappear in a more nearly homogeneous matrix. This would seem to support the view that localization of function is a secondary acquisition, derived from a primitive equipotentiality. With certain important qualifications, this is probably true; and if we follow in phylogeny the differentiation of local centers and their connecting tracts in correlation with types of function performed, the significance of localization appears. The problems of cerebral localization have usually been attacked in mammals and especially in man, where clinical applications are vitally important. Let us approach the subject from the other end of the phyletic series and look for the inception of localization patterns in primitive animals.

In the simplest known organisms localization of function is minimal and transient. In ameba any part of the cytoplasm may on occasion perform any function of which the organism is capable. There is a local differentiation of nucleus from cytoplasm, but in some bacteria even this localization disappears. A surface-interior pattern is always present, but the physical substance may shift from one to the other of these zones. In primitive multicellular species, ectoderm and entoderm were early differentiated—a specialization which persists throughout the animal kingdom as manifested in the basic distinction between somatic and visceral organs, a structural differentiation that has physiological meaning. Further specialization advanced more rapidly in somatic organs than in visceral, and in the former more rapidly on the sensory side than on the motor side. This again has physiological significance because the acute problems of subsistence involve adjustments to surroundings. The primitive motor responses are mass movements, but the surface of the body is exposed to a manifold of diverse stimuli which must be analyzed, and, accordingly, diverse organs of sense were locally differentiated. The course of this differentiation followed this rule: from the generalized and equipotential to the special and local. Thus in some primitive eyeless forms the entire skin is sensitive to light, in others only the most commonly exposed surfaces of it; and in the leech, Clepsine, Whitman ('92) found that a single animal exhibited all transitions from a series of segmentally arranged sense organs of generalized function in the posterior part of the body to well-formed eyes at the anterior end.

As I have elsewhere pointed out ('29), in the most simply organized vertebrates, the hagfishes, as described by Jansen ('30) and Conel ('29, '31), the brain is organized around two dominant sensory systems—olfactory and cutaneous—and the other special senses are in various stages of arrest or degeneration in correlation with a semiparasitic habit. Without eyes, jaws, or limbs, the visible behavior is reduced to a simple system of mass movements. Within the brain there is little local differentiation except for the primary sensory and motor fields directly connected with the peripheral end-organs, and yet this brain is the adjusting mechanism of a very rigid system of simple movements.

Larval Amblystoma is similar, though here the specialization of tissue is further advanced, and there is progressive advancement in representatives of later phylogenetic stages. The principle of progressive transformation from the general and dispersed to the specific and local applies throughout phylogenetic development; it is a general principle of embryogenesis (Weiss, '39, p. 288); it is clearly exemplified in human development, as has been demonstrated on the physiologic side by Hooker ('44) and on the structural side by Humphrey ('44, p. 39); and it appears in the course of conditioning reflexes (Coghill, '30).

In the phylogenetic history of vertebrates the basic pattern of sensory equipment was apparently laid down very early, with no radical changes except at the transition from aquatic to terrestrial life. And at this period of transition from fish to tetrapod the neuromotor apparatus experienced even more radical transformation, with elaboration of local reflexes which supplement and largely replace the more primitive mass movements.

In amphibian development this history is recapitulated in the long period which culminates at metamorphosis; and during this period the texture of the brain undergoes two divergent lines of differentiation of tissue in correlation with the expansion of two types of activity, the analytic and the synthetic, as described in the preceding section. The structural arrangement of the analytic apparatus is, in its main features, predetermined in the hereditary organization; it is stable and approximately the same in all members of the species. The intervening synthetic and integrating apparatus, on the other hand, is less rigid and is more labile in function. The pattern of its performance will vary from moment to moment in adjustment to every change in sensory and motor activity and every fluctuation of central excitatory state. But in even the most primitive vertebrates some cross-connections between sensory and motor zones, which are interrupted in the intermediate zone of correlation, are laid down in the stable, heritable structure. These serve the standardized ("instinctive") patterns of behavior, and the arrangement of these connections is determined more by the motor equipment of the animal than by the sensory equipment (Crosby and Woodburne, '38; Woodburne, '39).

From these considerations it follows that the concept of localization of function must not be formulated in static terms. It is localization of action, and the spatial pattern of this localization reflects every change in the character of the action. The structural pattern of this localization is more stable at the afferent and efferent endpoints of the system, and it becomes less so as we pass inward from these fixed points. The apparatus of standardized behavior like reflex is more rigidly localized than is that of more labile individually modifiable behavior.

Two types of structure which have just been contrasted may be characterized as unspecialized or *generalized*, and locally differentiated in *specific* stable and heritable patterns. The second was probably derived phylogenetically from the first, and in higher animals both of them exhibit progressive differentiation of structure in divergent directions. Some examples of these two types of cerebral architecture will next be cited, beginning with the second, which has been investigated in more detail.

SPECIFIC LOCALIZED STRUCTURE

An early stage in the evolution of localized conductors of specific sensory systems is illustrated by the connections of the lemniscus systems described in chapter xi. These tracts of Amblystoma are not well separated, and in the aggregate they comprise a rather dispersed collection of ascending fibers loosely assembled in several tracts, which are distinguished more by their general fields of origin and termination than by the functions which they serve. It is to be borne in mind that this low level of functional specificity of the secondary tracts is not correlated with a corresponding generalized structure and function of sense organs and related peripheral nerves. These organs, though different from those of higher animals, are highly specialized and as sharply localized. The correlation, on the contrary, is with the generalized character of the motor responses evoked. In higher animals with a wider range of motor capacities the functional specificity of the lemniscus tracts becomes more precise.

During the process of differentiation of these more specific tracts they retain collateral connections along the entire course, so that they continue to perform integrative functions similar to those of the less specialized ancestral pattern. In this connection we quote a passage from Dr. Papez ('36):

"In tracing the central connections of any one of the main afferent systems in the vertebrates one gains the impression that there is a progressive phyletic tendency of each system to enter into connections with all the important segments of the central organ. In this way there arises a totally integrated anatomical pattern common to all the receptorial systems in spite of the wide diversity of the receptors, their individual pathways and their interpolated centers. This central integration is not essentially of a reflex nature and cannot be appropriately described as a chain of reflex connections insomuch as each level has a highly individual structural organization designed primarily for the production of distinctive organic functions."

Papez appropriately emphasizes the integrating action of these long conductors; it seems to me, however, that his conception of "a progressive phyletic tendency of each system to enter into connections with all the important segments" in the interest of integration is a reversal of the actual course of phyletic history. These collateral connections are more numerous and more dispersed in lower forms than in higher. The integration is primary, and the analysis is secondary. It is true that the primary integration is not subordinated in the course of phylogeny; it is accentuated; but the apparatus employed is radically changed. Dispersed nonspecific connections are progressively replaced by localized specific structures, which are so interrelated as to work together harmoniously in the performance of standardized patterns of behavior. And, in addition to this, higher centers are elaborated, notably in the cortex, which progressively acquire dominant control of the total action system.

The phylogenetic history of the differentiation of the visual-motor system also illustrates the principle just stated. In most vertebrates the eyes are the dominant organs concerned with the orientation of the body and its members in space. The visual apparatus within the brain, accordingly, exhibits the most precise localization of function, and the refinement of this localization increases progressively in the phyletic series.

In Amblystoma the fibers of the optic tracts are widely spread in the brain stem, in marked contrast with those of other sensory systems, which tend to converge into a single receptive field. There is no evidence that this dispersal of fibers of retinal origin to the tectum, pretectal nucleus, thalamus, hypothalamus, and cerebral peduncle is correlated with any specificity of visual function. The explanation of this central spread of retinal fibers is to be sought on the motor side of the arc, that is, it is determined primarily by the nature of the response to be evoked.

In the Amphibia all these visual areas are centers of correlation, for in all of them optic terminals are mingled with those of other systems. Within this class, however, as we pass from generalized urodeles to specialized anurans, there is a conspicuous trend toward segregation of some terminals of the optic nerve in the tectum and lateral geniculate body of the thalamus. This trend culminates in mammals and is correlated with the differentiation of the visual area of the cerebral cortex, until in primates, as pointed out by Clark ('43), the retinal-geniculate-cortical pathway provides a very precise point-to-point projection of the visual field upon the cerebral cortex, and "there is no possibility that these impulses can be disturbed and modified 'en route' by other, unrelated, types of nervous impulse. . . . . In other words, the cerebral cortex receives retinal impulses in a remarkably pure and unadulterated form."

The highly specialized optic tecta of some lower vertebrates exhibit two types of specific localized structure, which differ in form and physiological significance. There is, first, an arrangement of sensory terminals spread superficially in mosaic pattern. This provides for point-to-point projection of retinal loci upon the tectum and perhaps for other forms of sensory localization. In the second place, there is a pattern of lamination at different depths from the surface. These laminae differ somewhat in their sensory connections, and the sensory influence is stronger in the more superficial members of the series. The arrangement of the deeper layers seems to be determined primarily by the directions taken by their efferent fibers. The mosaic pattern is primarily in terms of sensory analysis, the lamination pattern in terms of sensory correlation superficially and of motor analysis in the deeper layers. The cerebral cortex of mammals also exhibits both mosaic and laminated patterns of localization and in far more complex designs (Huber and Crosby, '33, '34).

Specific structure of this analytic type, with well-defined localization in both gray and white substance, increases in amount as we pass from lower to higher animals in the phyletic series; and this increment progresses from the sensory and motor periphery inward toward the upper cerebral levels, where the apparatus of integration and synthesis is most elaborately developed. In submammalian brains the amount of myelin present at successive levels of the brain stem is a rough indicator of the relative mass of tissue of this analytic type.

GENERALIZED STRUCTURE

This type of tissue, as pointed out above, predominates in the most primitive vertebrate brains. With advancing differentiation, we observe specialization of this tissue in three directions. (1) Some of it retains its primitive generalized structure with little change. In urodeles this is true of a large proportion of it; in mammals it survives in the periventricular system of cells and fibers of the diencephalon and mesencephalon and in some other regions. (2) A progressively larger proportion of this tissue is transformed into specifically localized structures, as just described. (3) Another large proportion of it is transformed into the relatively unspecialized tissues of the intermediate zone of correlation and its highly elaborated derivatives in the suprasegmental apparatus of the cerebellar and cerebral cortex.

Doubtless all parts of the body participate in the total integration and the determination of general attitudes and types of response, but the brain exercises dominant control over overt behavior and orders it in the interest of the welfare of the body as a whole. The apparatus of these totalizing functions evidently includes many diverse components, of which one of the most obvious is the neuropil, which in primitive vertebrates pervades the entire brain, so that activity in any part of it may affect the whole fabric, as elsewhere described. This dispersed tissue is not homogeneous, and it is not equipotential. It is doubtless always active and in diverse ways in different places at different times. Such localization of function as it exhibits can best be conceived in dynamic terms, that is, in terms of what intercurrent nervous volleys act upon it in momentarily changing places, rhythms, and intensities. We are dealing here with an equilibrated dynamic system comprising many activated fields in interaction, and this interplay is in patterns quite different from those of the stable, locally differentiated centers and tracts. It is more labile, and the patterns of performance are not stereotyped. Nevertheless, these fields are not structurally identical, and each one has distinctive physiological properties correlated with these histological differences.

Each field of neuropil differs from others in internal texture, in the source of afferent fibers, and in the distribution of efferents. A local field may be sharply segregated, as in the ventrolateral neuropil of the peduncle and the ventral interpeduncular neuropil, or it may interpenetrate tissue of the specific localized systems, as in the corpus striatum and optic tectum. The pattern of this localization is different structurally and physiologically from that of the specific systems of cells and fibers, and the two patterns of localization may both be present in the same block of tissue in primitive brains. In higher vertebrates these local differences are accentuated, the segregation of the synthetic apparatus is carried further, and its tissue is locally differentiated in a radically different way from that of the analytic apparatus, as is best exhibited in the associational tissue of the human cerebral cortex.

The "field" concept has been much exploited of late in several contexts, and it is fruitful, as applied, for instance, by Weiss ('39, p. 289) in general embryology and by Agar ('43, chap. ii) in general biology. As applied psychologically in Gestalt it has been difficult for the structurally minded neurologist to transfer the dynamic formulations into the biological frame of reference; but there is a structural "ground" within which every "configuration" of experience is set, and the primitive neuropil, together with its specialized derivatives, is one important component of the organic substrate of the "gestalt qualities." The "field" as here conceived is an organized living structure in action, some components of which we recognize as stable architecture and some as fluctuations in the excitatory state of the structural fabric. This structure has visible organization, and its properties are open to investigation anatomically and physiologically ('34a; '42, p. 293).

TWO CONCEPTS OF LOCALIZATION

The two kinds of structure which have just been considered perform functions which are localized according to different principles, a distinction which has been generally ignored. The specialized analytic structures have stable arrangement in space, and their functions have corresponding localization in three-dimensional mosaic patterns. The functions of the generalized structures are, in the main, synthetic rather than analytic, and they are usually described in dynamic terms in which a fourth dimension—a time factor—plays an important part. Yet this equilibrated dynamic system is not disembodied, and its component parts have locus in space and time. These loci, or fields, may have some degree of permanence with characteristic structural organization, as in the so-called "association centers" of the cortex, or they may fluctuate as ever changing patterns of linkage of the operating neurons. As Papez ('44) expresses it: "The anatomical structures are stable, the function is labile, depending on numbers of cells and their excitable or refractory states at any particular time." Many of these synthesizing patterns are repetitive, as in habit and memory. Though the actual structures involved may not be identical in successive repetitions, the pattern of performance is similar, that is, it recurs in conformity with an enduring "schema" (to employ Henry Head's term), or the engram of conventional terminology. Doubtless the engram has a structural (or chemical?) counterpart, but we do not know what it is.

The stable localization of the structural fields is contrasted with the evanescent localization of the pattern of their combination at each repetition of the schema. It is possible to find out where the tissue is that yields these dynamic schemata and to delimit it; but these limits cannot be circumscribed on the surface of the brain in simple mosaic patterns. The manifestation of any schema at a particular time is always a function of a configuration of nervous elements, which has location in space. But a very similar schema may at another time be exhibited by a different structural configuration, whose locus in space is by no means identical with the first ('30a).

It has recently been shown by Lashley and Clark ('46) that cortical structure is variable to a degree not hitherto appreciated in different individuals of the same species of monkey, and it is probable that the range of this variability will be found to be still greater in any human population. They conclude that "marked local variations in cell size and density among individuals of the same species may constitute a basis for individual difference in behavior"; but they challenge the validity of the criteria in current use for parcellation of the cortex into functionally specific areas, except for rather large areas of projection. This is supported by the experiments of Murphy and Gellhorn ('45) and the observations of Bailey and von Bonin ('46).

CHAPTER VII

THE ORIGIN AND SIGNIFICANCE OF CEREBRAL CORTEX

THE PROBLEM

THE human brain is the most complicated piece of mechanism that we know, and the products delivered by this thinking machine are, for us, the most interesting. Detailed descriptions of it and some of its operations have been written, but where and how it fabricates its unique wares is still the basic problem of science and philosophy. Thinking is a part of our living, and apparently we think all over, just as we live in all parts of our bodies. Yet it is evident that some parts of us play crucial roles in our mental life, just as other parts do in our movements, our digestion, and so on. That the cerebral cortex as a whole is a specific organ of much of our mental life is as well established empirically as anything in biology, but how thinking is done and where the critical processes are carried on are still mysterious.

Study of origins and early stages of embryologic and phyletic development has resolved many biological and psychological problems, but essential features of the inception of cerebral cortex remain obscure. Interest in this theme instigated my program of research upon the amphibian nervous system. All reptiles possess well-organized cortex, that is, superficial laminae of gray substance in the pallial part of the cerebral hemisphere, simple in pattern but obviously comparable with, and ancestral to, the human cortical complex. They also exhibit enormous enlargement of some subcortical parts of the hemisphere, notably the strio-amygdaloid complex. In birds these subcortical areas are still further enlarged and complicated, with reduction in the amount and specialization of the cortical tissue. In mammals the subcortical parts of the hemisphere are relatively smaller, some of their functions apparently having been taken over by the progressively expanding cortex ('21, p. 452; '26, p. 122). Correlation of these structural peculiarities with the characteristic modes of life of these several classes of animals gives some clues to the significance of the cortex in the vital economy.

As already pointed out, structural differentiation of cortex is incipient in the hippocampal sector of the pallium in all Amphibia; more clearly defined primordia of cortex are seen in adult lungfishes and in embryonic stages of some other fishes (Rudebeck, '45); but well-differentiated cortex of typical structure first appears in reptiles. The problem set is: What morphogenic agencies were operative during the emergence of cortex from a noncorticated matrix?

Early in the attack upon this problem it became evident that the key factors were to be sought, not in the pallial field, but in its environs. What comes into this field and what goes out of it at successive stages of morphogenesis and how are these factors related one to another in both subpallial and pallial territory? This requires in the upshot a histological analysis of the entire nervous system directed toward the physiological interpretation of all visible structure. In my *Brains of Rats and Men* ('26) the available evidence regarding the origin of cerebral cortex was surveyed. Since that time additional evidence has been recorded, and in the present work the results of a renewed examination are summarized. The problems centering in corticogenesis have not been solved, but some progress has been made. The conclusions reached can, at best, be only tentative, pending physiological control; and for such experiments exact information about the anatomical arrangements of parts is indispensable.

MORPHOGENESIS OF THE CEREBRAL HEMISPHERES

There is reason to believe that in the early ancestors of vertebrates the central nervous system was a simple tubular structure comparable with that of living Amphioxus. Accompanying enlargement of the dominant sense organs—nose and eye—the anterior part of this tube was expanded in four places, which became cerebral hemispheres, hypothalamus, epithalamus, and tectum opticum. In most lower vertebrates the olfactory system is very large and has played a dominant role in the earlier stages of the morphogenesis of the hemispheres. In fishes this differentiation took a wide variety of forms, some of which were surveyed in two papers ('21, '22a). These diverse forms and patterns of internal structure were shaped in adaptation to various modes of life which employed different equipment of sensory and motor organs.

It has been mentioned that the more sluggish fishes, and especially mudfishes living in stagnant water, have enormously enlarged and highly vascular chorioid plexuses, the ventricles are dilated, and the walls of the brain are thin. This insures a supply of oxygen to the brain which is adequate for their quiescent existence. The more active fishes living in well-aerated water lack these features and cannot survive in stagnant water. Their brains have solid masses of tissue in great variety of forms and are sensitive to asphyxiation.

The fossil record shows that amphibians were derived from generalized fishes similar to the living lungfishes. These are mudfishes with enlarged chorioid plexuses and thin-walled widely evaginated hemispheres. There is geological evidence that in early Devonian time, when the amphibians emerged from the water, there was general continental desiccation. The lakes and streams were drying up, and over extensive areas the freshwater fishes were faced with the alternative of adaptation to drought or extinction. The more highly specialized species perished, but some generalized and sluggish types of mudfishes made successful adjustment. Their extensive chorioid plexuses and widely expanded thin-walled hemispheres had survival value, for so they were tided over the critical period of oxygen deficiency during phyletic metamorphosis. These characteristics persist today in all urodeles, most of which have retained the ancestral mode of life.

In later evolutionary stages the expanded hemispheric vesicles had the further advantage that space is available for further differentiation, and especially for spreading out the correlation tissue in thin sheets, an arrangement which seems to be requisite for refinement of adjustment to the spatial relations of things and for high development of labile, individually modifiable behavior as contrasted with stable, heritable behavior of instinctive type. The interested reader will find further details about the origin and significance of the evaginated form of cerebral hemispheres in the two papers cited ('21, '22a) and in chapter xvi of my *Neurological Foundations* ('24c).

From the beginnings of differentiation of the evaginated hemispheres, their medial and lateral walls have shown striking and interesting differences. The telencephalic connections are variously arranged in the different groups of fishes; but in the Amphibia the cerebral hemispheres have acquired the definitive form with connections that are not fundamentally changed in the higher groups. There is no evidence that this arrangement is due to differences in the quality of the olfactory impulses transmitted from the olfactory bulb, except for the possibility suggested (p. 54; '21a) that specific impulses pass from the vomeronasal organ to the accessory bulb and amygdala. The nervus terminalis is more evidently specific, but of its functions nothing is known. All other descending olfactory impulses seem to be physiologically homogeneous. From this it follows that local differentiations in the hemispheres are due mainly to differences in the distribution of the various systems of ascending nonolfactory fibers or to differences in the destination of descending fibers arising from various parts. Evidently both factors are involved.

The connections between the hemispheres and the lower levels are assembled in the basal forebrain bundles and the stria medullaris thalami. The former are roughly comparable with the subcortical components of the mammalian extrapyramidal systems plus large numbers of ascending fibers, and the details of their connections are given in chapter xx. The analysis of these systems of fibers provides the key to the interpretation of the hemispheres of Amblystoma; and similar analysis in reptiles, birds, and mammals is essential for an understanding of the history of cortical evolution.

The fibers of the stria medullaris system are efferent or commissural, passing from all parts of the hemisphere to the habenula or decussating in the habenular commissure and returning to the hemisphere. The arrangement of the components of the stria is shown in figure 20, and this is essentially similar to that of mammals. The stria medullaris, in fact, is remarkably similar in all vertebrates; and these fibers evidently belong to a different category from those of the basal bundles, which contain ascending and descending fibers in arrangements that vary widely from species to species as we pass from lower to higher vertebrates. These variations seem to be especially significant for interpretation of morphogenesis. In this section, accordingly, attention will be directed to the latter systems, and the reader is referred to chapters xiv and xviii for the interesting details of the habenular connections.

The basal bundles comprise three groups of fascicles, the descending components of which are shown diagrammatically in figure 6: (1) dorsally and laterally the lateral forebrain bundles (*f.lat.t.*); (2) ventrally and medially the medial bundles (*f.med.t.*); (3) the olfactopeduncular tract (*tr.ol.ped.*), lying between the two preceding and with connections which are intermediate between them at both ends. Fibers ascend in these bundles from two sources: (1) from the somatic sensory field of the dorsal thalamus, by way of the thalamo-frontal tract (figs. 19, 111, *tr.th.f.*) in the lateral forebrain bundle, and (2) from the visceral field of the hypothalamus, by way of the medial bundle (fig. 113). The first of these bundles connects with the lateral wall of the hemisphere, the second with the medial wall. In conformity with this, the descending path from the lateral wall goes by way of the lateral forebrain bundle to the somatic motor field in the peduncle, and the path from the medial wall goes by way of the medial bundle to the visceral field in the hypothalamus (figs. 6, 11, 111, 112, 113). These connections inaugurated the different types of differentiation seen in lateral and medial walls of the hemisphere, a difference in type which becomes more pronounced in higher animals. The amygdala is of intermediate type, with somatic and olfacto-visceral connections of both afferent and efferent fibers (fig. 19).

All parts of the hemisphere are under olfactory influence, with olfacto-visceral correlations effected medially and olfacto-somatic laterally. The motor responses are radically different, and this difference is probably the basic determining factor in shaping the structural plan, not only of the olfactory connections, but of the organization of the hemisphere as a whole.

The hypothalamus is disproportionately large in lower vertebrates as compared with higher, and the olfacto-visceral functions of the hemisphere are correspondingly magnified. This doubtless accounts for the fact that differentiation in the pallial field is further advanced on the medial (hippocampal) side than elsewhere (fig. 99) and also for the fact that efferent projection fibers from this part of the pallium (fornix system, p. 254) appear in large numbers very early in phylogeny.

Fibers ascend from the dorsal thalamus to the hemisphere in fishes and in all higher animals. In Amblystoma these thalamo-frontal fibers arise in the generalized nucleus sensitivus, ascend in the lateral forebrain bundle, and end in the amygdala and middle part of the corpus striatum (figs. 19, 111, *tr.th.f.*). They are comparable with thalamo-striatal fibers of mammals and are precursors of mammalian sensory projection fibers, though in urodeles none have been seen to reach the pallial part of the hemisphere. This tract is small, and there is no evidence that different sensory systems are separately localized within it; its terminal nucleus, the corpus striatum, is correspondingly small and simply organized. The thalamo-frontal system is larger in reptiles and birds, where the striatal complex is magnified; but only in mammals are the several functional systems well segregated and separately localized, and here this localization is correlated with the differentiation of the sensory projection areas of the neopallium and the related nuclei of the dorsal thalamus. The course of the evolutionary differentiation of the cerebral hemisphere has been determined, in its main features, by the penetration of these nonolfactory fibers into its medial and lateral walls and the elaboration of related centers for the reception and correlation of these sensory impulses and appropriate motor discharge.

The entire ventrolateral wall of the hemisphere of Amblystoma is an olfacto-striatum. Its anterior end is under strong olfactory influence and is regarded as the primordium of the head of the caudate nucleus (figs. 98, 99). Posteriorly, the large amygdala (fig. 97) also receives many olfactory fibers. The middle sector receives fewer olfactory fibers and is a terminal station of the thalamo-frontal tract. From it fibers descend to the ventral thalamus and peduncle. It is, accordingly, regarded as paleostriatum, or somatic striatum, primordium of the mammalian lentiform nucleus.

The striatal gray of the middle sector is obscurely divided into dorsal and ventral nuclei, separated by a shallow sulcus striaticus (fig. 99, *s.st.*). Small cells of putamen type and larger cells of globus pallidus type are mingled in both nuclei, so that these structures are not separated as in mammals; yet their connections, as described in chapter xx, suggest that the ventral nucleus is the precursor of the globus pallidus. Both these nuclei receive fibers from the overlying piriform area, some of which descend into the lateral forebrain bundle (figs. 6, 111). The ventral nucleus is continuous with and intimately connected with the primordial caudate ('27, p. 298), and its efferent fibers go chiefly to the cerebral peduncle. The dorsal nucleus receives thalamo-frontal fibers and is in intimate relation with the amygdala and the piriform area. Its descending fibers have wide distribution to the tectum, thalamus, dorsal part of the peduncle, and dorsal, isthmic, and bulbar tegmentum as far back as the V nerve roots (fig. 6, *f.lat.t.d.*). In Necturus (fig. 111; '33*b*, p. 197; '33*e*) there is a complicated system of associational connections between the striatum and the piriform area; these are present also in Amblystoma, but the details have not been described.

Most of the white substance of the striatal field is occupied by a very dense and sharply circumscribed intermediate neuropil of peculiar texture (p. 53 and figs. 98, 99, 108, 109, 113; '27, p. 300; '42,

p. 202; Necturus, '33b, p. 149). There is a web of interlaced branches of ependymal elements and among these a still more densely woven tangle of dendrites and slender, contorted, unmyelinated axons. The thicker descending axons of the dorsal and ventral fascicles of the lateral forebrain bundle are assembled within this neuropil (figs. 111, 113). I have seen similar texture in Golgi impregnations of the head of the caudate in the opossum ('24d, p. 342), a structure adapted for diffusion and summation of all nervous impulses entering it.

The urodele type of striatal structure could be transformed into that of mammals by reduction of the olfactory component; further differentiation of the caudate nucleus and amygdala; segregation of the large efferent neurons in the ventral nucleus, which becomes the globus pallidus; and segregation of the smaller elements in the putamen. In Amblystoma these changes are only dimly foreshadowed. The ventral fascicles of the lateral forebrain bundle connect with the primary motor field of the ventral thalamus and peduncle and are comparable with those of the ansa lenticularis. The connections of the dorsal nucleus suggest relations with the reptilian neostriatum and the mammalian putamen. Here are probably to be found the earliest indications of those formative agencies which in later phyletic stages led to the differentiation of neopallium. In the still more primitive hemisphere of Necturus, these indications were recognized and discussed ('33e).

The strio-amygdaloid complex as a whole is the highest center of dominance in the control of the skeletal musculature, a role which is enormously enlarged in reptiles and birds. In mammals, parallel with the elaboration of cortex, the part which the striatum plays in the patterning of behavior is progressively reduced, but it retains important functions of co-ordinating and stabilizing motor performance. Just as the cerebellum is added to the sensori-motor systems for facilitation of muscular co-ordination, so in mammals the striatal complex is interpolated in the efferent cortical systems as an accessory facilitating mechanism.

In the Amphibia we find a critical stage in the morphogenesis of the cerebral hemisphere. The definitive major subdivisions are here blocked out in recognizable form, and the pallial part is incompletely segregated from the stem part; yet the most distinctive feature of the pallium—its superficial cortex—has not yet appeared. We want to know more about the agencies which are in operation here to initiate the separation of pallium from stem and what further changes led to the migration of the pallial gray from deep to superficial position and its subsequent complications. A beginning has been made, and I have at various times reported progress in this analysis and some discussion of its meaning ('24c, chaps. xv, xvi; '24d, p. 354; '26; '27, p. 315; '33a; '33b; '33e; '34a). Yet much remains to be done before we can fill in those finer details of structure which the physiologist needs to know in order to plan crucial experiments. Frogs are probably better adapted for such experiments than are salamanders, and to this end a more detailed analysis of the histological structure and connections of the forebrain of the frog is urgently needed. Sufficient knowledge of this structure is now available to enable the physiologist to explore the instrumentation of some components of the behavior pattern, as illustrated by a recent study by Aronson and Noble ('45).

THE CORTEX

More than thirty years ago I published some reflections under the title given to this chapter ('13a). Though parts of that paper require revision in the light of subsequent research, yet it sketches the background of the present discussion. Attention was called to Dewey's ('93) concept of the organic circuit as a substitute for the classical formulation of the reflex arc, as mentioned in the preceding chapter. Some illustrations of these organic circuits were given in my article. Here we need only to emphasize the fact that all behavior is the resultant of their interplay, for which provision is made in the cerebral mechanisms, such as are described in this and other works devoted to neuroanatomy. These are all circular reactions between receptor and effector organs or the related internal adjustors. In the course of phylogeny, cerebral cortex has been differentiated as the culminating member of a series of progressively more complicated integrating mechanisms adapted to make more efficient use of the preformed circuits of the brain stem in the interest of more flexible behavior in terms of individually acquired experience, as contrasted with the stereotyped patterns of the stem (for a convenient summary of the human connections see Papez, '44).

In this connection two sentences may be quoted from von Bonin ('45, p. 52): "It is of the essence of cortical organizations that sensory and motor areas become divorced more and more from each other—are pulled farther apart as it were—as evolution proceeds. .... As we ascend the evolutionary scale, the cortex assumes increasingly a structure which may be interpreted as leading to increased degrees of indeterminacy."

A fundamental feature of cortical functions is that they are delayed reactions (p. 78; '24c, p. 271); there is, first, the arrest or inhibition of some lower and more primitive patterns of behavior of reflex or instinctive type. This allows time for cortical reorganization of the component factors of the situation, conditioning of reflexes, or other modifications of the stereotyped patterns of response. The cortex, accordingly, is lifted up away from the lines of through traffic in the brain stem, and the dorsal convexity of the evaginated cerebral hemisphere is conveniently located, with ample space for indefinite enlargement.

During the phylogenetic development of cortex, ascending and descending pallial projection fibers are added to the pre-existing systems of the underlying stem. They do not entirely supplant them, for even in mammals, where cortical projection systems are highly developed, the subpallial parts of the hemisphere retain their own diencephalic connections.

Primitively, as in cyclostomes, the entire cerebral hemisphere is little more than an olfactory bulb and a secondary olfactory nucleus. In Amblystoma the hemispheric evagination is more extensive, and there is a large increment of ascending nonolfactory fibers; yet here the pallial part of the hemisphere receives the largest olfactory tracts, and all of it is essentially an olfactory nucleus.

The olfactory reflexes seem to be adequately provided for in the stem portion of the hemisphere. I have suggested ('33) that here, and especially in the corticated mammals, the olfactory sense, lacking any localizing function of its own, co-operates with other senses in various ways, including a qualitative analysis of odors (desirable and noxious) and also the activation or sensitizing of the nervous system as a whole and of certain appropriately attuned sensori-motor systems in particular, with resulting lowered threshold of excitation for all stimuli and differential reinforcement or inhibition of specific types of response. The olfactory cortex (and its predecessors in lower vertebrates) may, then, serve for nonspecific facilitation of other activities, in addition to its own specific olfactory functions. This facilitation may involve both general excitatory action and general inhibition. That the latter is present is indicated by the observation of Liggett ('28) that anosmic rats are more active than the normal controls. The organization of the olfactory system as a whole in all animals seems consonant with this interpretation; and in Aronson and Noble's study of the sexual behavior of frogs this facilitating action of the olfactory field is clearly demonstrated.

In primitive vertebrates the dominance of the entire anterior end of the brain by the olfactory apparatus implies more physiological homogeneity than in higher brains, where this tissue is invaded by larger numbers of nonolfactory fibers with more diverse specificities. The case is somewhat like the invasion of a hitherto isolated continent with homogeneous and primitive population by immigrants of numerous other races with very diverse cultural standards. When European peoples colonized North America, in some regions the newcomers intermarried with the natives and the two races amalgamated; in other places the indigenous population was driven farther and farther back or exterminated altogether. Something analogous to these processes has taken place during the invasion of the olfactory area by nonolfactory functional systems. In some regions there is blending of the old and the new, as in the amygdala, septum, and olfactory tubercle; in other places the indigenous olfactory system has more nearly retained its unmixed character, as along the margin of the olfactory bulb; and in other extensive regions of the hemisphere the indigenous elements have been almost entirely displaced by nonolfactory systems, as in part of the corpus striatum and the neopallial cortex. In the Amphibia this invasion of the olfactory field by nonolfactory systems is extensive, but the invading forces are not sufficiently diversified and localized to invoke the differentiation of cortical tissue in the pallial part of the hemisphere. This is probably correlated with the fact that amphibian behavior, by and large, is mass movement, with relatively little refined analysis into partial patterns.

The primitive differentiated cortex of reptiles has three well-defined sectors. These are spread, respectively, on the dorsomedial, dorsolateral, and dorsal convexities of the hemisphere. The first is archipallium, the precursor of the mammalian hippocampal formation. The second is paleopallium, or piriform cortex, represented in man by a relatively small area at the lower border of the temporal lobe and including the uncus and some adjoining areas. The third sector, the dorsal cortex, is of uncertain relationships. It occupies the position of the neopallium, which comprises the larger part of the human cortex, and probably is its precursor, though this apparently is not its only relationship.

In all amphibian brains these sectors of the pallial field can be identified, though no superficial cortical gray is present in any of them, as illustrated in figures 96–99. On the lateral aspect the piriform area (figs. 85, 86, 111, *p.pir.*) shows no evidence of cortical differentiation; it is, in fact, the chief secondary olfactory nucleus (*nuc.ol.d.l.*); nevertheless, its location and connections identify it unmistakably as the primordium of the piriform cortex of reptiles and lower mammals. Its neurons are small, simple, and similar to those most commonly seen in the brain stem (figs. 98, 99, 105). On the dorsal convexity there is an undifferentiated and poorly defined field (*p.p.d.*), which is the precursor of the reptilian dorsal cortex.

The medial sector—primordium hippocampi (*p.hip.*)—shows a first step toward cortical differentiation, for here the compact central gray layer is dispersed by outward migration of the cells throughout the thickened wall, and these are imbedded in dense neuropil. These neurons vary from small to quite large and, in general form, resemble those of other parts of the pallium, though they are evidently more specialized (figs. 97, 98, 99, 105). One to several thick and thorny dendrites arise from the cell body and spread widely, some reaching the external limiting membrane. The axon may arise from the cell body, but usually from the base of one of the dendrites. It may divide, sending one branch into the dorsal pallium and one to the septal nuclei or medial forebrain bundle (fornix). Some axons are short, branching freely within the area of spread of the dendrites (fig. 105; '39b, fig. 44), but most of them send one or more long fibers from this arborization into tracts which leave this field. Close to the surface are a few tangential neurons which at one time I regarded as precursors of the reptilian cortical cells. Similar cells are found also throughout the brain stem, and in the pallium (fig. 98) their axons take short courses as correlation fibers. They are more numerous in the frog (P. Ramón y Cajal, '22, figs. 6, 7). The differentiation of the hippocampal cells is further advanced in Amblystoma than in Necturus ('33a, p. 183) and less so than in anurans. The hippocampal neuropil increases in density and complexity as we pass from Necturus to Amblystoma and the frog.

The four layers of neuropil characteristic of the brain stem (p. 30) are very unequally developed in the amphibian hemisphere (Necturus, '33b, p. 176). The periventricular neuropil is everywhere abundant. The deep neuropil of the alba contains an elaborate system of association fibers, which has been described in detail (figs. 111, 113; '33b, p. 194; '33e). The intermediate neuropil contains many of the recognizable long tracts, and in the striate area it is elaborately developed, as already described. Superficially of the striate neuropil is a strio-amygdaloid neuropil, which is continuous dorsally with the piriform and dorsal neuropil. This sheet as a whole is evidently the synaptic field of the pallial associations. It receives the dendrites of the underlying gray substance but contains no cell bodies. In higher animals this synaptic zone seems to exert a neurobiotactic influence, so that in embryonic stages all the neurons of the pallial field migrate outward and are incorporated within it, thus producing the laminate cortex.

In the amphibian primordium hippocampi, this movement has begun but is not consummated. The deep gray layer has been broken up, and its elements are dispersed. The periventricular neuropil of the grisea and the neuropil of the alba merge, so that the entire area is pervaded by a dense entanglement of dendrites and axons, within the meshes of which the cell bodies are imbedded. This neuropil is denser in two places—rostrally and ventrally, where subpallial connections predominate, and dorsally, where pallial associations predominate. In the reptiles, with differentiated cortex, the corresponding two parts of the hippocampal cortex are structurally different.

In Necturus the most rostral fascicles of the strio-pallial association go far forward and dorsally above the posterior end of the olfactory bulb to reach the dorsolateral sector of the anterior olfactory nucleus and territory adjacent to it (fig. 111). This is clearly a secondary olfactory nucleus of subpallial type, adjacent to the olfactory bulb and traversed by the great dorsolateral olfactory tract, from which it receives numberless terminals and collaterals. Its principal discharge is backward into the primordium piriforme, a pallial area (fig. 111, *tr.ol.pal.l.*). These connections would, perhaps, have no special significance in themselves, but comparison with reptiles shows that there the corresponding region exhibits remarkable peculiarities. In urodeles the area in question is one of the least differentiated parts of the hemisphere, except for the strong fascicles of the strio-pallial association, and perhaps this lack of specialization favors the role which it seems to play as germinative tissue for neopallial cortex.

Cortex of simple pattern is present in turtles in each of the three pallial fields seen in Amphibia; and the dorsal and lateral cortex (general cortex and piriform cortex) are related to a massive subcortical thickening of the lateral wall of the hemisphere, which was called the "dorsal ventricular ridge" by Johnston and "hypopallium" by Elliot Smith. The thalamic radiation, comparable with the amphibian tractus thalamo-frontalis, is large in turtles. It ends chiefly in the rostral part of the hypopallium, but some of these fibers pass through without synapse into the dorsal or general cortex. The latter fibers are true thalamic sensory projection fibers with connections of neopallial type. In front of this region there is a "pallial thickening," from which motor cortical projection fibers go out to the cerebral peduncle—again a neopallial type of connection—and this part of the cortex is electrically excitable (Johnston, '16).

In the alligator the topographic relations are very different. The dorsal, or general, cortex has no contact with the hypopallium except at its rostral end in the region of the pallial thickening of turtles, which Crosby ('17, pp. 358, 381, figs. 5, 6) calls "primordial general cortex." This primordium she regards as the germinal tissue or focal point in the differentiation of the general cortex, and she gives a clear statement of the factors which probably were operative in the differentiation of this general cortex.

The dorsolateral sector of the anterior olfactory nucleus of Necturus, together with some adjoining tissue, is just such an area of basal, i.e., subpallial, type as Crosby postulated; it is in the exact position with reference to other parts of the hemisphere as her primordial general cortex; and it receives especially strong fascicles of the strio-pallial association, which turn far forward to reach it. It is significant that in the alligator this primordium is the only region where both projection fibers of the lateral forebrain bundle and shorter fibers from the hypopallium and corpus striatum can connect with the general cortex (Crosby's figs. 5–9, 12–19, 37). Bagley and Langworthy ('26) have shown that in the alligator this area and parts of the cortex adjoining are electrically excitable, thus furnishing experimental proof that true motor projection fibers of neopallial type arise from it. The underlying hypopallium was tested and found to be unexcitable. Ariëns Kappers ('29, p. 140) accepts Crosby's interpretation of the reptilian primordial neopallium and states that in the lizard, Varanus, a small number of thalamic projection fibers ascend directly to this area. This, he says, is the source of the neopallium of mammals, not the more differentiated general cortex of the dorsal convexity of the hemisphere.

The dorsolateral sector of the anterior olfactory nucleus and the rostral end of the primordial piriform area of Necturus may, accordingly, be regarded as critical points in further search for the earliest primordium of the neopallium. This region is related with the primordial general cortex of reptiles, which may be regarded as the precursor of the subiculum and other transitional fields rather than of neopallium, *sensu stricto*. The preceding account of the probable history of cortical evolution is drawn largely from Crosby's graphic and discerning analysis published in 1917.

The structure of the pallial field of Amblystoma and its connections were described in 1927 and subsequently in greater detail in several papers devoted to Necturus ('33b, '33e, '34, '34a). In these papers and some earlier publications I commented on the fact that the first well-differentiated cortex appears in reptiles in three clearly defined areas; and the opinion was expressed that a prerequisite for this differentiation is the penetration into the pallial field of thalamic projection fibers in separately localized tracts with different physiological properties. This minimal localization of function in the projection systems goes hand in hand with local differentiation in the pallium and amplification of the cortical associational connections of these areas. This process of local cortical differentiation continues to advance in complexity of pattern in proportion as the systems of thalamic projection fibers are amplified and diversified.

This principle of cortical morphogenesis receives its first and clearest exemplification in the obvious difference in the subpallial connections of the medial and lateral parts of the pallial field, hypothalamic connections predominating medially and thalamic connections laterally. In Ichthyopsida the hypothalamic influence is much stronger than the thalamic, a relation which is strikingly reversed in higher vertebrates. These diencephalic influences are not sufficient to cause cortical differentiation in the amphibian pallium, though there are some local differences in the three recognizable pallial areas and Söderberg ('22) found clearer evidence of cortical incipience in some early larval stages. Holmgren ('22) found evidence of cortical differentiation in developmental stages of selachians and some other fishes, and in adult lungfishes a primitive cortex is clearly delaminated externally of the central gray (Rudebeck, '45).

In none of the fishes and amphibians do we find so well-differentiated cortex as in reptiles. Why is this? The answer seems to be that in all Ichthyopsida the entire forebrain is dominated by the olfactory system to such an extent as to retain a measure of physiological homogeneity, which is not favorable for cortical differentiation. The basic feature of cortical function is the association of diverse components of the action system with separate localization of the functional systems involved. In fishes and amphibians this localization of function in the forebrain is incipient, but it is not sufficiently advanced to evoke cortex of definitive type. In reptiles, on the other hand, the great increase in the system of somatic sensory exteroceptive thalamic radiations is correlated with enlargement of the corpus striatum complex, including an extensive area quite free from olfactory and hypothalamic connections and the extension of some of the fibers of the thalamic radiation to the dorsal pallial field without interruption in the striatum. Thus the pallial field is subdivided into three well-circumscribed areas, each with a physiological specificity different from those of the others and one of which is emancipated from dominance of the olfactory system. Now for the first time in phylogenetic history the pallium possesses an intracortical system of association fibers adapted for the specific cortical type of function ('26, pp. 78, 123; '27, pp. 315 ff.; '33e; '34a).

In a survey of the history of cortical evolution the birds occupy an anomalous position. They are much more highly differentiated than reptiles, but in an aberrant direction, with no mammalian affinities. In most of them the olfactory system is reduced almost to the vanishing-point, and the optic system is greatly enlarged. There is extensive local differentiation of thalamic nuclei, but not in the mammalian pattern. The system of ascending thalamic projection fibers is larger than in reptiles, and most of these fibers end in the enormously enlarged and complicated corpus striatum. Correlated with the latter point is the striking fact that, despite the great increase in thalamic projection fibers, the cortex of many birds is scarcely more extensive than in reptiles and in some species is less well differentiated (Craigie, '40). Birds are more highly specialized in both structure and behavior than are the lower mammals, and yet their cerebral cortex is rudimentary in comparison with even the most primitive mammals. The explanation for this is that the bird's more diversified behavior is largely stereotyped in instinctive patterns, adequately served by subcortical apparatus, while the patterns of mammalian behavior, even in the lowest members of the class, are in larger measure individually learned. And enhancement of learning ability goes hand in hand with cortical differentiation.

It is impossible to define a primordial boundary between pallial and subpallial territory ('27, p. 316; '33a). There is apparently no primitive (palingenetic) distinction between the pars pallialis and the pars subpallialis of the cerebral hemisphere; cortical types of structure and function may be differentiated out of such raw materials as are available, and the locations of these indefinite boundaries will vary from species to species. Even in the human brain the boundary is in some places obscure and controversial.

This, in outline, seems to be the history of the origin and evolution of the cerebral cortex. The details have not yet been filled in, and this can be done only by experimental methods, checked and controlled at every step by accurate histological analysis of the tissue operated upon. When the facts about the sequences of the evolution of cortical structure and function are colligated with experimental studies of behavioral capacities of the animals in question, we shall have a secure foundation upon which to build a sound comparative psychology, and this, in turn, will clarify much that is now obscure in human experience.

PHYSIOLOGY AND PSYCHOLOGY

Returning, now, to a general survey of the factors involved in the differentiation and normal operation of the central nervous system, we find that these fall into two categories. Some of them conform perfectly with well-known laws of traditional mechanics of the inorganic realm, as formulated in the Newtonian system and its modern derivatives. Others have not been successfully fitted into this frame of reference. From the beginning of inquiry into this problem, there has been a tendency to set these refractory components apart from the natural order in some mystic realm of vitalism. To the naturalist this solution is not acceptable, for the two classes of phenomena are empirically indissociable.

That the operations of nature are not bound by the man-made rules of Newtonian mechanics is now evident. It has been shown that the formulas of Newtonian mechanics, Euclidean geometry, and Aristotelian logic are not universals. They are valid in a restricted field but not in the realm opened up by current conceptions of relativity and quantum mechanics. In view of this situation, the field of neurological inquiry is immeasurably enlarged and complicated.

In our search for the laws of growth and normal action of the nervous system, we naturally and properly look, first, for those features which can be fitted into the conventional formulations of inorganic mechanics. Up to the present time the science of neurology has been concerned almost exclusively with this aspect of the problem, with eminently successful results. Mechanical stress and tension, pressure and movements of fluids, local chemical action, surface tensions and permeabilities, electrical phenomena of many kinds—these and other physical factors now under investigation are bringing to light many basic principles of nervous action. In a wider field D'Arcy Thompson's great work, *On Growth and Form* ('44), does not trespass beyond these boundaries, for he says (p. 15): "When we use physics to interpret and elucidate our biology, it is the old-fashioned empirical physics which we endeavour, and are alone able, to apply." A useful general summary of *Medical Physics*, edited by Otto Glasser ('44), has recently been published. But this line of attack sooner or later reaches limits beyond which it has not yet been possible to go.

In human neurology the major problem of all times has been the relation of these physicochemical processes to the conscious experience which emerges from them. The normal subjective life is not disorderly, but the laws of this order as revealed by introspective psychology seem to be incommensurable and disparate with those of the underlying physicochemical system as known objectively. This gap has not been bridged by any acceptable formulation in terms of Newtonian mechanics, and more and more of the experts in this field are coming to believe that this cannot be done. This does not imply any appeal to mysticism. Quantum mechanics takes its place in the order of nature along with Newtonian mechanics, and it may well be that the solipsistic qualities of that "private" conscious experience which is not objectified are related to the events of the objectively known "public" physicochemical world in accordance with principles as different from those of Newtonian mechanics as the latter are from quantum mechanics. If so, these principles of the mind-body relationships have not yet been formulated, and we live in hope that some day this will be done. Just as quantum mechanics has added a time dimension to the three Cartesian co-ordinates of space and additional dimensions beyond our range of experience in theoretic mathematical physics, so it may well be that introspective experience and objective or extraspective experience are related in terms of dimensions not yet recognized and given scientific expression, defining a dimension as "any manifold which can be ordered," as Reiser ('46, p. 89) does in an interesting discussion of this problem.

The theories of mind-body relationships are mentioned here because I regard mentation as a vital process as truly as are muscular contraction and glandular secretion. The laws of operation of these three processes are different, their products are different, and the apparatus employed is different. All these operations have locus in place and time. The organs which perform them have been slowly differentiated and matured during embryonic and phyletic history. We have reason to believe that mind is not an exclusive prerogative of mankind. Mental capacity has developed parallel with the growth of its organs. The genetic and phylogenetic approach to the mind-body problem has already yielded significant data; and, as this study is carried backward toward earlier stages (embryonic and phyletic), the unresolved problems of human psychophysics do not disappear.

Something akin to the mental as we experience it may be a common property of all living things and even of the cosmos as a whole, as some suppose. Or it may be that, just as life emerged on our planet from the nonliving in some as yet undiscovered way, so mind appeared as an emergent at some unknown stage of organic evolution. If so, the naturalist must assume that the emergence in both cases occurred in lawfully ordered ways within the frame of the natural. In the present state of knowledge an open-minded skepticism on this question is the only safe attitude.

More intensive study of the properties of the nervous tissues seems to be the most promising approach to these unsolved problems. In the past, escape from mystery has too often been sought through verbalisms and mysticism. Rigid adherence to scientific method—accepting as evidence not wishful thinking but verifiable experience—will avoid this pitfall. Obviously, conventional methods of inquiry must be pushed to the limit of their availability, and in the meantime new formulations of problems must be sought with all the resourcefulness that scientific imagination can command, not neglecting the possibility that some of these formulations may lie outside the frame of current Newtonian and quantum mechanics.

CHAPTER VIII

GENERAL PRINCIPLES OF MORPHOGENESIS

THROUGHOUT the preceding descriptions and interpretations, some general morphological principles are implicit. Since these are moot questions, it is fitting that these assumptions be explicitly stated and clarified. Several years ago in a discussion of the morphogenesis of the brain ('33a), some of these principles were critically examined, and a part of what follows is condensed from that essay.

MORPHOGENIC AGENCIES

A century and a half ago the German *Naturphilosophie* elaborated the mystic and poetic conception of an archetypical form, which was popularized by Oken and Goethe and culminated with Sir Richard Owen and Louis Agassiz. As Professor Owen wrote in 1849: "The archetypal idea was manifested in the flesh under diverse such modifications, upon this planet, long prior to the existence of those animal species that actually exemplify it." These *geistige Kräfte* were conceived as enduring morphogenic agencies which shape the course of all animal and plant differentiation.

After the publication of Darwin's *Origin of Species* in 1859, a new school of morphologists arose under the leadership of men like Gegenbaur, Haeckel, and T. H. Huxley, the guiding principle being a search for phylogenetic relationships as key factors in morphogenesis. The mystic enduement of the earlier archetype was replaced by a sound biological principle of progressive change effected by verifiable internal and environmental agencies.

This conception of morphology as the visible record of phylogenetic history has stood the test of time. It is dynamic, not static; and our search is for the natural agencies which have operated to produce the observed modifications of form and correlated behavior during the course of evolutionary change. This century of progress has, however, witnessed a curious relapse, with resurgence of the ancient predilection for rigid categories and artificial systems of logical analysis, which yield in the end a formulation of inflexible, and therefore obstructive, morphological principles. Ancient and heritable patterns, such as metamerism, germ layers, and so on, were formalized, and the tendency was to regard them as stable and immutable factors in morphogenesis. The problems were thus simplified in terms of misleading logical categories, resulting in a static, rather than a dynamic, analysis of development and evolution. These formal and rigid concepts have too often retarded, rather than facilitated, a true understanding of the structure.

The dialectic of some current "form-analytic" programs of research on morphogenesis of the brain seems to postulate a predetermined primordial pattern of the neural tube, which is preserved throughout all stages of differentiation and which can be recognized by the arrangement in space of cellular areas and their relations to one another independently of their functional connections or of any dynamic agencies in morphogenesis other than cellular proliferation. The result is an oversimplified, logically consistent, morphological schema uncontaminated by functional or other complicating factors; but this has little interest for the working anatomists and physiologists, for it has no obvious relation with the structural forms with which they are practically working. Nerve fibers are quite as important as cell bodies in cerebral organization and as elements in cerebral forms. By what right does the morphologist ignore them in his study of form? We have ample evidence that the growth of nerve fibers and the migration of cells may be determined by functional requirements, which differ from species to species in correlation with different modes of life. Are forms which have physiological meaning of no significance in morphology, and can they safely be ignored by morphology?

My paper of 1933 cites several illustrations from my own experience and that of others of the seductive influence of rigid categories of morphological concepts, which simplify analysis by neglect of other significant factors in morphogenesis. In modern morphology the search is for genetic relationships, and homologies are defined in terms of such relations. In all phylogenetic study we must constantly keep in the foreground of attention two main classes of morphogenetic agencies. These are, first, the conservative factor of stable genetic organization and, second, the more labile influence of the specific functional requirements. Both factors are always present, and one important task of the morphologist is the analysis of his material so as to reveal the parts played by each of them. A sound and fruitful morphology will take both into account. For the practical purposes of descriptive anatomy and experimental physiology an analysis in terms of functional efficiency is indispensable, and a morphology which ignores the dynamic factors of tissue differentiation in terms of physiological adaptiveness lacks something and is sterile, as I have repeatedly emphasized ('08, '10, '13, '22b, '25a, '33a).

This does not imply that in either embryonic or phylogenetic development the pattern of structural differentiation is determined primarily by peripheral influences, which, acting directly on the germ plasm, produce heritable changes in its structure. Whether such inheritance of acquired characters ever occurs is in controversy. Certainly it is the general rule that adaptation of structure to physiological requirements is effected by indirection—natural selection or some other principle—but that it is acquired in some way is evident, and these physiological requirements are key factors in morphogenesis. This is more evident in the nervous system, perhaps, than in other organs of the body, for one of the prime functions of nervous tissue is adjustment of the body to its environment.

The analysis of morphogenic factors is best accomplished experimentally, as is well illustrated by Holtfreter's recent study ('45). This paper sets out in sharp relief the contrast between growth and differentiation. The relation between intrinsic and extrinsic factors and between nuclear and cytoplasmic influences in morphogenesis is discussed in two recent books by W. E. Agar ('43, chap. v) and R. S. Lillie ('45, chap. x). The reciprocal relations between genes and cytoplasm are now under active investigation.

There is decisive evidence that in embryogenesis the pattern of differentiation of both sensory and motor systems is determined largely by intrinsic agencies and that it proceeds more or less autonomously up to functional capacity. These structural patterns are laid down in the inherited organization, and that organization has been elaborated in the course of phylogenetic development in adaptation to the physiological needs of the species in question. Morphology here ties in with ecology.

The last point carries with it the necessary implication that the intrinsic agencies which initiate and shape the course of embryonic differentiation are not strictly autonomous and that even the hereditary factors have arisen *ab initio* as responses of the protoplasmic organization to environmental influences. It is this point of view which has been emphasized by Child ('41) in his search for more general morphogenic agencies which antedate the specialized features of ontogenetic differentiation. The old controversies about preformation versus epigenesis are outmoded and may well be ignored, for both factors are present in all development. It can no longer be claimed that "the basis of developmental pattern is an inherent property of protoplasm and therefore continuously present and independent of external conditions" (Child, p. 6), for at no period in the life of any organism is it independent of its environment. The vital process is fundamentally an interaction between the protoplasmic organization and its surroundings—respiration, nutrition, and all the rest. But when a given heritable pattern of internal organization has once been established, then the several components of this organization may acquire a large measure of autonomy in both further development and adult function. This specialization has obviously taken place; and when, in the course of development of vertebrate neuromuscular organs, they are approaching functional capacity, the patterns of this performance, though not independent of internal and external environment, clearly are not directly determined by any specific influences acting upon them from receptor organs. In this restricted sense these patterns are initiated intrinsically, and their performance is autonomous. But throughout development there is a complicated interaction of the inherited factors with one another, and this intrinsic interplay of morphogenic agencies is an important feature of developmental mechanics.

The widely current belief that heritable variations sometimes occur in progressive series set in a definite direction rather than always in accordance with the probability curve of chance deviations around fixed unit characters has been strongly supported by many independent lines of evidence. It has been pointed out that the progressive senescence of tissue in both ontogenetic and phylogenetic series involves a fixation or stabilization of originally undifferentiated plastic tissue into permanent structural patterns. So far as this differentiation is heritable and irreversible, the future course of evolution is thereby intrinsically determined, for variations will be distributed around the new pattern as a mode in accordance with a different frequency curve than would be shown if the inherited structural pattern were different. The process of differentiation is therefore itself a natural cause of limitation of the future course of evolution within boundaries set by the efficient working of the established pattern. The nervous systems of arthropods, teleosts, reptiles, birds, and mammals furnish illustrations of the effect of such irreversible differentiation on the course of animal evolution. The significance of this principle in the evolution of the nervous system was briefly discussed in my paper on orthogenesis in 1920, and more fully subsequently ('21, '22a, '24c, '26, '33a, '48).

In the embryo, as soon as connection is made between the sensory zone and the motor zone in the central nervous system, specific peripheral influences begin to operate to a much greater degree, and these may modify the subsequent course of development of the already fabricated sensory and motor apparatus and "inflect" the pattern of performance. The nature and degree of this modification by use is, however, limited to the range permitted by the pre-existing intrinsic organization (Coghill, '29, p. 86; Weiss, '41, p. 59). That this range is very restricted in the Amphibia has been made clear by much recent experimental work. The intrinsic growth factors which predetermine appropriate patterns of nervous organization in embryological development are operative also in the regeneration of peripheral nerves in the adult animal (see the experiments of Sperry cited on p. 228).

With progressive increase in the complexity of adjustment to the external and internal environment, there has been a corresponding differentiation of structure. The genetic organization determines the primary pattern, and this pattern is modified by use and the personal experience of the individual. The first of these factors yields that stable organization of tissue which is common to all members of the species and which forms the subject matter of most of the literature of neuroanatomy. The structural changes effected during growth by the second factor are harder to recognize, and this critical field is largely unexplored. In our examination of Amblystoma we are searching for primordia of both these kinds of tissue—the stable heritable structure and the more labile, individually modifiable tissue involved in conditioning of reflexes and other adjustments to personal experience. Both components are at a low level of differentiation in this animal, but their characteristic structure is recognizable, and the successive steps of their further evolutionary development can be followed.

Most species of urodeles apparently are reversals from more highly differentiated ancestral forms. This, however, in its main features is not a dedifferentiation but an arrest of development, so that adult characteristics of the descendants resemble larval features of their ancestors. Such reversals and the accompanying instances of actual degeneration doubtless are brought into being by changes in the mechanism of heredity, involving rearrangement of the complicated pattern of the genes (Emerson, '42, '45). Evolution is an irreversible process, with divergent ramifications in inconceivable variety and many instances of approximate parallelism. The apparent repetitions, when viewed in time, are not circular but spiral in form; and apparent retrogressions are never exact reversals of the preceding historical sequence. When a mammal returns to the water, it does not become a fish. The amphibians present conspicuous illustrations of recapitulation in individual development of some features of the phylogenetic history, and most of the apparent retrogression is really an arrest of development, which does not go back so far as the piscine ancestry. Their limbs are not fins, and their gills are not fishlike.

There is another functional factor in morphogenesis to which too little attention has been given—a greater or less capacity for so-called "spontaneous" activity, that is, behavior initiated internally and manifested in patterns determined by the intrinsic structure. All protoplasm is active as long as it is alive, and the essential vital properties are inherent in protoplasmic organization. This behavior is not a mere reflection of external influences, for the living stuff transforms the energies which impinge upon it and recombines them in original designs. Reserves are accumulated, and these are expendable on occasion in accordance with need, with or without external excitation. In the central nervous system this intrinsic "spontaneous" automaticity attains maximum potency, and it is exhibited by even isolated fragments of it in characteristic patterns, of which oscillographic records can be made. In the intact brain the interplay of these intrinsic activities is always going on, and as we pass from lower to higher animal types it becomes progressively a more and more important factor in determining patterns of behavior. The enormous reserves of potential nervous energy in the brain are evoked and manifested as stabilizing influences (cerebellum, corpus striatum, etc.) and also in that spontaneity, initiative, and inventiveness which culminate in cortically directed human behavior. These capacities are shown in some measure by all animals, and a search for the apparatus employed in even so lowly a creature as a salamander may be fruitful (see '48, chap. xv, for discussion of Coghill's contributions on automatism, spontaneity, and motivation).

It has been pointed out by von Bonin ('45) that the logical foundations of the concept of morphology based on phylogenesis as developed by T. H. Huxley, Gegenbaur, and others of their time are insecure. The mathematical argument need not here be examined, for it is based on certain restricted postulates, and in animal evolution there are many variables not embraced by these postulates. It certainly does not follow that "the task of understanding structures on the basis of their phylogenetic history" is "an insoluble problem." Though neither cultural history nor phylogenetic history has been reduced to mathematical formulation, there is general agreement that history, judiciously interpreted, is an accredited guide for understanding the present and prognosticating the future. Phylogenetic history is not a sealed book, and Dr. von Bonin assures me that he would be the last to deny a positive value to the historical approach to problems of morphogenesis and that such studies have actually contributed much toward an understanding of human cerebral structure and function. The positive paleontological evidence regarding the phylogeny of the brain is more extensive and illuminating than is generally recognized. Though fossilized brains are unknown, the very large number of casts of skull cavities, when skilfully interpreted, yield a surprising amount of reliable information about the nervous organs which once occupied those cavities, as illustrated, for instance, by Stensiö's studies ('27) of fossil ostracoderms. It must be freely granted, of course, that conclusions reached are tentative, to be accepted only as checked against other lines of evidence, particularly the known sequence of evolutionary history as revealed by fossilized skeletal remains.

No single mode of attack upon problems of morphogenesis is adequate. Experimental methods yield the most decisive evidence, and these require adequate knowledge of anatomical structure. This last is the contribution of comparative anatomy and comparative embryology, and both of these must be functionally interpreted to be fruitful. The anatomist should recognize the limitations of his method. His task is to lay foundations—stable and adequately broad—and to suggest fruitful working hypotheses. The Amphibia occupy a strategic position here for the same reason that the experimentalists find them so useful.

MORPHOLOGICAL LANDMARKS

Comparative study shows that some general features of structural plan run through the vertebrate series with remarkable constancy and that other features undergo amazing transformations. The discovery of the laws in accordance with which these transformations are effected is the goal toward which we are working. We are confronted with a similar, but not parallel, series of problems in the study of embryological development. In so far as we succeed in our search for these laws, we advance our understanding of fundamental vital processes.

The stable structural features of the nervous system are the most useful landmarks for the comparative anatomist. They are expressions of the conservative hereditary factor in morphogenesis; but this stability cannot safely be interpreted as the simple manifestation of some primordial archetypical pattern, for these features are retained during the course of phylogenetic history only in so far as the fundamental features of the peripheral connections and their internal relationships are constant, that is, because they are parts of an apparatus of adjustment to environment which is common to all vertebrates. Other parts of the brain are more variable because, with complication of the behavior pattern in higher species, more elaborate and diversified mechanisms of adjustment and integration are requisite.

In all vertebrate brains the most fundamental structural landmark is the transverse plane separating the spinal cord and rhombencephalon below from the cerebrum (as defined in the BNA) above. In Amblystoma this plane is marked externally by the fissura isthmi and internally by the sulcus isthmi. The zonal arrangement as described in chapter v is well defined in the rhombic brain and the midbrain, rostrally of which it is obscured by various secondary modifications which become more complicated as we pass from lower to higher members of the vertebrate series. A second important landmark is the transverse plane separating the diencephalon from the telencephalon, marked externally by the deep stem-hemisphere fissure.

These two planes also mark the positions of two strong flexures of the neural tube in early embryonic stages, caused by inequalities of growth of the dorsal and ventral zones of the neural tube. The first of these flexures to appear is a ventral bending of the neural tube in the mesencephalic region, caused by precocious enlargement of the tectum. This is followed by a flexure in the reverse direction at the di-telencephalic junction (p. 212). These flexures are less obvious in adult brains because they are somewhat straightened in later stages and masked by growth of interstitial tissues; but the site of each of them is a zone of transition between major divisions of the brain with distinctive physiological characteristics.

The foundations of the current anatomical analysis of the brain were laid by Wilhelm His in terms of human embryological development. The early neural tube was divided into a linear series of blocks separated by transverse planes, and a longitudinal sulcus limitans on each side marks the boundary between a dorsal sensory alar plate and a ventral motor basal plate. The adult derivatives of this embryonic mosaic are the primary anatomical units. The nomenclature derived from this analysis as officially adopted (the BNA), or modifications of it, is now almost universally employed, to the great advantage of human descriptive neurology. But this scheme has its limitations. Some features of it are quite inapplicable to the brains of lower vertebrates; for, though the embryonic neural tube is similar in most of them, its adult derivatives vary so widely in adaptation to diverse modes of life that no inflexible formula is applicable. Since the brain of Amblystoma is generalized, few of these difficulties arise here.

There is difference of opinion about where the sulcus limitans ends anteriorly. If the embryonic floor plate ends at the fovea isthmi (Kingsbury, '30), it is evident that the basal plate extends farther forward to include the mesencephalic cerebral peduncle and probably more or less of the adjoining parts of the hypothalamus and ventral thalamus. The remainder of the mesencephalon and diencephalon (including the retina) and the whole of the telencephalon are derived from the expanded anterior part of the alar plate and the related neural crest. The adult derivatives of alar and basal plates include much tissue that is specifically neither sensory nor motor; and this intercalated associational fabric crosses the boundaries of the primitive embryonic mosaic in ways which differ from species to species. Each species must be analyzed in terms of its own mode of life and distinctive action system.

C. von Kupffer ('06) put special emphasis upon two deep transverse sulci in the ventricular wall of the neural tube in early embryonic stages of a series of lower vertebrates, including Necturus and Salamandra. These were termed sulcus intraencephalicus anterior and posterior. Study of later stages of developing urodele brains shows that this emphasis was well placed, for these sulci mark the positions of the two transitional sectors of the brain to which reference was made above— the first, the diencephalic, and the second, the isthmic sector.

In our specimens of Amblystoma, von Kupffer's anterior sulcus in early motile stages (Harrison's stages 33–36) is a sharply defined groove, which extends dorsally from the lateral optic recess in front of the chiasma ridge to the region of the velum transversum. Its dorsal part is variable, but clearly the primary course is into the posterodorsal initial evagination of the hemispheric vesicle, as described by von Kupffer and by Rudebeck ('45). This is clearly the case also in A. jeffersonianum, as shown by Baker and Graves ('32) in their five stages from 5 to 17 mm. long. My published references to this sulcus and its adult derivatives have been successively modified, as more material was examined ('10, pp. 419, 432; '27, p. 238; '33b, p. 240; '35a, p. 252; '38, p. 212; '38b, pp. 401, 402; '39a, p. 262). These differences in interpretation are doubtless due in part to the natural variability of the specimens and in part to lack of a sufficiently close series of well-preserved stages to reveal the actual sequence of the changes. In our Amblystoma material the ventral part of this sulcus seems to shift its position and to be transformed directly into the sulcus preopticus, and it was so described ('39a, p. 262). Rudebeck finds in dipnoans, Necturus, and Triturus that the sulcus intraencephalicus anterior passes from the lateral optic recess dorsalward to the posterodorsal hemispheric ventricle and that the definitive sulcus preopticus arises as a secondary outgrowth from this primary groove. Our specimens of Amblystoma have not revealed this secondary origin of the sulcus preopticus but resemble that of the anuran, Pelobates, as described by Rudebeck ('45, p. 53).

In Amblystoma the posterior intraencephalic sulcus of von Kupffer persists as the sulcus isthmi of the adult. In the 6-mm. Ammocoetes (von Kupffer's fig. 47) it extends transversely from the plica rhombo-mesencephalica to a point in the floor a short distance spinalward of the tuberculum posterius, i.e., to the fovea isthmi, and it is similar in several other species figured. During larval development of Amblystoma it and the related external fissura isthmi shift their relative positions (p. 179). As emphasized above, this plane of separation between cerebrum and rhombencephalon, whether or not it is marked by a visible sulcus in the adult brain, in all vertebrates is the boundary between the two chief subdivisions of the brain.

The distinction between these subdivisions is conspicuous in prefunctional and early functional stages. Coghill ('24, Paper IV, p. 97; '31, Paper X, p. 162) reports that at all stages of development of Amblystoma from premotile to swimming the rate of proliferation of cells and differentiation of neuroblasts is rapid in rhombencephalon and spinal cord, on the one hand, and in the cerebrum, on the other hand; but "there is a distinct gap between fields of both differentiation and proliferation at the isthmus. Such a gap does not appear at any other level in the brain or cord." Diencephalon and telencephalon appear to be about equally involved in this process, and so do rhombencephalon and cord; but during these stages growth appears to be initiated independently in these two major divisions of the nervous system.

The experiments of Detwiler ('45, '46), to which reference has already been made (p. 62), show that ablation of the cerebral hemispheres and visual organs of Amblystoma in prefunctional stages results in no demonstrable change in size or weight of the medulla oblongata. He finds no evidence that the hemispheres or visual organs exert any morphogenic influence upon the medulla oblongata up to a larval age of 48 days (33 mm. in length), and these mutilated larvae are capable of performing all the ordinary feeding reactions.

The most primitive and fundamental patterns of total behavior are organized in the spinal cord and lower medulla oblongata, particularly at the bulbo-spinal junction. Very early these come under the control of the vestibular apparatus (Coghill, '30, p. 638) and midbrain, and this control must be maintained throughout life if the primitive mass movements are to retain their efficiency (Detwiler, '45; '46). The isthmus, interpolated between these regions, seems to be concerned mainly with the organization and control of local reflexes (partial patterns) of the medulla oblongata under the influence of both ascending and descending systems of correlation fibers.

For convenience of description the rhombic brain is here arbitrarily divided into four regions, each with characteristic structure and functions: (1) the bulbo-spinal junction, (2) medulla oblongata, (3) cerebellum, (4) isthmus. In these regions the functionally defined zones are unevenly differentiated, and the cerebellum as a "suprasegmental" apparatus is in process of emergence from the sensory and intermediate zones. Similarly, in the forebrain the pallial field exhibits an early prodromal phase leading toward cortical differentiation, as seen in reptiles.

The rhombic brain receives all sensory components of the cranial nerves except the olfactory and the optic. In lower vertebrates, in which the auditory apparatus is at a very low level of differentiation, the sensory components of the rhombic nerves are relatively unspecialized, in sharp contrast with the highly specific optic and olfactory systems; and the physiological dominance of the two systems last mentioned in the control of behavior is the determining factor which gives to the cerebrum unique properties that are in marked contrast with those of the rhombencephalon.

The isthmus is a transitional sector, within which the patterns of all bulbar activities are ordered and integrated. It bulks larger in lower vertebrates than in higher, in which the cerebral cortex has taken over the larger part of this control. Above it the cerebellum was differentiated, not as part of the apparatus which patterns performance but as an ancillary mechanism on the efferent side of the arc, to reinforce and regulate the execution of movements.

Within the cerebrum the two primary centers of dominance—optic and olfactory—are separated by a similar transitional sector in the diencephalon. This is plastic tissue, not dominated by any single sensori-motor system; it is the meeting place of ascending and descending sensory paths. In noncorticated vertebrates we find here the apparatus of a type of adjustment from which influences pass forward into the hemispheres and there act as morphogenic agencies in the elaboration of cortical structure.

It appears, then, that the loci of some characteristic features of the vertebrate brain were fixed by their peripheral connections in the earliest members of the series and that some of these have remained essentially unchanged throughout the phylogenetic history. Others have emerged very gradually from a nonspecific matrix which is diffusely spread throughout a wide field. The search for primordia of the latter type in the lower forms as parts of a mosaic pattern with rigidly defined boundaries cannot be successful. In each animal species the tissue requisite for successful adjustment to the mode of life adopted is fabricated out of such raw material as is available, and nature is not bound by our formal rules of logical consistency. The major subdivisions of the brain were thus defined very early in vertebrate phylogeny, and they retain their general characteristics throughout the series, but there is no apparent limit to the range of modifications which these sectors may undergo in adaptation to specific physiological requirements.

In premotile stages of Amblystoma, Coghill mapped several areas in the walls of the neural tube, characterized by distinctive proliferation and differentiation. Adopting a modification of his scheme, I gave arbitrary numbers to twenty-two such areas from the olfactory bulb to the cerebellum ('37, p. 392), and their development can be followed up to the adult stage. These units of the mosaic pattern of the premotile embryo undergo remarkable shiftings of position during larval development and an equally remarkable diversity in patterns of differentiation and fibrous connections. When the comparative embryologist surveys the vertebrate series as a whole, he recognizes a striking similarity in the early stages of differentiation of the neural tube of all of them. In later stages of development this similarity gives way to wide diversity in the progress of differentiation of these primordial units of structure, and practically all this divergent specialization can be seen to be directed toward adaptive modifications of structure, correlated with differences in the action systems of the several species. Some limits to the range of this modifiability are set by the inherent qualities of the genetic organization so that some general principles of morphological pattern can be recognized everywhere. Yet the structure when viewed phylogenetically is remarkably plastic, and the available materials are adapted to a wide variety of uses in diverse combinations and interconnections in all the different phyla. The most alluring feature of these comparative studies lies in our ability to sort out of this apparent confusion of detail those strong threads of ancestral influence which are interwoven in ever changing designs under the influence of adaptive adjustment to different modes of life.

THE FUTURE OF MORPHOLOGY

During the past half-century, morphology has seemed to be declining in favor, its problems submerged in the more attractive programs of the experimentalists. Nevertheless, activity in this field has not abated, and now there is a renaissance, the reasons for which are plain. Conventional methods of anatomical research have laid a secure factual foundation, but the superstructure must be designed on radically different lines. Several centuries of diligent inquiry by numerous competent workers have produced a vast amount of published research on the anatomy and physiology of the nervous systems of lower vertebrates; but most of this literature is meaningless to the student of the human nervous system, and, as mentioned at the beginning of this book, its significance for human neurology has until recently seemed hardly commensurate with the great labor expended upon it. The last two decades have inaugurated a radical change, in which we recognize two factors.

In the first place, technical improvements in the instrumentation and methods of attack have opened new fields of inquiry hitherto inaccessible. To cite only a few illustrations, new methods for the study of microchemistry and the physical chemistry of living substance, radical improvement in the optical efficiency of the compound microscope, the invention of the electron microscope, and the application of the oscillograph to the study of the electrophysiology of nervous tissue are opening new vistas in neurology, which involve quite as radical a revolution as that experienced a few centuries earlier when microscopy was first employed in biological research.

A second and even more significant revolution is in process in the mental attitudes of the workers themselves toward their problems and toward one another. A healthy skepticism regarding all traditional dogmas is liberating our minds and encouraging radical innovations in both methodology and interpretation. And, perhaps as a result of this, the traditional isolationism and compartition of the several academic disciplines is breaking down. The specialists are now converging their efforts upon the same workbench, and cooperative research by anatomists, physiologists, chemists, psychologists, clinical neurologists, psychiatrists, and pathologists yields results hitherto unattainable. What is actually going on in the brain during normal and disordered activity is slowly coming to light.

Here the comparative method comes to full fruition, and comparative morphology acquires meaning, not as an esoteric discipline dealing with abstractions but as an integral and indispensable component of the primary task of science—to understand nature and its processes and to learn how to adjust our own lives in harmony with natural things and events, including our own and our neighbors' motivations and satisfactions.

The objective toward which we are directing our efforts is a better understanding of human life and its instrumentation. Our mode of life has been achieved through eons of evolutionary change, during which the conservative and relatively stable organization of the brain stem has been supplemented and amplified by the addition of cortical apparatus with more labile patterns of action, resulting in greater freedom of adjustment to the exigencies of life. In all behavior there is a substrate of innate patterns of great antiquity, and in practical adjustments these primitive factors are manipulated and recombined in terms of the individual's personal experience. Memory and learning are pre-eminently cortical functions, but these cortical capacities have not been given to us by magic, and we want to know how they have been developed and the roots from which they have grown.

The incentives which motivate research in comparative neurology are the same as those of all other science, pure and applied, and of all truly humanistic endeavor in other fields—to find out what is good for humanity and how to get it. This implies, as I have recently exhorted ('44), that the humanistic values of science must always be acknowledged and cultivated.

PART II
SURVEY OF INTERNAL STRUCTURE

INTRODUCTORY NOTE

THE first part of this work includes a schematic outline of the organization of the amphibian brain and discussion of some morphological and physiological principles suggested by this inquiry. In Part II, evidence is presented in sufficient detail to document the conclusions and principles summarized in Part I. This material is arranged by topographic regions, as these are listed in chapter iv. These descriptions are supplementary to those in Part I, and each topic should be read in connection with the corresponding passages of the preceding text.

CHAPTER IX

SPINAL CORD AND BULBO-SPINAL JUNCTION

THE SPINAL CORD AND ITS NERVES

THOUGH the spinal cord is not included in this survey, some features of its upper segments must be considered here because of their connections with the brain and particularly with the complicated structure at the bulbo-spinal junction. The spinal cords of urodeles have not been adequately described, and our material is not suitable for this purpose. Early stages of development of Amblystoma have been described by Coghill, and the cord of larval Salamandra by van Gehuchten ('97). A wide variety of experimental studies of development involving the spinal cord have been reported by others.

THE SENSORY ZONE

The slender dorsal gray column is continuous with the nuclei at the bulbar junction, to be described shortly. These cells are activated by the dorsal spinal root fibers and the spinal V and spinal vestibular roots. A few of them in and near the mid-plane are a spinal continuation of the commissural nucleus of Cajal and are probably visceral sensory in function. This supposition is supported by Sosa's description ('45) of similar cells in the septum dorsale of mammals and birds, which he regards as spinal representatives of Cajal's nucleus.

Most of the dorsal root fibers immediately upon entrance bifurcate into descending and ascending branches, and the latter in the upper segments comprise most of the massive dorsal funiculi, with which some other fibers are mingled, notably those of the descending vestibular root and bulbar correlation tracts a and b of Kingsbury. The large spinal V root lies ventrally of these funiculi, and below this is a dorsolateral funiculus, containing fibers of the spinal lemniscus, spino-cerebellar tract, and other fibers of spinal and bulbar correlation. Many of these dorsolateral fibers decussate, descending from the dorsal gray as internal arcuates.

THE INTERMEDIATE ZONE

The intermediate zone is not clearly defined, its gray substance being continuous with that of the motor zone. In the alba the neuropil of the reticular formation is less well developed than in the medulla oblongata.

THE MOTOR ZONE

The motor zone of the cord is continuous with that of the medulla oblongata, with no recognizable boundary; and throughout these regions the peripheral motor neurons are mingled with co-ordinating neurons, and they often resemble the latter so closely that they cannot be distinguished unless their axons are seen to enter the motor roots.

THE SPINAL NERVES

The upper spinal nerves are modified. The first usually has no sensory root or ganglion. The arrangement of the motor roots of the first and second pairs is exceedingly variable. The nervus accessorius and the nervus hypoglossus are not separately differentiated. The primordia of the former are represented in the lowermost vagal rootlets, which emerge from the lateral aspect of the medulla oblongata; and the primordia of the latter are in the first and second spinal nerves, the ventral roots of which emerge at the ventral surface. In one specimen the lowest vagal root was seen to emerge at the level of the calamus scriptorius, but usually it is at a more rostral level. The lowest root of the first spinal nerve usually emerges in the region of the calamus, and the first root of this nerve at variable distances rostrally.

The first spinal nerve of Salamandra as described by Francis ('34, p. 159) agrees with that of Amblystoma. On page 161 he quotes Goodrich, who has shown that the urodele hypoglossus innervates muscles derived from the ventral outgrowths of the second, third, and fourth myotomes and that "the hypoglossus of Amphibia and Amniota may certainly be considered as homologous, although not necessarily composed of the same segmental nerves."

The neurons of the ventral horns of gray include tegmental elements, and motor cells which give rise to peripheral fibers ('44b, fig. 10; van Gehuchten, '97). Both types may have very large, much-branched dendrites, which in the larva ramify through almost the entire cross-sectional area of the cord and may cross to the other side in the ventral commissure. In the adult animal, internuncial connections within the dorsal and ventral zones and between these provide for co-ordinated spinal reflexes; but all movements of the trunk and limbs are subject to further control from bulbar and other higher centers. The details of the structural apparatus by means of which these ordered movements are effected remain obscure. At the inception of motility in the embryo the first neuromotor responses to stimulation are mass movements, and the apparatus of local reflexes matures later (Coghill, '29). This implies that integrative functions of total-pattern type mature earlier than do the partial patterns of the local reflexes. Coghill's studies revealed a transitory system of peripheral and central connections in the early stages of the development of motility when mass movements prevail, followed by radical changes as the action system becomes more complicated.

Before the spinal ganglia are functionally mature, a series of transitory giant ganglion cells (Rohon-Beard cells) within the cord send peripheral processes out to skin and myotomes and central processes, which effect connection with the neuromotor elements. The transitory cells subsequently disappear and are replaced by the more specialized elements of the spinal ganglia.

Intramedullary cells of sensory type were observed by Humphrey ('44) in the spinal cords of human embryos. Two types of bipolar sensory cells appear in embryos of 5 mm., one of which is transient and is regarded as homologous with the Rohon-Beard cells. The other type persists to functional stages, and at the beginning of motility (22.5 mm.) many of these are changing to a unipolar shape and resemble cells of the spinal ganglia. These intramedullary unipolar cells are found in embryos of from 16 to 144 mm. in length, and are regarded as functioning components of the dorsal roots in the early stages of motility. Youngstrom ('44) also reports the occurrence of sensory cells within the spinal cords of human embryos of from 19 to 63 mm. These cells are in the mantle layer and resemble those regarded by Humphrey as comparable with Rohon-Beard cells. Similar intramedullary cells of sensory type have been seen by many others in embryos and adults to accompany root fibers of spinal and cranial nerves (see Pearson, '45, for illustrations); and it is probable that these are all derived from the neural crest, like the mesencephalic nucleus of the V nerve (as Piatt, '45, has demonstrated).

On the motor side of the arc two types of peripheral neurons were described by Coghill ('26, Paper VI) and Youngstrom ('40): (1) The thick primary fibers appear first in ontogeny and course for long distances in the ventral funiculus before emergence. The first ventral root fibers arise as collaterals of these longitudinal axons. (2) Thinner secondary fibers, which appear later, pass out from the gray of the cord more nearly transversely. The Rohon-Beard cells are centrally so connected with the primary motor cells as to evoke mass movement of the musculature of the trunk in response to adequate stimulation of any kind. In subsequent stages central connections between spinal ganglion cells and secondary motor neurons are made, and these are regarded as provision for execution of local reflexes. The primary motor neurons persist in adult Amblystoma. They occur in larval anurans but disappear at metamorphosis (Youngstrom, '38). Humphrey ('44) describes cells in the spinal cords of very young human embryos, which she believes are surviving vestiges of primary motor neurons of amphibian type.

In our sections of the adult the ventral spinal roots contain fibers of primary and secondary type. The primary root fibers are thick and heavily myelinated centrally and peripherally. Some of the thinner secondary fibers are well myelinated, and many of them seem to lose their myelin as they emerge from the spinal cord. Coghill ('26, Paper VI, p. 135) reported that in early swimming stages "a single fiber may innervate an entire myotome, and branches of these same fibers form the earliest nerves to limbs and tongue." At this early stage, however, the musculature of the limb bud is still an undifferentiated primordium. Youngstrom confirmed these observations and expressed the opinion that the limb musculature, like that of the trunk, has a double innervation of both primary and secondary fibers; but no details of the distribution of these fibers in the definitive limb musculature are given. More recently, Yntema ('43a, p. 331) says of the primary fibers in larvae of from 12 to 19 mm. in snout-anal length that "typically, they supply the myotomic musculature. In addition, fibers of this kind run to muscles of the extremities"; but again details of their distribution in the limb are lacking. In a personal communication he adds: "I have found evidence for the distribution of primary motor fibers to at least some muscles of the girdles of larvae, and have seen larger fibers which appear to be primary in the limbs themselves."

In the frog, with an action system very different from that of Amblystoma, the development of these nerves shows corresponding differences, for, as mentioned above, Youngstrom ('38) found that larval frogs have primary and secondary fibers like those of Amblystoma; but in the adult frog the primary fibers have completely disappeared. The opinion expressed by Taylor ('44) that in frog larvae the primary fibers do not enter the limbs may have no bearing on the innervation of limbs in Amblystoma because of the radical difference in the neuromuscular apparatus of these species. The primary fibers evidently are concerned with massive movements of the trunk musculature. The significance of the two sorts of fibers in the innervation of the limbs is still obscure. In Amblystoma the number of primary motor fibers is not markedly reduced by removal of the early undifferentiated neural crest, while secondary fibers are, as a rule, greatly reduced in number, the growth of the latter being dependent on the presence of sheath cells and the former not (Yntema, '43a). It will be of interest to learn whether the primary motor fibers have functions in the embryogenesis of Amblystoma comparable with those postulated for the "pioneer motor neurons" observed in the bird by Hamburger and Keefe ('44, p. 237).

THE BULBO-SPINAL JUNCTION

Little need be added here to the general description of this important region in chapter iv and to the details of structure and connections recently published ('44b). The topography as seen in transverse Weigert section is shown in figure 87. If the calamus scriptorius is taken as the arbitrary boundary between spinal cord and brain, this junctional region in Amblystoma may be considered to comprise the segments of the first and second pairs of spinal nerves, the second below the calamus and the first above. The entire length of the first spinal segment overlaps the lower vagus region of the medulla oblongata.

In the sensory zone the somatic sensory systems of the neurons of the dorsal gray columns are somewhat enlarged to form the nucleus of the dorsal funiculi, which extends far forward in the lower vagus region. Medially of this is the much larger collection of compactly arranged smaller cells of visceral-gustatory function—the commissural nucleus of Cajal. This nucleus extends downward from the calamus for a distance of about one spinal segment, below which visceral sensory function is represented by scattered cells in the dorsal median raphe. Above the calamus the commissural nucleus merges insensibly with the nucleus of the fasciculus solitarius.

The funicular nucleus is regarded as comparable with the external cuneate nucleus of mammals rather than with the nuclei of the f. gracilis and f. cuneatus, since Amblystoma has no medial lemniscus ('44b, p. 318). The arrangement and connections of the commissural nucleus are similar to those of man.

Secondary fibers from the funicular nucleus (many of them myelinated) pass downward as internal arcuate fibers to the spinal cord and medulla oblongata, some uncrossed and some decussating in the ventral commissure. Other crossed fibers join the tractus spino-cerebellaris and the spinal lemniscus (fig. 3). The secondary fibers from the commissural nucleus are unmyelinated. Some of them are internal arcuates, which distribute to neighboring parts of the spinal cord and medulla oblongata of the same and of the opposite side; and some pass directly laterally to the pial surface, where they turn rostrad in tr. visceralis ascendens (fig. 8; '44b, figs. 10, 11, 12, *tr.v.a.*) to reach the superior visceral nucleus in the isthmus and the ventro-lateral neuropil of the peduncle. Some further details about the connections of these nuclei are in the next two chapters.

The region of the calamus scriptorius is evidently an important center of correlation and integration of general somatic and visceral-gustatory sensibility of the entire body, with efferent discharge directly to the motor zone and also to higher centers of sensory correlation. Here root fibers of cutaneous and deep sensibility from the head, trunk, and limbs; of vestibular and lateral-line sensibility; and of gustatory and visceral sensibility converge into a common pool, which is the first integrating center of these functional systems to mature in ontogeny.

CHAPTER X

CRANIAL NERVES

DETAILS of the peripheral distribution of the several systems of nerve components have been recorded for a considerable number of amphibian species, notably in many important papers by H. W. Norris. The first of this series was Strong's paper ('95) on the larval frog, which was followed by Coghill's description of the cranial nerves of Amblystoma, published in 1902. Their arrangement here may be regarded as typical for the vertebrate phylum as a whole, with no extreme specialization of any system. The constancy of the arrangement of these components at the superficial origins of the nerve roots in all vertebrates is remarkable, in view of the extreme diversity of both peripheral and central connections of their fibers and of the enormous differences in the number of fibers represented in the several systems among the various species. Except for the specific differences just mentioned, the chief departures from uniformity of composition of the nerve roots are the suppression in all Amniota of the large lateral-line components of the Ichthyopsida and the correlated differentiation of the cochlear apparatus in the higher classes.

The central connections of the olfactory and optic nerves and the nervus terminalis are described in the chapters relating to the forebrain and the midbrain. The other functional systems are discussed in chapters iv and v, and to those general statements some additional details of their arrangement in Amblystoma are given here.

DEVELOPMENT

Some peculiar features of the development of the somatic motor roots were mentioned in the preceding chapter. The development of the visceral motor roots was described by Coghill, though many details remain to be filled in. The early development of the sensory systems of root fibers was studied by Coghill ('16, Paper II) and Landacre ('21 and later papers). In Landacre's paper of 1921 the embryos studied were identified as Plethodon glutinosus, but they subsequently proved to be Amblystoma jeffersonianum (Landacre, '26, p. 472). Older stages were described by Kostir ('24).

The embryological studies just mentioned were based on series of normal embryos. The conclusions reached have been checked experimentally, extended, and in some details corrected by Stone ('22, '26) and by Yntema ('37, '43), so that we now have very accurate information about the sources of the nerve cells of each sensory component of the ganglia of the V to X cranial nerves. The ganglion of the trigeminus is derived chiefly from neural crest, which also contributes some cells to the ganglia of the VII, IX, and X nerves (Landacre, '21, p. 15). Yntema ('37) found no neurons of neural-crest origin in the facial ganglion; but, since there is a small general cutaneous component of this nerve in adult Amblystoma, it is probable that some cells of neural-crest origin are present, as is known to be the case in some other animals. Part of the trigeminal ganglion (profundus ganglion of Landacre, '21, p. 23) is delaminated from the lateral ectoderm. The lateral-line ganglia are derived exclusively from the dorsolateral placodes of the ectoderm, and the ganglion of the VIII nerve from the auditory vesicle. The visceral ganglia of the VII, IX, and X nerves arise from epibranchial placodes. Landacre derived only the special visceral (gustatory) component of these ganglia from these placodes, but Yntema has shown that the larger part of the general visceral component also is of placodal origin. According to Yntema's analysis, epibranchial placodes give rise to general and special visceral components of the cranial ganglia, dorsolateral placodes to lateral-line components, the auditory placode to the VIII ganglion, and neural crest to general cutaneous and general visceral components. The mesencephalic nucleus of the V nerve is derived chiefly from a portion of the neural crest which is incorporated within the neural tube, though it is not certain that this is the exclusive source of these cells (p. 141, and Piatt, '45).

SURVEY OF THE FUNCTIONAL SYSTEMS

GENERAL SOMATIC SENSORY NERVE ROOTS

Here are included cutaneous sensibility of several modalities—touch, temperature, pain, and, in aquatic animals, refined chemical sensitivity. Associated with these nerves are those of deep pressure. The nervous apparatus of these various qualities of sense has not been successfully analyzed in lower vertebrates. Their fibers are mingled peripherally and also centrally, except for those of the mesencephalic V root. It is not improbable that some peripheral fibers may serve more than one of the modalities of sense as centrally analyzed.

The peripheral fibers of this system are usually described as the general cutaneous component of the nerves, though some of them are distributed to deeper tissues. Most of them enter the brain in the trigeminus root and smaller numbers in roots of the VII, X, and (probably) IX nerves. The vagal fibers of this system have wide peripheral distribution (Coghill, '02), including the ramus auricularis and other vagal branches and also anastomotic connections with branches of the IX and VII nerves. The peripheral distribution of the VII fibers has not been described. A few fibers of the sensory IX root have been seen (rarely) to descend in the spinal V fascicles. Most urodeles are said to lack a general cutaneous component of the IX nerve, though there is some evidence of it in Necturus ('30, p. 22). If the presence of these fibers is confirmed, they probably join the general cutaneous component of the vagus peripherally.

Many of the trigeminal fibers divide immediately upon entering the brain into the thick descending branches of the spinal V root and thinner ascending branches (fig. 40) of the cerebellar root. The longest course that can be taken by one of these bifurcated fibers is shown in figure 3. Some of these fibers take deeper courses, penetrating the spinal V root to enter the fasciculus solitarius (p. 148). Some of the mesencephalic V fibers also divide near their exit from the brain, with descending branches arborizing in the reticular formation of the upper medulla oblongata (p. 141 and fig. 13).

The spinal V root is large and well myelinated. It can readily be followed through the length of the medulla oblongata and for an undetermined distance into the spinal cord (figs. 87–90). Each of its fibers for its entire length is provided with a fringe of short collaterals (fig. 38), which are directed inward into a neuropil which is continuous dorsally with that similarly related with the VIII and lateral-line roots, the whole forming a common pool for the reception of all somatic sensory components (fig. 9). In the calamus region these collaterals mingle with collaterals and terminals of spinal root fibers of the dorsal funiculus and the spinal vestibular root, bulbar correlation tracts a and b, and the dorsolateral funiculus. This axonic neuropil is permeated by dendrites of the nucleus of the dorsal funiculus and commissural nucleus of Cajal.

In many of our Golgi preparations the central courses of the sensory V fibers are electively impregnated, often with no other fibers visible in their vicinity. Some of these from the larva have been illustrated ('14a, figs. 48–51, 54; '39b, figs. 42, 46, 47, 57–61, 67, 77). Figures 27–32 show V roots as seen in horizontal Cajal sections of the adult brain. Figure 32 passes through the two motor V roots and their nucleus; figure 40 includes three impregnated neurons of this nucleus and the bifurcating fibers of the sensory root, which are also shown in figure 38. Woodburne ('36, p. 451) saw thick root fibers of the trigeminus entering the cerebellar root; but, in the absence of Golgi sections, the thinner collaterals of the spinal root were not demonstrable.

The superior or cerebellar root of the trigeminus is much smaller than the spinal root, and at the ventrolateral border of the auricle it joins the spino-cerebellar tract (figs. 30, 31, 91). Many fibers of both tracts end here with open arborizations in a neuropil which is the primordium of the superior (chief, or pontile) nucleus of the mammalian trigeminus; but some fibers of both tracts pass through this neuropil and continue dorsomedially into the body of the cerebellum, where they end, some on the same side and some decussating in the commissura cerebelli (figs. 31–34, 37, 91). These commissural fibers are joined by others arising from cells in the vicinity of the superior trigeminal neuropil, and many of the decussating fibers, after crossing, reach the superior neuropil of the other side, thus forming an intertrigeminal commissure. We here confirm, in the adult, Larsell's description ('32, p. 413) of the cerebellar commissure of the larva as composed chiefly of trigeminal and spinal components. In Amblystoma the superior sensory nucleus of the trigeminus probably is concerned chiefly with the proprioceptive aspects of cutaneous sensibility (deep proprioception being provided for in the mesencephalic V root). This cerebellar connection persists in man, but here the chief V nucleus has also acquired refined types of sensibility which Amblystoma lacks.

Neither the superior nor the spinal nucleus of the trigeminus has well-defined boundaries. The central cells which engage terminals and collaterals of the sensory V fibers may also have synaptic contacts with terminals of all other sensory systems that enter the medulla oblongata. There are, however, certain lines of preferential discharge for each group of sensory systems, and the segregation of local nuclei and secondary pathways for each functional system is incipient.

LATERAL-LINE, LABYRINTHINE, AND COCHLEAR SYSTEMS

These special somatic sensory systems are closely related genetically, structurally, and physiologically, but much remains obscure about their relationships. The labyrinthine apparatus seems to be at the focus of these systems. It is very conservative, except for the cochlear part, showing relatively little change in structure and function from lowest to highest vertebrates; moreover, its physiological properties have been thoroughly explored. The lateralis system attains its maximum in fishes, persists in larval amphibians and adults of some urodeles, and disappears entirely in all higher groups, both embryonic and adult. Organs of hearing are poorly developed in fishes. Auditory functions seem to be performed by the vestibular apparatus and also (for slow vibration frequencies) by the organs of the lateral line, which undoubtedly have other functions also.

The peripheral end-organs of all these systems are specialized epithelial structures, in contrast with the free nerve endings of the general somatic system. The vestibular end-organs of the internal ear resemble the end-organs of the lateral lines, in that in both cases there are specialized epithelial cells which are the receptive elements. The epithelium is thickened, and among the slender elongated supporting elements there are shorter ovoid cells with ciliated outer ends. These specific nerves have thick myelinated fibers, the branched unmyelinated terminals of which closely embrace the cell bodies of the specific receptive elements (Larsell, '29; Chezar, '30; Speidel, '46).

The lateral-line organs of Amblystoma are papillae, some of which are depressed in pits but are not inclosed in canals as in most fishes. Their arrangement conforms with the general pattern in fishes, with rows above and below the eye, on the lower jaw, and extending into the trunk as far back as the tail. The related nerves comprise one of the largest systems of the larva, which is reduced but not lost at metamorphosis. These thick and heavily myelinated fibers enter the brain in two large roots spinalward of the VIII roots and three or four which enter dorsally and slightly rostrally of the VII roots. They are conventionally assigned to the VII and X pairs of nerves, though they are more properly aligned with the VIII roots.

The arrangement of these roots is shown in figures 7, 9, 89, 90. Most of their individual fibers bifurcate immediately upon entrance into the brain into ascending and descending branches with numerous widely spread collaterals. These root fibers are arranged in fascicles, which span almost the entire length of the medulla oblongata (fig. 7), except for the most dorsal of the three or four lateralis VII roots, which ends in a "dorsal island" of neuropil ("cerebellar crest" of Larsell) at the level of entrance (figs. 7, 33, 45). Lateral-line fibers have not been seen to descend into the spinal cord. Anteriorly, they enter the auricle and end here (fig. 91); none have been traced into the body of the cerebellum, though secondary lateralis fibers after synapse in the auricle enter the com. vestibulolateralis cerebelli in company with vestibular fibers (figs. 32, 33, 34, *com.cb.l.l.*).

The exact functions served by the lateral-line organs are still imperfectly understood. The organs of the lateral lines and those of the internal ear have many similarities in embryological development, structure of the receptive apparatus, and central connections. They probably have had a common evolutionary origin from a more generalized form of cutaneous sense organ similar to the so-called "sensillae" of some invertebrates. This may be the explanation of the intimate association in the human ear of sense organs of such diverse functions as the cochlea for hearing and the semicircular canals for equilibration, both being highly refined derivatives of primitive tactile organs. The sense organs of the lateral lines are probably intermediate in function between tactile sensibility of the skin and the auditory and equilibrating functions of the internal ear. In fishes they have been shown to be sensitive to mechanical impact, slow vibrations, and currents in the water (Parker and Van Heusen, '17; Parker, '18). Hoagland ('33) and Schriever ('35) have investigated the functions of lateral-line nerves of fishes with the aid of oscillograph records of their action currents. Hoagland finds that these organs are in a state of continuous activity and that the nervous discharge is increased by application of pressure, by ripples and currents in the water, by movements of the trunk muscles, and by temperature changes.

In Amblystoma larvae Scharrer ('32) found evidence that the lateral-line organs may participate in the snapping reaction when moving prey is seized; and, subsequently, Detwiler ('45) reports that the lateral-line organs of these larvae constitute an adequate receptor apparatus for the detection of food in motion after extirpation of the eyes and nasal organs. The central connections of these nerves suggest that they play an important part in proprioception, and this is supported by Hoagland's experiments.

The VIII nerve of Amblystoma carries fibers from the membranous labyrinth, the structure of which resembles those of fishes plus a recognizable rudiment of the cochlea. These fibers enter the brain by two closely associated roots, dorsal and ventral, each of which contains many rather fine myelinated fibers, with some very coarse fibers mingled with them. Each fiber has a T-form division within the brain, the branches ascending and descending through the entire length of the medulla oblongata (figs. 7, 87–90). The dorsal and ventral roots remain separate as far forward as the V root and backward as far as the second root of the vagus. Beyond these limits the two roots merge. It is evident that fibers of the ventral fascicle take longer courses within the brain than do those of the dorsal fascicle, but the significance of the separation of vestibular fibers into two roots has not been determined. Some of these fibers descend for a long and undetermined distance into the spinal cord, mingled with the more ventral fibers of the dorsal funiculus and those of correlation tract b. The ascending fibers enter the auricle (figs. 29, 30, 31, 91), where many of them end. Others continue into the body of the cerebellum, decussate in the vestibulo-lateral cerebellar commissure, and terminate in the vestibular and lateralis neuropil of the auricle of the opposite side (figs. 32, 33, 34).

Within the medulla oblongata the collaterals and terminals of the vestibular fibers arborize in the common pool of neuropil, which also receives terminals of the V and lateral-line roots. Most of the neurons of the second order in the acousticolateral area spread their dendrites within this neuropil so as to engage terminals of several of these fascicles of root fibers of different physiological nature (fig. 9).

There is ample physiological evidence that salamanders exhibit vestibular control of posture and movement similar to that of other animals, and this implies that there is some central apparatus that is selective for the specialized end-organs of the internal ear. Sperry ('45a) has shown that in the case of the frog this specificity is preserved after section of the VIII nerve and its subsequent regeneration and that the precision of restoration of vestibular function is quite as exact as it has been shown to be in the case of regeneration of the optic nerve (p. 229). Since the specific functions of the several vestibular end-organs are not visibly localized in the medulla oblongata of the salamander, some other method of selection must be employed. Sperry's experiments on frogs lead him to favor the supposition that there are physicochemical axon specificities and selective contact affinities between the different axon types and neurons of the vestibular centers, a supposition which accords with much other evidence (p. 79). Differences in threshold and the time factor in the transmission rhythm may act selectively at the central synapses.

Though Amblystoma has no recognizable cochlear root of the VIII nerve, there is a primordium of the pars basilaris cochleae, which is better developed in the frog. This rudiment is lacking in Necturus, so that among the Amphibia successive stages in the early differentiation of the cochlear apparatus can be observed.

The fibers of the dorsal lateral-line VII root are shorter than those of the others, all ending in the "dorsal island" of neuropil at the posterior border of the lateral recess of the ventricle (figs. 33, 45; '44b, fig. 14; Larsell, '32, fig. 57). These fibers, like those of the other lateralis VII roots, come from lateral-line components of all three chief peripheral branches of the lateral-line VII nerves ('14a, p. 357). The dorsal island appears to be a remnant of the dorsal neuropil (cerebellar crest) of the lobus lineae lateralis described by Johnston ('01) in fishes.

In very young larvae of the frog (Larsell, '34) the relations are similar to those of urodeles, but soon a dorsal branch of the VIII nerve (derived from the primordial cochlea) enters this area dorsally of the dorsal lateral-line root. The dorsal island of neuropil retains its individuality to the time of metamorphosis, meanwhile becoming entirely surrounded by cells which proliferate from the dorsal lip of the area acusticolateralis. The dorsal VIII root is greatly enlarged to become the cochlear nerve; it terminates in relation with the cells surrounding the dorsal island, which now constitute the cochlear nucleus. After metamorphosis is complete, all lateral-line fibers degenerate, so that the gray of the larval area acusticolateralis becomes in the adult the cochlear nucleus dorsally and the vestibular nucleus ventrally.

Without here going into the further details of this differentiation, it is evident from Larsell's studies that the dorsal gray of the area acusticolateralis of urodeles and larval anurans is, during the metamorphosis of the frog, transformed directly into cochlear nuclei.

There is no degeneration of these cells and replacement by others. The same neurons which in the tadpole are activated from lateral-line organs lose their lateral-line connections in the adult frog and receive their excitations from the auditory apparatus, with a radical change of function. That which I at one time regarded as improbable ('30, p. 60) is exactly what happens in ontogeny, and doubtless the phylogenetic history is similar, as Ariëns Kappers has long maintained. Parallel with the differentiation of the cochlear nerve and nucleus in anurans, the related lateral lemniscus is enlarged and specialized in its definitive form.

In Necturus, which has no recognizable cochlear primordium, the functions of the dorsal cells of the area acusticolateralis evidently are related exclusively with lateral-line organs. The dorsal lateralis VII root terminating in the dorsal island does not differ physiologically from the other lateral-line roots of the VII nerve, so far as known. Like them, it receives fibers from lateral-line organs distributed over the entire head ('30, p. 21). But there is an obscure indication of a lateral lemniscus. Why do the fibers of the dorsal lateral-line root end in the restricted area of the dorsal island instead of extending through the whole length of the acousticolateral area like the other lateralis roots? The answer is probably to be sought in the phylogenetic history of the extinct ancestors of living urodeles. In fishes the lobus lineae lateralis of this region is covered by a neuropil, which has been termed the "cerebellar crest" and which extends forward into continuity with the superficial neuropil of the cerebellum. Larsell ('32, p. 410) regards the neuropil of the dorsal island as a survival of the cerebellar crest of fishes. This is the region within which the dorsal cochlear nucleus of anurans has been differentiated; and, if, as is generally believed, the living urodeles are descendants of more highly specialized ancestors with better organs of hearing, the preservation of their dorsal island may be regarded as a vestigial record of an ancestral history now lost.

The amphibian auricle (pp. 20, 44) receives terminals of trigeminal, lateral-line, and vestibular fibers. The connections of these fibers and their secondary pathways make it clear that this area contains primordia of two quite distinct mammalian structures. One of these is the terminal station of lateral-line and vestibular root fibers, and this tissue in higher animals is incorporated within the cerebellum and becomes the flocculus, as described by Larsell. The other primordium is trigeminal, and this in Amblystoma is probably concerned chiefly with proprioceptive functions of the skin and deep tissues of the head, as indicated by its strong cerebellar connection. This connection persists in mammals but is relatively insignificant here because, as mentioned above, the enlarged mammalian superior V nucleus is concerned chiefly with refined functions of the skin that Amblystoma does not possess.

MESENCEPHALIC NUCLEUS AND ROOT OF THE TRIGEMINUS

The system of the mesencephalic nucleus and root of the trigeminus is here well developed in typical relations, with some instructive special features. Its thick, well-myelinated fibers go out with branches of the V nerve. The details of their peripheral courses in Amblystoma have not been described. Experiments by Piatt ('46) indicate that the majority of the fibers of this system, which go out from the tectum opticum, are distributed to the jaw muscles. The more caudal cells of the mesencephalic nucleus probably have other connections. No evidence has been found for supply of any eye muscles from this nucleus.

Unlike other sensory systems, the cell bodies of these neurons lie within the brain. Their arrangement and the courses of the fibers arising from them have been described in the larva ('14a, p. 361) and in the adult ('36, p. 345) and are shown diagrammatically in figure 13. These cells vary in size, cytological structure, and number. Most of them are very large and of so characteristic appearance that they are easily recognized. They are sparsely distributed throughout the tectum in all layers of the gray substance, somewhat less numerous anteriorly near the posterior commissure, and densely crowded within and adjoining the anterior medullary velum. Occasionally, they are seen in the body of the cerebellum and in the nucleus cerebelli. Ten large larvae of A. punctatum had an average of 159 of these cells, the extremes being 76 and 208 (Piatt, '45). A subsequent count by Piatt ('46) of the total number of these cells in ten larvae of 45 mm. gave an average of 261 cells, equally divided on right and left sides. Of these cells, 86 on each side are in the tectum opticum and 45 in the nucleus posterior tecti and velum medullare anterius. Individual variations in numbers of cells are large, but approximate bilateral symmetry is quite consistently present. These cells are unipolar, the single thick processes accumulating near the outer border of the tectal gray and here acquiring myelin sheaths. These fibers are directed posteroventrally in loosely arranged dorsal and ventral fascicles, which converge toward the V nerve roots.

The ovoid cell body has a smooth contour, with no processes except the single thick fiber. It is imbedded in dense neuropil and closely enveloped by a web of these fibers. Every contact of the fibers of the neuropil with the cell is a synaptic junction. This is doubtless the explanation of the wide dispersal of these cells in all parts of the tectal gray, and they are so arranged that the entire extent of the deep tectal neuropil may be simultaneously activated by excitation of the mesencephalic V system.

The striking resemblance of the cells of the mesencephalic V nucleus with those of the semilunar and spinal ganglia and with the transitory Rohon-Beard cells of the spinal cord (Coghill, '14, Paper I) has often been commented upon and is well illustrated by the excellent photographs published by Piatt. They have, accordingly, been generally regarded as derivatives of the embryonic neural crest that have remained within the neural tube. This hypothesis has been tested experimentally by Piatt ('45), with the conclusion that neural crest is at least one source of these cells, though a possible origin from other sources is not excluded. Many observers have reported the presence of intramedullary cells of sensory type along the courses of roots of spinal and cranial nerves (Pearson, '45, cites instances), and some of these cells also may be of neural-crest origin. Others may be of autonomic type, migrating out from the brain (Jones, '45), though this is controverted.

Just as the typical unipolar cells of the sensory ganglia of spinal and cranial nerves have a single process, which divides into peripherally and centrally directed branches, so the mesencephalic V fibers (or some of them) divide shortly before emergence from the brain into peripheral and central branches (fig. 13). The central branches descend as far as the level of the IX nerve roots. My earlier statement ('14a, p. 362) that these fibers "arborize among the dendrites of the motor VII neurons" is misleading, for these terminals are spread widely in the intermediate zone between the levels of the V and the IX roots.

This bifurcation of the root fibers and the fact that the bodies of the cells of the mesencephalic V nucleus are in synaptic contact with all the deep neuropil of the tectum suggest that afferent impulses transmitted by these fibers may take either or both of two courses: (1) They may pass upward to the tectum, where they activate the deep neuropil diffusely and here are in relation with terminals of the optic and lemniscus systems; or (2) they may descend into the reticular formation of the medulla oblongata, where they act directly upon the motor nuclei and also upon the apparatus of bulbar neuromotor co-ordination (including the cells of Mauthner). These descending branches are accompanied by fibers of the spinal V root and by other fibers from the tectum and tegmentum, which end in the same field of the reticular formation (fig. 13, *tr.t.b.p.* and *tr.teg.b.*). In larvae of early feeding stages, thick uncrossed fibers, which descend from the tectum and subtectal areas into the bulbar reticular formation, are especially clearly seen, and also others which take similar courses after decussation in the ventral commissure. Some of these fibers arise from neurons of the isthmus, which are in synaptic connection with terminals of the secondary visceral-gustatory tract.

These connections of mesencephalic V fibers seem well adapted to facilitate the feeding reactions, a conclusion which is supported by observations on the cat by Corbin ('40) and the literature which he cites. In Amblystoma the field of reticular formation within which the movements of the mouth and pharynx are organized receives the descending mesencephalic V fibers, collaterals of V fibers, and fibers of correlation from the nucleus of the f. solitarius, isthmic visceral-gustatory nucleus, tectum, and the underlying dorsal tegmentum.

PROPRIOCEPTIVE SYSTEMS AND CEREBELLUM

Control of the course of muscular movement in process is insured by a variety of sensory end-organs, including those in muscles, tendons, joints, and the overlying skin. At the beginning of motility in the embryogenesis of Amblystoma a single peripheral sensory element (the transitory Rohon-Beard cells) may perform both exteroceptive and proprioceptive functions (Coghill, '14, Paper I, p. 199), and this may be true of some spinal ganglion cells in the adult, though here special proprioceptive apparatus also is provided. The Rohon-Beard cells are believed to be derived from a portion of the neural crest which is incorporated within the neural tube; and the mesencephalic nucleus of the trigeminus, as just described, has a similar origin. The latter cells survive in the adults of all vertebrates, in the service apparently of co-ordination of movements involved in the feeding reactions.

In the head the membranous labyrinth is the dominant organ of this system, with participation of nerves of cutaneous and deep sensibility, and probably the lateral-line organs also. Some proprioceptive control is doubtless organized in the reticular formation of the cord and bulb, but we have little information about how this is done. From the entire sensory zone of these regions, proprioceptive influence is filtered off and directed to the cerebellum, which is the general clearing-house for these functions. Many vestibular root fibers and a smaller number of trigeminal fibers go directly to the cerebellum, and secondary fibers from the sensory zone enter it by way of the spino-cerebellar tract and bulbar correlation tracts a and b (p. 159; '44b). That exteroceptive and proprioceptive functions are not completely segregated in these brains is shown by the fact that many fibers of the spinal lemniscus (tractus spino-tectalis) send collaterals into the cerebellum ('14a, p. 376). Within the cerebellum the general somatic sensory and vestibular components of the proprioceptive system are locally segregated, the former in the body of the cerebellum and the latter in the auricle, and this localization is a primary feature of the cerebellum in all vertebrates, as Larsell has shown. This author ('45) has also made it clear that cerebellar function includes much more than proprioception, or else the concept of proprioception must be redefined in more inclusive terms. The second alternative, I think, is better, as I have suggested in an article ('47) on the proprioceptive system, from which some of the following paragraphs are taken, by courtesy of the editor of the *Journal of Nervous and Mental Disease*.

Sherrington ('06, p. 347) defines the cerebellum as the head ganglion of the proprioceptive system, taking as the basis for his classification of receptors "the type of reaction which the receptors induce." In his exposition of this idea he makes it clear that the proprioceptive system is segregated from other sensory systems, not in terms of the receptors involved but because the system as a whole exerts regulatory control over the action of all skeletal muscles. The criteria employed here are applied in the efferent, not the afferent, side of the arc. In view of present knowledge of cerebellar function, Sherrington's original concept of proprioception should be emphasized and amplified.

It has long been recognized that in the cerebellum of lower vertebrates the sensory inflow is of two kinds, which are separately localized, viz., (1) the vestibular and lateral-line systems in the lateral part and (2) the spinal and trigeminal systems in the median body. The second category traditionally comprises deep sensibility of several sorts, notably that of muscle spindles, tendons, joints, and some other internal end-organs. Current physiological research requires radical revision and broadening of this traditional analysis. It has been shown that in mammals different cutaneous areas, vibrissae, audition, and vision have local representation in the cerebellum, as do also various systems of synergic muscles. In lower vertebrates the cerebellum has a broad connection with the hypothalamus, implying representation in the cerebellum of olfactory sensibility also.

In brief, cerebellar control of muscular movement employs practically all modalities of sense represented in the action system of the animal. The function of the cerebellum as the "head ganglion of the proprioceptive system" is not to pattern the muscular response (for these functions are localized elsewhere) but to facilitate its execution; and this facilitation employs all available sensory experience. Many organs of sense perform simultaneously both exteroceptive and proprioceptive functions. Sherrington's fruitful analysis of the action system into interoceptive, exteroceptive, and proprioceptive components was not based upon the specificities of the receptive organs, considered either anatomically or physiologically; but, on the contrary, the distinction was drawn in terms of what the animal does in response to sensory excitations. The interoceptive systems are defined in terms of internal adjustments, chiefly visceral. The exteroceptive systems are those which evoke adjustments of the body or its members to events in the external world. The proprioceptive systems are ancillary to the activities of the musculature in maintenance of tonus, posture, and regulation of the action of synergic groups of agonist and antagonist muscles in appropriate strength and sequence.

Proprioception, therefore, must be defined not in terms of the modalities of sense employed but in terms of the results achieved. Cerebellar proprioceptive control is accomplished by the application of all relevant types of sensory inflow to specific and successive muscular activities which may be in process from moment to moment; and the definition of "proprioception" must be formulated in terms of the motor response rather than of the sensory systems involved. The proprioceptive system, accordingly, includes all peripheral end-organs and nerves and all central adjustors in the spinal cord, brain stem, cerebellum, and cerebral hemispheres that collaborate in the co-ordination and synergizing of muscular activity in process. In a recent conference with Dr. Larsell he suggested to me that, in view of the inadequacy of current conceptions of the true nature of the proprioceptive system and the faulty connotations of the term in present usage, it might be better to avoid the word hereafter and replace it by the more inclusive name, "proprius system."

The preceding comments on the proprioceptive system apply, *mutatis mutandis*, to Sherrington's exteroceptive and interoceptive systems also. In his original definitions of these terms, Sir Charles was careful to insist that each component of each of these three subdivisions of the total pattern of behavior must be viewed in its entirety as a unitary act and that the significance of these acts can be understood only in terms of their reciprocal relationships with one another and with the total action system of the animal. The critical feature of each of these acts is the end-result, the actual behavior exhibited. The names originally given to these three classes of functions put the emphasis on the receptive organs, where it does not belong. Some obscurity and confusion may be avoided if the unity of these several components of behavior is recognized in their nomenclature. The exteroceptive systems, viewed in their entirety, are *somatic*, the interoceptive systems are *visceral*, and the proprioceptive systems are ancillary to all muscular activity and, accordingly, may be termed *proprius*. Sherrington's terms, "exteroceptors," "interoceptors," and "proprioceptors" are suitable names for the receptive organs, with the qualification that the same organ may, on occasion, activate somatic, visceral, or proprius responses.

VISCERAL SENSORY AND GUSTATORY NERVE ROOTS

General visceral sensory fibers of wide peripheral distribution enter the brain by the vagus roots, and the IX and VII roots contain smaller numbers of similar fibers from the mucous surfaces of the mouth and pharynx. Taste buds are widely distributed in these mucous surfaces, and the gustatory fibers are indistinguishably mingled with the general visceral fibers peripherally in the roots of the VII, IX, and X nerves and centrally in the f. solitarius, more of them entering the brain anteriorly than posteriorly. This mixed group of peripheral fibers, as a whole, is quite distinct from all other functional systems and it was termed by the earlier students of nerve components in lower vertebrates the "communis system" because all its fibers converge into a single central bundle, the f. communis (Osborn, '88). This we now know is homologous with the mammalian f. solitarius. The peripheral and central arrangements of the chemoreceptors illustrate some general principles which will next be examined.

CHEMICAL SENSIBILITY

The peripheral terminals of the sensory fibers of the V to X cranial nerves take three forms: (1) The fibers of the general somatic sensory and visceral systems have free nerve endings within or beneath epithelium or widely spread in deeper tissues. (2) The end-organs of the special somatic sensory systems are differentiated epithelial structures of the internal ear or lateral lines with receptive hair cells, which are shorter than the surrounding supporting cells. (3) The chemoreceptors of the gustatory (special visceral sensory) system are budlike epithelial structures, which resemble the naked lateral-line organs but differ from them in that the specific receptive cells are slender, elongated elements, which span the entire thickness of the epithelium. The fibers which innervate them are generally thinner than those of lateral-line organs and are less myelinated or unmyelinated.

Those species of fishes which have taste buds abundantly distributed in the outer skin and also naked organs of the lateral lines not inclosed in pits or canals present both morphological and physiological problems of great difficulty ('03, '03a, b). In the earlier literature all these cutaneous organs were termed indiscriminately "terminal buds," with resulting confusion which was not clarified until the nerve fibers which supply them were found to belong to different functional systems. The fibers supplying lateral-line organs, wherever situated, converge centrally into the acousticolateral area, and fibers supplying taste buds, whether in mucous surfaces or in the outer skin, converge into the f. solitarius and its nucleus. The separation of the gustatory from the lateral-line system of cutaneous sense organs by the anatomical method has been confirmed by physiological experiments performed by the writer, G. H. Parker, and others.

Though this distinction is perfectly clear in some species of fishes, in others there are transitional forms of "terminal buds," and much remains obscure about the functions of these various types of receptors. The problem is complicated by the fact that in fishes the skin is everywhere very sensitive to a large variety of chemical substances (Sheldon, '09; Parker, '12, '22; Ariëns Kappers, Huber, Crosby, '36, chap. iii; for a more general discussion of the chemical senses see Moncrieff, '44). The skin is sensitive, in general, to different substances from those which activate the olfactory organ and taste buds, but there are some puzzling exceptions.

For instance, the gurnard fishes (Prionotus, Trigla) have three rays of the pectoral fin which are modified to serve as "feelers" in the search for food on the floor of the sea. Somewhat similar filamentous pelvic fins of the gourami, codfish, and several other teleosts are abundantly supplied with taste buds with the usual functions and nervous connections ('00, '03; Scharrer, Smith, and Palay, '47); but the free pectoral fin rays of the gurnards have no taste buds, and yet it has been shown (by the authors last mentioned) that these fin rays are sensitive to the same sapid substances as are the cutaneous taste buds of other fishes and that the reactions also are similar. These authors, in tracing the central courses of the large nerves which supply these free fin rays, find that these fibers have central connections similar to those of the pectoral fins of other fishes, belonging, that is, to the general cutaneous system. They do not enter the f. solitarius. They present evidence also that some secondary fibers from these general cutaneous centers connect centrally with the superior gustatory nuclei of the isthmus and hypothalamus, just as do the true gustatory fibers arising in the nucleus of the f. solitarius. This is interpreted to mean that these nerves of general cutaneous chemical sensibility are so specialized that they can serve typical gustatory reactions, though they do not connect peripherally with taste buds.

These observations seem to show that some peripheral fibers of the general cutaneous system, without specialized receptive end-organs, may acquire functions substantially identical with those of cutaneous taste buds and that such fibers have central connections similar to those from taste buds. It is evident that no rigid categories can be recognized here in terms of our conventional classification of "the senses" or of their organs, a principle illustrated also by Whitman's ('92) observations on the cutaneous sense organs of the leech, Clepsine, to which reference is made on page 84. Nature is not bound by our rules of logical analysis.

Taste buds within the mouth are interoceptors, but similar buds in the outer skin of fishes are typical exteroceptors, used in the selection and location of food, as are also the free nerve endings of the general cutaneous nerves that respond to chemical excitants. It is evident that all these nerve endings co-operate with the nerves of ordinary tactile sensibility in the normal process of finding food. That this co-operation is intimate and in some cases indispensable has been shown by Parker ('12) in the case of the catfish, Ameiurus. In this fish the skin has general chemical sensitivity to acid, alkali, and salt, a sensitivity which is served by general cutaneous nerves. In the skin there are also innumerable taste buds which are innervated by fibers which enter the f. solitarius. The general chemical sensibility is preserved if the taste buds are denervated, but the specific gustatory function of the taste buds is lost if the general cutaneous innervation of the surrounding skin is eliminated. A similar relation prevails with taste buds within the mouth, for these have a double innervation; and in man the gustatory function is abolished if the trigeminal innervation of the tongue is surgically destroyed, even though the specific innervation of the buds remains uninjured; this loss, however, is temporary, and after a few weeks gustatory function returns (Cushing, '03).

Amblystoma has no cutaneous taste buds, but the mouth cavity is abundantly supplied with them. They are especially numerous on the palate among the vomerine teeth, and these buds have a peculiar accessory innervation—a compact skein of circumgemmal fibers of uncertain origin ('25b; Estable, '24). These fibers separate from a plexus related with the ramus palatinus and may be derived from a trigeminal anastomosis; but this has not been demonstrated.

The tactile, general chemical, and gustatory systems are as intimately related centrally as they are peripherally. All taste buds of all animals, wherever found, are supplied by fibers which discharge centrally into the nucleus of the f. solitarius or its derivatives. In those fishes which have cutaneous taste buds with exteroceptive functions the central connections of these buds differ from those of buds within the mouth which have interoceptive functions. These details need not be given here; the interested reader is referred to a recent paper ('44b) and references there given. These differences are explained by the fact that stimulation of interoceptive taste buds evokes visceral responses, but excitation of exteroceptive buds is followed by somatic movements for capture of food.

In Amblystoma, as in man and all other vertebrates, all fibers from taste buds enter the f. solitarius. Most fibers of all modalities of general cutaneous sensibility of the head enter the sensory V nucleus; but a small number of them pass through this nucleus to enter the f. solitarius, thus providing for integration of general somatic sensory and gustatory sensibility. The prefacial f. solitarius carries gustatory impulses forward into the neuropil of the superior trigeminal nucleus in the auricle. A third and much more extensive provision for bringing general cutaneous and both general and special visceral sensibility into physiological relation is at the bulbo-spinal junction (chap. ix).

The preceding analysis illustrates the intimate physiological relationship which exists among the various modalities of sense which may be concerned with the resolution of mixed sensory experience in the interest of securing the appropriate responses. The integrating apparatus is spread from the peripheral end-organs throughout the central nervous system. Within this machinery for conjoint action there have been differentiated the specific sensory and motor systems, that is, the analyzers. The first step in this analysis is the separation in the central adjustors of the visceral from the somatic systems. Thus the fibers of taste and general visceral sensibility converge into the f. solitarius, well separated from all the somatic sensory systems which are assembled more superficially. This segregation obviously has arisen because of the radical differences in the courses taken by the efferent fibers from visceral and somatic receptive fields to visceral and somatic effectors, respectively.

In general, the gustatory fibers tend to end near their entrance into the brain, and the general visceral fibers to descend toward the lower end of the system. In elasmobranchs the nucleus of the f. solitarius is locally enlarged, with a separate lobe for each of the nerves of the gills. These enlargements, which show as a beadlike row in the wall of the fourth ventricle, are probably chiefly gustatory. In the carp and some other teleosts with enormous numbers of taste buds, there are separate enlargements of this nucleus known as facial, glossopharyngeal, and vagal lobes. These are known to be largely gustatory in function. In other species of teleosts there are various modifications of these arrangements. In some birds with very few taste buds the f. solitarius is clearly double. A very slender medial bundle carries the few gustatory fibers and the much larger lateral bundle, the general visceral fibers (Ariëns Kappers, Huber, Crosby, '36, p. 370). In Amblystoma, as in mammals, none of these specializations have occurred, and the visceral sensory system as a whole retains its primitive characteristics.

SOMATIC MOTOR NERVE ROOTS

These roots in Amblystoma comprise only fibers for the extrinsic muscles of the eyeball in the roots of the III, IV, and VI nerves. As previously mentioned, all peripheral motor neurons are mingled with those of the motor tegmentum and are usually indistinguishable from them except in cases where their axons can be followed into the nerve roots. The cells of the nuclei of the eye-muscle nerves are fairly clearly segregated, and in some reduced silver preparations they react specifically to the chemical treatment (fig. 104); but even here their dendrites are widely spread and intertwined with those of tegmental cells, so that both kinds of neurons would appear to be similarly activated by the neuropil within which they are imbedded. The oculomotor nucleus lies in the posteroventral part of the peduncular gray (figs. 6, 18, 22, 24, 30, 31, 104). The nucleus of the IV cranial nerve lies about midway of the longitudinal extent of the isthmic tegmentum and far removed from the oculomotor nucleus (figs. 61, 104). The thick IV root fibers (most of them myelinated) ascend along the outer border of the gray to decussate in the anterior medullary velum in the usual way.

As mentioned in chapter xiii, the sensory zone of the isthmus contains cells of the mesencephalic V nucleus and others which send axons peripherally to meninges and chorioid plexus. Some of the latter go out with the IV nerve roots to unknown destinations. It is possible that some of these cells are secondarily displaced neurons of the motor IV nucleus, similar to those described by Larsell ('47b) in cyclostomes. There is no definite evidence that this condition exists in urodeles; and, indeed, all connections of the cells lying within and adjoining the superior medullary velum require further study.

The floor plate of Amblystoma throughout its length contains a special type of ependymal elements and the cell bodies of some neurons. These neurons do not invade the floor plate from the basal plate, but they develop within this plate intrinsically, as was first pointed out by Coghill ('24, Paper III). In the adult medulla oblongata there are few of them at any one level, but they constitute a definite nucleus raphis, which is enlarged in some places, notably so in the interpeduncular nucleus. Most of the cells of the nucleus of the VI nerve are median, as in Necturus ('30, p. 14). These cells in ordinary preparations cannot be distinguished from others of the nucleus raphis except by observation of axons emerging in the VI nerve root. They are distributed sparsely in the ventral raphe and adjacent to it between the levels of the VII and IX nerve roots ('44b, fig. 2). They emerge usually by two widely separated roots, though more rootlets are sometimes seen.

GENERAL VISCERAL EFFERENT NERVE ROOTS

Preganglionic fibers for unstriated muscles and glands leave the brain in the III, VII, IX, and X roots, probably with the IV nerve and its environs (p. 181) and perhaps with the parietal nerve (p. 235). In other vertebrates, fibers of this system have been described as leaving the brain with the optic nerve, with the nervus terminalis, and independently from other regions of the brain for the meninges. Our material is inadequate to reveal satisfactorily either the central connections or the peripheral courses of any of these fibers, so that this topic remains to be clarified. The large unmyelinated hypophysial nerve belongs in this system, as described on page 244.

SPECIAL VISCERAL MOTOR NERVE ROOTS

The striated muscles related with the visceral skeleton of the head —jaws, hyoid, branchial arches, and their derivatives in higher animals—belong in a special category (p. 69). These muscles are visceral in phylogenetic and embryologic origin and primitively in function, but in all craniate Chordata they have acquired the same striated structure as somatic muscles, as well as various degrees of somatic function. They occupy, accordingly, an ambiguous position and are sometimes termed "special somatic muscles" ('22, '43). Their innervation is of similarly intermediate character. In Amblystoma these motor fibers are thick and well myelinated and arise from large cells of motor type which are more or less clearly segregated in separate nuclei. These nuclei lie laterally of the somatic motor nuclei and well separated from them.

This system of fibers is represented in the V, VII, IX, and X pairs of cranial nerves. Their peripheral distribution has been described by Coghill ('02). Their large motor nuclei are imbedded in the tegmental gray, with no clear boundaries. Their approximate positions in the larva are shown as projected upon the floor of the fourth ventricle in 1914a, figure 1; but, as seen in other figures of that paper, their dendrites ramify widely among those of the tegmental neurons. This indicates that both kinds of cells are activated from the same sources.

The motor V nucleus lies more laterally than the other members of this group. A Golgi impregnation of three of its elements is shown in figure 40. From it two roots arise, one from its posterior end and one farther forward. There are two well-separated motor nuclei and roots of the VII nerve, the roots emerging ventrally of the VIII root. The cells of the anterior nucleus are scattered among similar cells of the tegmentum near the level of exit from the brain. The posterior nucleus is farther spinalward and is the anterior part of a well-defined column of cells, which also includes the motor IX and X nuclei. The motor VII root fibers pass forward dorsally of the f. longitudinalis medialis (fig. 89) and turn laterally to emerge from the brain, as in Necturus ('30, p. 15). There is no visible boundary between the posterior motor VII and the motor IX and X nuclei.

CHAPTER XI

MEDULLA OBLONGATA

THERE is no systematic description of the medulla oblongata of adult Amblystoma comparable with my paper on Necturus ('30). Coghill's papers give many details of the early stages of development, but no comprehensive description has been written. The 38-mm. larva was described in 1914, and subsequent study convinces me that the conditions found there and in Necturus are not fundamentally different in adult Amblystoma, though the latter is considerably more specialized. This is evident upon comparison of the figures of adult Amblystoma recently published ('44b) with those of the two papers just cited, to which the reader is referred for general discussion. At metamorphosis the changes are far less in Amblystoma than in Salamandra and some other urodeles. Descriptions of the medulla oblongata and cerebellum of Cryptobranchus, Siren, Proteus, Salamandra, and some other urodeles are found in the papers cited on page 11. Some details of the schematic outline given in chapters iv and v will now be filled in.

SENSORY ZONE

The sensory zone forms the massive dorsolateral wall of the fourth ventricle from the calamus scriptorius forward, to and including the thick auricle under the cerebellum (p. 44). This zone is reduced in size but not suppressed in the floor of the lateral recess of the ventricle spinalward of the auricle. In the remainder of the medulla oblongata the gray substance forms a low ridge bordering the taenia of the fourth ventricle and projecting into the ventricle, the acousticolateral area, which is smaller in the adult than in the larva (figs. 9, 89, 90). Ventrally of this area is a narrower ridge, which expands posteriorly —the visceral lobe (figs. 9, 88). This contains the nucleus of the fasciculus solitarius.

The white substance of the sensory zone is composed chiefly of the fascicles of sensory root fibers and the associated neuropil. There are, in addition, secondary arcuate fibers and two longitudinal bundles of correlating fibers—tracts *a* and *b* of Kingsbury (figs. 7, 9, 87–90).

The sensory root fibers are related peripherally with end-organs which seem to be physiologically as specific as those of man, though the specificities are of different types; but at the first synapse this specificity is no longer preserved in terms of localized gray centers or pathways of conduction. A neuron of the second order may have functional connection with afferent fibers of tactile, vestibular, lateral-line, or gustatory roots, singly or in any combination.

Since most of the sensory root fibers span the entire length of the medulla oblongata and have numberless collateral branches throughout this length, evidently there is little provision for localization of function in terms of fore-and-aft relations in space. Mass responses of the entire musculature are readily evoked by excitation of any or all components of this sensory complex, but how are selective local responses effected? When the arrangement of the secondary connections of these sensory root fibers is examined, it is evident that this question cannot be answered in terms of any mechanism of switchboard type. The axons of these secondary neurons do not take random courses. They tend to be assembled in bundles of fibers, which have a common direction and destination. Such bundles as ascend to higher levels (the lemniscus systems) evidently have some kind of specificity, for they retain their identity up to termination in definite and different places. This specificity may be determined by the source of their excitation—the modality of sense and the location in the body of the receptive end-organs—or it may be determined by the nature of the response to be evoked—muscular contraction, secretion, and the location of the effector organs activated. Probably both these factors are operative in establishing the pattern of arrangement of the higher centers of adjustment and their connecting tracts of fibers.

The problem of the apparatus employed in making discriminative responses to specific types of sensory excitation is not simplified by these observations. In higher animals the specificity of the various modalities of sense is preserved in central pathways of conduction leading up to higher centers of adjustment in the thalamus and the cerebral cortex. But somewhere in the course of this transmission these diverse, localized, functional systems are brought into relation with one another and are integrated in such a way as to result in appropriate responses to the total situation. In these amphibians the process of integration begins at the first synapse. The organization of the neurons of the second order is such as to facilitate activation of large masses of musculature by sensory excitation of any kind, that is, reactions of total-pattern type as defined by Coghill. This is the only kind of response which can be made by Amblystoma in the earliest stages of development of motility; and in the adult animal it is clear, from both the structural organization of the nervous system and the overt behavior, that the action system as a whole is predominantly of total-pattern type. One of the coarser features of the structural mechanism of this integrated behavior is seen in the way the neurons of the second order act as collectors of excitations of diverse modalities and widely distributed peripheral origin.

Within this primordial apparatus of integrated mass action, as development advances, local areas are differentiated with progressively more restricted functional connections. This is manifested anatomically by contraction of the spread of the dendrites of the secondary elements so as to have synaptic contact with a smaller number of fascicles of root fibers. In midlarval stages some of these elements spread their dendrites through the whole extent of the white substance of the sensory zone, thus having contact with terminals of all fascicles of root fibers (fig. 9, no. 1). Some of these dendrites also reach downward into the zone of correlation, the reticular formation (fig. 9, no. 4). Other elements are in functional relation with a few or only one of the root fascicles (fig. 9, nos. 2, 3). In the adult the dendritic spread is not so wide, and a larger proportion of the neurons are in synaptic relation with one or a few contiguous fascicles of root fibers; yet even here there are few neurons of the second order which may be activated by root fibers serving a single modality of sense.

There is, accordingly, as development advances a progressive restriction and local specialization of the central apparatus of analysis of the manifold of sensory experience, and this is an important factor in the acquisition of the local reflexes which are individuated within the larger total patterns of behavior as this process has been described by Coghill. Furthermore, this process of local specialization of structure within the sensory field, as exhibited in ontogeny, has a parallel in phylogenetic history, as illustrated, for instance, by comparison of the organization of this field of Amblystoma with that of the frog, as described by Larsell ('34, p. 504), and with that of the specialized fishes ('44b).

In the tracts of fibers which ascend to higher levels of the brain stem from this sensory field—the lemniscus systems as described below—there is a similar lack of segregation of separate pathways for the several modalities of sense. The general bulbar lemniscus (*lm.* of the figures) carries fibers which may be activated from any or all of the sensory nerve roots. The segregation of some specific sensory pathways is incipient in the lemniscus systems, and this is seen with especial clarity in the secondary ascending visceral-gustatory path (*tr.v.a.*).

INTERMEDIATE ZONE

The reticular formation of the spinal cord is continued forward into the medulla oblongata (fig. 9; '14a, p. 378; '30, p. 59). In Weigert sections (figs. 89, 90) it may be recognized as a lighter field between the bulbar lemniscus (*lm.*) and the spinal lemniscus (*lm.sp.*). Here the spino-bulbar, spino-cerebellar, and spinal lemniscus tracts are assembled in the alba.

The neuropil of this field receives dendrites from the underlying gray and from the adjoining sensory and motor zones (fig. 9; '14a, figs. 22–43; '44b, figs. 7, 8). Some of these cells are small, with local connections; others are very large, with wide dendritic spread and long axons (most of them myelinated), which descend as internal arcuate fibers to the ventral funiculi. Many of them decussate in the ventral commissure; and both crossed and uncrossed fibers, before or after crossing, bifurcate into ascending and descending branches. The descending bulbo-spinal fibers and the ascending bulbo-tegmental fibers are part of the neuromotor apparatus of bulbar and spinal reflexes. The lemniscus systems, on the other hand, arise chiefly in the sensory zone, though some of their fibers may be axons of neurons of the intermediate zone, a distinction of no great significance, in any event, because many neurons have extensive dendritic spread in both zones. The lemniscus systems also differ from the neuromotor apparatus just described, in their terminal connections, passing not to the motor zone but to higher levels of the sensory zone.

Among the large cells of the intermediate zone the two giant cells of Mauthner are of special interest. They lie at the level of the vestibular roots and have large dendrites directed outward among terminals of these root fibers ('14a, fig. 12). In the 12-mm. larva they are of enormous size ('14a, fig. 53), with thick dendrites, which ramify throughout the entire area of the white substance. In the adult the dendritic spread is less extensive, yet it is sufficiently wide to embrace the greater part of the alba of the intermediate and motor zones and especially the terminal area of vestibular fibers. The thick myelinated axons immediately decussate and descend in the f. longitudinalis medialis through the entire length of the spinal cord. These two cells are collectors of nervous impulses from many sources and are part of the apparatus of control of mass movements of the musculature of the trunk. Their specific functions have not been fully explained, though they have been much studied. They are magnified and highly specialized examples of the elements of the nucleus motorius tegmenti of the medulla oblongata.

We lack sufficient knowledge of the structural and physiological properties of this tegmentum to frame satisfactory hypotheses of the actual mechanism of the integrative and co-ordinating functions which seem to be operating here. The intermediate tegmentum is structurally so intimately interwoven with the tegmentum of the motor zone that both probably act in these operations as a functional unit.

At the anterior end of the medulla oblongata the trigeminal tegmentum merges with the isthmic tegmentum, and dorsally of this junction the nucleus cerebelli occupies the position of the intermediate zone. In higher animals the gray of this nucleus is incorporated within the body of the cerebellum as the deep nuclei. The definitive cerebellum, accordingly, is a derivative of both sensory and intermediate zones.

MOTOR ZONE

Little can be added here to the general description in chapter v. The larger cells of the motor field may spread their dendrites through the whole of the motor and intermediate zones and also upward into the sensory zone and across the ventral commissure to the opposite side ('14a, figs. 29, 30, 31, 37, 41, 42; '44b, figs. 7, 8). This type of structure is extended forward into the isthmus, where it is more specialized and the small and large cells are segregated, though not completely so. Most axons of the large cells are myelinated in the adult. In midlarval and late larval stages, thick unmyelinated axons of the large cells take curious courses. Some descend uncrossed in the ventral or ventrolateral funiculi. Others decussate in the ventral commissure. The latter may cross transversely as internal or external arcuate fibers, or they may take long ascending or descending courses before crossing obliquely and then turning spinalward. Before or after crossing, these fibers may divide into long ascending and descending branches, and they may give long branched collaterals along the entire course. The terminals of an individual fiber may reach a large part of the motor field of the medulla oblongata of both sides. How far forward the ascending branches may extend has not been determined; some of them certainly pass beyond the isthmus. Some illustrations of these fibers have been drawn ('39b, figs. 42, 46, 51, 57, 58, 61, 66, 68). In the medulla oblongata they are mingled with similar fibers of the thalamo-bulbar, pedunculo-bulbar, and tegmento-bulbar systems ('39b, fig. 23). Relatively few of them enter the f. longitudinalis medialis. The extensive and diffuse spread of the terminals of these fibers makes it evident that, if they are employed in any definitely local reactions, the localization is effected by some device other than the arrangement in space of their terminals.

In mammalian neurology the transverse segment of the brain stem, which lies under the cerebellum, is named the "pons" after its most conspicuous external feature. This name is obviously inappropriate here, where there is no pons or any recognizable primordium of it.

FIBER TRACTS OF THE MEDULLA OBLONGATA

The white substance of the medulla oblongata contains many compactly arranged, well-myelinated fibers. Some are very coarse and some thinner. Mingled with these, especially in the sensory and intermediate zones, there are many unmyelinated fibers. In most preparations the several systems are not clearly fasciculated, so that analysis is difficult in the adult. In larvae elective Golgi preparations have revealed many details, though much remains obscure. Most of the ventromedial fibers are descending. Laterally within and surrounding the reticular formation are fibers of local correlation and the ascending lemniscus systems; and dorsolaterally there are the fascicles of sensory root fibers, together with some correlating systems. Most of the latter appear as arcuate fibers, which are very numerous at the outer border of the gray, with some at intermediate and superficial depths of the alba (figs. 87–90).

Sensory fibers of the second order are of five sorts: (1) reflex connections by arcuate fibers with the bulbar motor zone of the same and the opposite side, either directly or with synapse in the intermediate zone; (2) bulbo-spinal connections, crossed and uncrossed, descending in the ventral and ventrolateral funiculi; (3) bulbo-cerebellar connections (see chap. xii); (4) correlation tracts a and b (of Kingsbury), intrinsic to the sensory zone; (5) the lemniscus systems, passing between the sensory zone and higher levels of the same zone, most of them decussating in the ventral commissure. The fibers of all these kinds are widely dispersed, and their analysis is still incomplete.

Arcuate fibers are everywhere present in the medulla oblongata, and each of the five groups of secondary connections mentioned above is represented in these arcuates. There are also deep arcuates from the intermediate and motor zones to the ventral commissure. These are doubtless part of the apparatus by which the patterns of performance of synergic groups of muscles are organized, but the details of actual operation of this apparatus are unknown. It seems to be well established that the primary organization of these motor patterns is intrinsic to the motor and intermediate zones. These intrinsic mechanisms may be activated (1) in a nonspecific way from the sensory zone, with resulting mass movement; (2) more specifically from parts of the sensory field dominated by one or another system of peripheral sense organs, with local motor response; (3) from higher centers of control which co-ordinate all motor activity in the interest of integrated behavior of the body as a whole. The second and third of these types of behavior are recognizably present in Amblystoma, though at a very primitive level of specialization.

SENSORY CORRELATION TRACTS *a* AND *b*

All peripheral somatic sensory fibers which enter the medulla oblongata terminate in a common pool of neuropil, which pervades the sensory zone within which the various qualities of sense seem to be more or less completely merged. The secondary fibers which leave these fields are, however, not strictly equipotential. There is evidently an incipient localization of function among these fibers, but the apparatus by which this specialization is achieved is not clear. It seems probable that part of this apparatus is to be sought in two longitudinal tracts of correlation, which were first described by Kingsbury ('95) in Necturus. To these he gave noncommittal names, "tracts *a* and *b*," one lying above, the other below, the fascicles of lateral-line nerve roots (figs. 9, 87–90). I have written a general description of these tracts of Amblystoma ('44*b*) and, in particular, of their relations posteriorly with the dorsal funiculus of the spinal cord and the nucleus funiculi and anteriorly with the cerebellum. Some additional features may now be added.

Tract a.—This tract follows the taenia of the fourth ventricle for most of its length, anteriorly passing below and through the dorsal island of neuropil, under the floor of the lateral recess of the fourth ventricle, and then arborizing in the lateral vestibular area of the auricle (figs. 7, 32, 33; '44b, fig. 14). Posteriorly, at the calamus scriptorius, it converges with its opposite fellow, and the two tracts enter, respectively, the medial fascicles of the dorsal funiculi of the spinal cord. Tract *a* is regarded as a mixed collection of fibers of correlation, related primarily with the lateral-line nerve roots, a supposition which is supported by Kreht's observation ('30, p. 316; '31, p. 422) that in adult Salamandra, in which the lateral-line roots are atrophied, tract *a* also disappears. It is connected by arcuate fibers with the intermediate and motor zones of both sides, and probably some of these fibers are commissural, connecting the acousticolateral areas of the two sides.

The nucleus of the dorsal funiculi extends for a considerable distance anteriorly and posteriorly of the calamus scriptorius. Fibers of the slender median fascicle of the dorsal funiculus of the cord terminate in this nucleus, and some of them continue forward for an undetermined distance in tract *a*. The nucleus funiculi is connected through tract *a* with the lateral-line area farther forward by fibers, some of which are probably descending and some ascending. The latter may go as far as the cerebellar primordium in the auricle, though there is no demonstration of this. In Triturus, as described by Larsell ('31, p. 48), unmyelinated fibers descend into tract *a* from the auricle, thus providing for cerebellar discharge into the lateral-line area.

Tract b.—This tract is a mixed bundle of myelinated and unmyelinated fibers, believed to be related primarily and perhaps specifically with the vestibular roots (figs. 7, 31). The literature contains several descriptions of tract *b* of Amblystoma ('14a, p. 373; '39b, p. 604; '44b, p. 317; Larsell, '32) and other urodeles. In the aggregate, these give an incomplete account of it. From the data at hand it is concluded that it contains a variety of crossed and uncrossed, ascending and descending, fibers connecting the vestibular areas of both sides with each other, with the underlying tegmentum, with the cerebellum, and with the spinal cord. Kreht ('31, p. 423) describes in Proteus a probable connection with the f. longitudinalis medialis.

Anteriorly, tract *b* converges with tract *a* into the vestibular neuropil at the lateral border of the auricle (fig. 91). This is cerebellar territory, the primordium of the flocculus of mammals. Silver impregnations show that many of these fibers end in free arborizations in the auricle ('39b, figs. 43, 67, 77, 98). This connection is clear in

Triturus also (Larsell, '31, p. 50) and is comparable with the secondary vestibulo-cerebellar tract of mammals. Fibers descending from the auricle into tracts *a* and *b* are probably ancestral to the mammalian f. uncinatus of Russell.

At the bulbo-spinal junction, tract *b* merges with the spinal root of the vestibular nerve, and its fibers descend for an undetermined distance in the lateral fascicles of the dorsal funiculus of the cord. There is, accordingly, a dorsal vestibulo-spinal connection by both peripheral and secondary fibers, a connection which puts the vestibular apparatus into especially intimate relation with the neuropil of the nucleus funiculi. It is probable that fibers ascend from the nucleus funiculi in tract *b*, and some of these may extend as far as the auricle. As I have pointed out ('44*b*), this cerebellar connection, if confirmed, would be the obvious precursor of the system of arcuate cerebellar fibers described in primates by Ferraro and Barrera ('35) as passing from the external cuneate nucleus to the cerebellum by way of the corpus restiforme. Tract *b* connects only with the vestibular field of the cerebellum, and this cerebellar connection of the nucleus funiculi (if present) would have a physiological significance quite different from that of the larger connection by way of the tractus spinocerebellaris (fig. 3), for the latter connects only with the nonvestibular body of the cerebellum.

In addition to the dorsal uncrossed vestibulo-spinal connection just described, there is an extensive crossed and uncrossed connection between tract *b* and the motor field by arcuate fibers. Many of these fibers bifurcate, with branches which ascend and descend in the ventral funiculus. These may correspond with the secondary vestibular fibers of the mammalian f. longitudinalis medialis, but here they are dispersed, and the details of their courses have not been described. In the absence of elective impregnations, our material does not reveal their courses. Their presence may be expected, in view of the fact that they comprise an important constituent of this fasciculus in mammals. An incomplete impregnation in an advanced larva gives some evidence of them. In figure 38 the contorted fiber directed medially from the region of entrance of the VIII root is probably a secondary vestibular fiber, which enters the f. longitudinalis medialis, and the two impregnated fibers seen in this fasciculus may belong to this system.

The inferior olive has not been identified in urodeles, but among the dispersed arcuate fibers are some with connections which suggest the presence of a primordium of this structure ('39b, p. 604). Some of the smaller tegmental neurons in the vicinity of the VIII roots send axons laterally into tract b, where they divide with ascending and descending branches. Similar fibers are seen to enter this tract at all levels between the V and X roots. Some of these come from the opposite side. The ascending branches of these fibers pass under the auricular gray and recurve dorsalward to arborize within the gray of the rostral face of the auricle. This may represent a vestigial remnant of the large olive of some fishes, here reduced to insignificant proportions because of the small size of the cerebellum (Ariëns Kappers, Huber, Crosby, '36, pp. 668–89).

THE LEMNISCUS SYSTEMS

The less myelinated field of alba of the medulla oblongata between the sensory and motor zones, which I term the "reticular formation," contains the ascending fibers of the lemniscus systems and a neuropil, which receives dendrites of neurons of the underlying gray and also of many neurons of the adjoining sensory and motor zones. This seems to be the field in which patterns of local bulbar reflexes are organized, but its texture and connections have not been satisfactorily analyzed. From the reticular formation and the overlying sensory gray, thick axons, some of which are myelinated, descend as internal and external arcuate fibers to the ventral funiculus. Many of these pass to the motor zone as part of the neuromotor apparatus of bulbar and spinal reflexes. These enter a mixed spino-bulbar and bulbo-spinal tract associated with the spinal lemniscus, as shown in figures 38 and 39. Others, usually thinner fibers, immediately decussate in the ventral commissure and then ascend in the lemniscus tracts to higher levels of the sensory zone.

In the lemnisci of mammals, fibers of the various sensory systems are segregated in separate tracts. In Amblystoma the arrangement is radically different. Segregation of functional systems is incipient but so little advanced that mammalian names are inapplicable. Under this heading there are included here all ascending fibers from the spinal cord and medulla oblongata, which terminate in the sensory and intermediate zones at higher levels. These are arranged in four groups which take separate courses: (1) the spinal lemniscus complex (*lm.sp.*), (2) the general bulbar lemniscus (*lm.*), (3) tr. bulbo-tectalis lateralis (*tr.t.b.l.*), and (4) the ascending secondary visceral-gustatory tract (*tr.v.a.*). The spino-cerebellar tract is closely associated with the first group. There is some evidence of a separate trigeminal lemniscus, as described below.

1. THE SPINAL LEMNISCUS (FIGS. 3, 9, 10, 11, $lm.sp.$)

The spinal lemniscus is an extensive system of fibers ascending from the spinal cord and here lying immediately ventrally of the spinal V root. In the cord these fibers arise as axons of cells of the dorsal gray column, which decussate in the ventral commissure. In the region of the calamus this tract receives extensive additions from the nucleus funiculi, the spinal V nucleus, and the commissural nucleus of the opposite side (figs. 3, 87). Some of these crossed fibers from the calamus region have thinner collaterals, which descend for a short distance in the ventral funiculus of the spinal cord. The ascending fibers, at first, lie in the ventral funiculus (figs. 41, 42) but soon turn dorsally to join those from lower levels of the cord ('44b, figs. 9–11). The most lateral and ventral fibers of this mixed bundle terminate in the reticular formation of the medulla oblongata —tr. spino-bulbaris (figs. 3, 88, 89, 90). The others at the level of the V nerve roots separate into spino-cerebellar and spino-tectal tracts. Many of the spino-cerebellar fibers are collaterals of the spino-tectal fibers (fig. 10; '14a, p. 376). Some fibers of the spino-tectal tract continue past the midbrain to end in the dorsal thalamus (fig. 34). The entire course of this complex is shown in a diagram ('39b, fig. 26) drawn from three adjoining sagittal sections of a larva, and this is confirmed by elective Golgi impregnations of the adult. Terminals of this tract in the dorsal thalamus are seen in figure 44.

2. THE GENERAL BULBAR LEMNISCUS ($lm.$)

This large fascicle, the general bulbar lemniscus, receives fibers from all parts of the sensory zone of the medulla oblongata (figs. 9, 11). These decussate in the ventral commissure and ascend medially of the reticular formation (figs. 89, 90). In the isthmus they turn dorsally and traverse the midbrain ventrally of the tectal formation, some fibers continuing forward into the dorsal thalamus (figs. 27–34, 91–94). These fibers may apparently carry nervous impulses activated by all kinds of sensory fibers which enter the medulla oblongata. Their terminals are spread quite uniformly throughout the entire extent of the tectum, and here they mingle with those of the spinal lemniscus (fig. 11; '39b, fig. 96). There is little evidence of physiological specificity in either of these tracts, save that one comes from the spinal cord and the other from the medulla oblongata and that the latter may be activated by a wider variety of sensory excitations, including the vestibular, lateral-line, and visceral-gustatory systems. This lemniscus is one of the largest tracts of the urodele brain and is very large in fishes also. In anurans and higher forms it can be recognized with difficulty because its fibers are dispersed or segregated in other specialized tracts like the trigeminal lemniscus. Amblystoma has nothing comparable with the medial lemniscus of mammals.

There is some evidence of an incipient trigeminal lemniscus, though most of the secondary general cutaneous fibers ascend in the general bulbar lemniscus ('39b, p. 606). From the region of the superior sensory V nucleus under the auricle a large number of arcuate fibers descend to the ventral commissure. Some of these enter the general bulbar lemniscus. Others are added from the entire length of the spinal V nucleus, and in the calamus region similar fibers join the spinal lemniscus. These connections were observed by Woodburne ('36, p. 455). From the region of the superior V nucleus, fibers ascend uncrossed as far as the isthmic tegmentum, where they end. This may be the precursor of an uncrossed trigeminal lemniscus. In the preparation from which figure 39 was drawn, no fibers of the lemniscus systems are impregnated. The visible external and internal arcuate fibers make local bulbar connections, descend to the spinal cord, and ascend only as far as the isthmus. Some of these fibers divide into ascending and descending branches before or after decussation. Anteriorly on the right, a compact fascicle of external arcuates descends from the superior trigeminal neuropil, turns forward, and decussates close to the ventral surface within the interpeduncular neuropil. Most of them descend in the bulbo-spinal tract, but some turn forward as tr. bulbo-isthmialis. This may be a precursor of a separate crossed trigeminal lemniscus. Their course is parallel with tr. bulbo-tectalis lateralis (primordial lateral lemniscus), but deeper.

3. TRACTUS BULBO-TECTALIS LATERALIS (FIGS. 11, 91, 92, *tr.b.t.l.*)

The tr. bulbo-tectalis lateralis is a mixed system of fibers, closely associated with the general bulbar lemniscus and of similar origin from the sensory zone, chiefly from its middle part in the vicinity of the vestibular nerve roots. In the isthmus it lies externally of the general bulbar lemniscus, and it terminates exclusively in the primordial inferior colliculus (nucleus posterior tecti) and the underlying isthmic neuropil. These fibers are mingled with those of a system which descends from the inferior colliculus—tr. tecto-bulbaris posterior (figs. 12, 13, 29–34, 89, 90, *tr.t.b.p.*). Their segregation from those of the general bulbar lemniscus in the urodeles seems to be determined primarily by their terminal distribution; they end in the inferior colliculus, while the larger bulbar lemniscus ends chiefly in the superior colliculus.

This tract is probably the precursor of the lateral lemniscus, which first makes its appearance in definitive form in anurans. A "lateral lemniscus" has been described in the brains of fishes by many authors, but this is a misnomer. That large tract of fishes is comparable with the general bulbar lemniscus of urodeles and has none of the distinguishing features of the mammalian lateral lemniscus, though it doubtless includes the primordium of that tract. Its current name, "f. longitudinalis lateralis," is more appropriate.

Both these bulbar lemniscus tracts are well developed in Necturus and the reader is referred to their description in that animal for discussion of the phylogenetic relationships ('30, pp. 51 ff.; Ariëns Kappers, Huber, Crosby, '36). Necturus lacks a cochlear rudiment, and hence the lateral bulbar lemniscus here probably has a very imperfect auditory function, if any. In Amblystoma and most other urodeles there is such a rudiment, and here audition of a primitive type may be represented, along with other functions, in this tract. In adult Anura there is a well-developed primordium of the cochlea, from which a cochlear division of the VIII nerve arises. Correlated with this, there are cochlear nuclei, from which a true lateral lemniscus passes to the inferior colliculus. Parallel with this differentiation, both general and lateral lemniscus tracts, as seen in urodeles, are radically reorganized. Larsell ('34, p. 521), after thorough examination of this question, accepts my interpretation of the lateral tract of urodeles as an incipient lateral lemniscus, citing my remark ('30, p. 58): "It may be that this incipience of an auditory lemniscus in Necturus is a central anticipation of a later peripheral specialization of the cochlear rudiment—that is, an apparatus for sorting out centrally the meager auditory component of the mixed functional complex of the undifferentiated VIII nerve. But it is more likely another evidence that Necturus is a degenerated or an arrested derivative of some more highly differentiated ancestor." A good brief summary of the evolution of the auditory apparatus has been given by Papez ('36). For further comments on this tract see pages 188, 214.

4. THE SECONDARY VISCERAL-GUSTATORY TRACT (*tr.v.a.*)

The fibers of the secondary visceral-gustatory tract are segregated in terms of physiological specificity more definitely than are those of the other lemniscus systems. They are axons of neurons of the nucleus of the f. solitarius and commissural nucleus, which ascend uncrossed ventrally of the spinal V root. Most of the visceral sensory fibers of the VII root are gustatory, and most of those of the vagus roots are general visceral. It is probable that the secondary fibers arising more anteriorly are activated mainly from taste buds, but these cannot be distinguished from those of general visceral sensibility. The central courses of these fibers have been fully described and illustrated in the larva ('14a), in the adult ('44b), and in Necturus ('30). A useful summary of the comparative anatomy of this system has been published by Barnard ('36). The adult arrangement is shown here in figures 7, 8, 9, 23, 30, 37, 38, 87–90.

One is impressed by the stability of the general plan of these arrangements throughout the vertebrate series, despite the most extreme modifications of the details in adaptation to different modes of life. This applies particularly to the peripheral and bulbar connections. The ascending connections are less well known, and they probably show more radical changes in the series from lower to higher vertebrates. Fishes and amphibians exhibit a common plan, with infinite variety of detail; the arrangement in Amblystoma appears to present this plan reduced to its simplest form, and this generalized plan may be taken as a point of departure from which the cerebral apparatus of visceral-gustatory adjustments of higher animals has been derived.

All the peripheral fibers enter the f. solitarius, which spans the entire length of the medulla oblongata. They are thin, and many of them are myelinated. Most of those from the geniculate ganglion, which form the visceral sensory VII root, have a T-form division, with ascending and descending branches as they enter the f. solitarius. Most (perhaps all) of the prefacial f. solitarius is composed of these ascending facialis fibers, which are less heavily myelinated than are their descending branches and, presumably, are gustatory in function. These fibers ascend to the auricle, where they end in the same neuropil as the ascending V root (figs. 7, 30, 38; '14a, p. 365). By this arrangement, correlation may be effected between visceral-gustatory and general somatic sensibility of the mouth cavity. The visceral sensory fibers of the IX and X roots are less myelinated than are those of the VII root. Most of them descend without division in the f. solitarius, but some divide with branches, which ascend for short distances (fig. 37; '44b, figs. 14, 17).

In Amblystoma, as in all other vertebrates, the two f. solitarii converge at the calamus scriptorius into the commissural nucleus, and part of their fibers decussate here in the dorsal commissura infima of Haller. Here at the bulbo-spinal junction there is an important field of correlation between the visceral-gustatory systems and the general somatic sensory systems of the entire body. The details of this structure in Amblystoma, the comparative anatomy of this region, and its strategic importance for fundamental physiological problems have recently been discussed ('44b) and are summarized here in chapter ix.

The nucleus of the f. solitarius contains some neurons with dendritic connections exclusively with this fasciculus, but most of these elements have other connections also. Some of their axons cross in the ventral commissure and are lost to view in the vicinity of the general bulbar lemniscus; others ascend uncrossed in the secondary visceral-gustatory tract to the isthmus, midbrain, and hypothalamus. Some of these secondary fibers bifurcate, with branches taking both these courses (fig. 9). The first of these pathways is probably more primitive, for Barnard ('36, p. 513) writes: "There is no differentiated, uncrossed secondary gustatory tract in the lamprey. All connections with higher centers are through the bulbar lemniscus system." It is believed that in Amblystoma the decussating fibers are concerned chiefly with bulbar reflexes; whether any of them ascend to higher levels in the bulbar lemniscus is not clear. In any event the uncrossed tract is evidently the chief pathway to higher centers. Rostrally of the V roots this tract divides, some of its fibers continuing forward to area ventrolateralis pedunculi and some turning dorsad to the superior visceral nucleus in the isthmus. We have many elective preparations of these fibers which have been cut in various planes, and the entire course is clear.

In the ganoid fishes (Johnston, '01; Barnard, '36, p. 517) the secondary visceral tract takes essentially the same course as in Amblystoma; but the superior visceral nucleus lies more ventrally and posteriorly, under the cerebellum. In those teleosts in which the gustatory system is enormously enlarged (carp and catfish), the secondary visceral-gustatory nucleus shows corresponding enlargement and is displaced forward and dorsalward in the isthmus to a position similar to that seen in Amblystoma. This makes it probable that in Amblystoma the fibers of the secondary tract that pass through the isthmus and go directly forward to the peduncle and hypothalamus are general visceral in function and that those that take the more dorsal course to the superior nucleus are chiefly gustatory in function, though this distinction may not be sharply drawn. In support of this conclusion, Barnard ('36, p. 595) mentions the fact that this superior nucleus disappears in birds, in which the gustatory apparatus is greatly reduced. If we may accept the suggestion of Fox ('41, p. 418) and others that the ventral tegmental nucleus of mammals is comparable with the superior visceral-gustatory nucleus of lower forms, it is evident that the mammalian nucleus has a more posteroventral position, similar to that of the generalized ganoid fishes.

The isthmic secondary visceral nucleus was first identified in larval Amblystoma ('14a, pp. 364, 373) and subsequently in Necturus ('17, p. 248; '30, p. 62). Its cells form a low ventricular eminence at the posterior dorsal tip of the isthmic tegmentum (fig. 2B). They are of medium size, scattered and clumped in an open neuropil (fig. 34), with outlying cells dispersed in the alba. As seen in transverse sections, they lie immediately ventrally of the posterior end of the isthmic sulcus ('25, fig. 19; '42, fig. 43). Their long dendrites are directed laterally and ventrally into the posterior isthmic neuropil, where they engage terminals of the secondary visceral tract, ascending root fibers of the trigeminus, tr. bulbo-tectalis lateralis, tr. bulbo-isthmialis (fig. 38), collaterals of tr. tecto-bulbaris rectus (fig. 37; '42, fig. 67), and terminals of the dorsal tegmental fascicles, notably those of group (9), as described in chapter xx, from the cerebral hemispheres. There is also a diffuse connection with the cerebellar formation, and this is much more intimate in some fishes ('05).

An extension of the secondary visceral-gustatory tract to the tectum has been described by Brickner ('30) in some teleostean fishes. In these fishes there is a large commissure connecting the two visceral-gustatory nuclei, which includes many decussating fibers of the secondary visceral-gustatory tract. A separate fascicle of the latter fibers turns forward, as tr. gustato-tectalis, to reach the anterior part of the tectum opticum. This connection has not been observed in Amphibia, but it may exist. It makes provision for correlation of visceral-gustatory with visual experience, and it is accompanied by a return path from the tectum to the visceral-gustatory nucleus by way of tr. tecto-bulbaris rectus—a typical reflex circle (p. 76).

At the outer border of the gray of this nucleus, in a few preparations there are large spindle-shaped neurons, with dendrites extending for long distances tangentially to the gray layer. One dendrite may ramify among terminals of the secondary visceral tract and dorsal tegmental fascicles of groups (7), (8), and (9) and the other in the deep neuropil and the f. tegmentalis profundus. One such element is illustrated in figure 101, showing an axon directed dorsally toward the decussatio veli, possibly a commissural connection between the two visceral nuclei similar to that described in the frog and in fishes.

These afferent connections suggest that this isthmic nucleus is a correlation center, where gustatory, general visceral, and a considerable variety of other types of experience are brought into relation, with the gustatory component dominant and all in the interest of reflexes concerned with feeding. The afferent fibers of the secondary gustatory tract terminate not only in the neuropil of the visceral-gustatory nucleus but also in that of the adjacent isthmic and dorsal tegmentum ('25, fig. 19; '42, fig. 43), where reflexes of the jaw and hyoid musculature are believed to be organized (p. 190).

This hypothesis is supported also by the courses taken by the efferent fibers. These go out in several directions. First, there is a dispersed group of fibers which spread in the underlying tegmentum (figs. 8, 13). This is a direct connection with the neuromotor apparatus of the mouth and pharynx. The remainder enter the tertiary visceral tract (fig. 8, *tr.v.t.*), which has rather wide distribution. This tract of Amblystoma is relatively small, and its fibers are dispersed and so intimately mingled with others that analysis is difficult. The diagrams show what has been clearly seen. In figure 8 the dotted component of *tr.v.t.* is highly probable, though not confirmed by elective impregnations.

In the description of the brachium conjunctivum (p. 176) this fascicle is shown to descend close to the gray in the posterior lip of the isthmic sulcus to reach the decussation. In this part of their course the cerebellar fibers comprise a large component of the f. tegmentalis profundus. As they emerge from the cerebellar formation, they pass through the superior visceral nucleus, and here they are joined by a smaller number of similar thin unmyelinated axons of cells of this nucleus. These two systems of fibers are mingled, and analysis is impossible except in electively impregnated material. We have few such specimens of either Amblystoma or Necturus, but those we have show that in the ventral commissure immediately spinalward of the decussation of the f. retroflexus some of the visceral-gustatory fibers turn forward in company with those of the ventral secondary visceral tract, and both groups of fibers arborize in the ventrolateral neuropil of the peduncle. Some of these fibers probably pass through this neuropil without synapse to reach the hypothalamus, as in fishes, in company with pedunculo-mamillary fibers and mingled with those of the mamillo-peduncular, mamillo-tegmental, and mamillo-interpeduncular tracts. At their decussation the tertiary visceral fibers spread in the alba of the isthmic tegmentum before and after crossing, and some of them probably reach the interpeduncular neuropil, though this needs confirmation. All these systems and others (including, perhaps, mamillo-cerebellar fibers) are mingled, and none of the tracts are closely fasciculated.

In some fishes these tracts are separately fasciculated, and the amphibian arrangement, so far as revealed in our material, conforms with the teleostean pattern. The tertiary visceral-gustatory tract is very large in some of these fishes ('05, pp. 420, 436, figs. 20, 23, 37, 38), passing from the secondary nucleus directly to the hypothalamus. This direct connection has been described by Larsell ('23, p. 109) in the frog; and in both Amblystoma and Triturus he saw some indications of it. In the frog these direct fibers do not take the deep course, as described above for Amblystoma, but are superficial, accompanying mamillo-cerebellar fibers. I find evidence of this superficial connection in Amblystoma also, but in the absence of elective impregnations no accurate description can be given. These fibers in some of our preparations are seen as a well-fasciculated tract. In the Cajal sections, from which figures 25–36 were drawn, it is not so clearly shown as in some other specimens, but its location can be identified. These fibers assemble at the lateral surface rostrally of *tr.v.a.* (fig. 34), in company with those of *tr.t.b.p.*, as drawn in that figure. Ten sections farther ventrally, they are drawn, but not labeled, in figure 33 superficially of *tr.t.b.r.* In some Golgi preparations, fibers in this position connect with the hypothalamus, and this is the course taken by the tertiary visceral tract in the frog (Larsell, '23, figs. 2, 4). This is the position also of the tr. mamillo-cerebellaris, which I provisionally identified in Necturus ('14, p. 8) and subsequently ('17, p. 250) questioned and the presence of which has been confirmed by Larsell.

In the light of such fragmentary evidence as we now have, it seems to me probable that in Amblystoma the ascending tertiary visceral-gustatory fibers take a deep course, accompanying the brachium conjunctivum, and that the superficial fascicle described above is a mamillo-cerebellar tract, which also connects with the isthmic tegmentum and the isthmic visceral-gustatory nucleus.

The bulbar apparatus of visceral-gustatory adjustment is simply organized in Amblystoma, as, indeed, it is in man, with minimal provision for local specialization. The division of the ascending path into dorsal and ventral moieties in the isthmus is similar to the divergence of the olfactory paths to epithalamus and hypothalamus. In both cases somatic correlations are effected dorsally and olfacto-visceral correlations ventrally. The dorsal gustatory nucleus may directly activate the skeletal musculature concerned with feeding, through its connections with the underlying tegmentum, or it may discharge into the hypothalamus, where all visceral adjustments are organized.

The swallowing of a morsel of food in response to excitation of taste buds is one of the simplest acts of which the body is capable. This may be done reflexly through the local activation of the bulbar connections of the f. solitarius; but, simultaneously with this local action, nervous impulses may be transmitted to higher centers which are so interconnected as to bring this simple act into relation with any other activities that may be in process in response to internal or external stimuli. The final result of any particular sensory excitation is dependent upon the central excitatory state of the central nervous system as a whole and of every local part of it. And the quality of this excitatory state is determined not only by the stimuli presently acting upon it but also by the past experience of the race and of the individual. This theme is amplified in a recent discussion of the apparatus of optic and visceral correlation ('44a).

CHAPTER XII

CEREBELLUM

IN URODELES the cerebellum is incompletely separable from the brain stem, and the entire rhombencephalon evidently acts as a closely integrated unit. Larsell's papers ('20, '31, '32, '45) on the cerebellum of Amblystoma and Triturus give good accounts of its development and structure.

Three of the primordia from which the mammalian cerebellar complex has been assembled are here clearly separate in relations easily recognized. The fourth mammalian component—the pontile system—is not represented in Amblystoma. The three components present are: (1) the vestibulo-lateralis system in the auricles, primordia of the floccular part of the flocculonodular lobes; (2) the median body of the cerebellum, which is ancestral to the larger part of the vermis and adjoining regions; and (3) the nucleus cerebelli, internal to the other two and in intimate relations with both of them. The topographic arrangement of these parts as seen in transverse sections is shown in figure 91. In higher brains these three components are variously merged, and the nucleus cerebelli is subdivided and incorporated within the cerebellar mass as the deep nuclei. The connections of the vestibular and lateral-line systems with the auricle have been described above. The afferent connections of the body of the cerebellum are shown in figure 10. These include primary and secondary sensory trigeminal fibers, spino-cerebellar fibers, tecto-cerebellar fibers, and probably a hypothalamo-cerebellar tract. Evidence from both comparative anatomy and embryology indicates that cerebellar differentiation began within the sensory zone, and in the adults of all vertebrates it retains some features characteristic of this zone, including termination within it of vestibular root fibers and (in many species) of trigeminal root fibers also. It is morphologically supra-segmental and physiologically supra-sensory, an adjustor of higher order, not primarily concerned with determining the pattern of performance but rather with facilitation and regulation of its execution. The motor zone plays no part in it genetically or functionally except as a sort of accessory after the fact. In later phylogenetic stages, with the appearance of the pons and cortico-pontile connections, this situation is changed; but even in mammals these connections, though topographically in the motor zone, are foreign to its intrinsic organization, for the pons and all its connections are dependencies of a supra-segmental apparatus and the cells of the pontile nuclei are embryologically derived from the sensory zone. In the most primitive vertebrates in which cerebellar differentiation is minimal (almost absent in myxinoids, Jansen, '30; Larsell, '47a) the sensory zone of the medulla oblongata abuts against that of the midbrain, which in these animals is the dominant adjustor of all somatic (nonvisceral) activities. Necturus presents a very early stage in the fabrication of the cerebellar complex ('14), scarcely more advanced than in cyclostomes. In Amblystoma the process is further advanced. It is evident that cerebellar differentiation began under the influence of several kinds of sensory excitations, viz., the vestibular, lateral-line, general cutaneous, and deep sensibility of the muscles, joints, etc. ('24; Larsell, '47, '47a). Connection with the visual system of the tectum was effected in early phylogenetic stages. With the elaboration of the cerebral cortex in mammals, another major component was added—the cortico-pontile system. This history has been thoroughly explored and documented by Larsell ('20–'47a). Brodal ('40) and Brodal and Jansen ('46) have greatly extended our knowledge of the connections between the inferior olive and the cerebellum in mammals, but we have little exact information about them in lower forms.

The isthmus is a critical junctional field. Immediately spinalward of it the sensory and intermediate zones of the medulla oblongata are enlarged to form the massive auricle, within which are terminals of the ascending vestibular and associated lateral-line roots (fig. 7). This is probably the first primary component of cerebellar architectonic. More medially the ascending sensory root of the trigeminus reaches its terminal (superior) nucleus. Some of these V root fibers, accompanied by secondary V fibers from the superior nucleus, continue into the median body of the cerebellum (figs. 10, 31–33, 91), where they are joined by fibers of the spino-cerebellar tract. Some fibers of the mixed bundle decussate in the commissura cerebelli (*com.cb.*). This is the second primary component of the cerebellum from which the median mass (corpus cerebelli) of larger cerebella has been derived and into which converge sensory influences of all sorts except those from the internal ear and the lateral-line organs. This median mass receives fibers from the mesencephalic tectum and (in lower-vertebrates) also from the hypothalamus. The tecto-cerebellar tract may be activated from the mesencephalic V nucleus and from terminals of the optic and lemniscus systems.

The very rudimentary cerebellum of Myxine has been the subject of much controversy. Holmgren ('46, p. 54) described in the embryo two possible primordia, one in the auricle of the rhombencephalon and one at the posterior end of the mesencephalic tectum. Larsell's comprehensive study ('47, '47a) of this region of cyclostomes has clarified the problem. Myxine has the most primitive cerebellum known, and, so far as its differentiation has gone, it conforms with the typical vertebrate pattern.

The spino-tectal terminals in the inferior colliculus are primitively contiguous with the vestibular and lateralis terminals in the auricle, and from the wedlock of these two systems the cerebellum was born. The primordial inferior colliculus, accordingly, was primitively concerned with proprioception. Upon this foundation the cochlear system of higher animals was built in much the same way that the bulbar cochlear nuclei emerged from the lateral-line nuclei of the medulla oblongata (p. 138). It is significant that in Amblystoma the primordial inferior colliculus contains dense collections of cells of the mesencephalic V nucleus, that fibers of the mesencephalic root of the V nerve penetrate cerebellar tissue, and that cells of the mesencephalic V nucleus are found sparsely scattered in the cerebellar gray (p. 140). It seems probable that the mesencephalic V nucleus was first differentiated at the posterior end of the tectum and that from this focus its cells spread forward into the optic tectum and, in smaller number, backward into the cerebellum.

In the gray of both the median body of the cerebellum and the lateral auricle some of the neurons are specialized as Purkinje cells, in contrast with smaller cells, which are precursors of the cerebellar granules. The largest and best developed Purkinje cells are loosely arranged at the outer border of the central gray, but they are not delaminated from the granular layers as in mammals. If cortex is defined as laminated superficial gray, Amblystoma has no cerebellar cortex, though its primordium is clearly evident. The Purkinje cells, though of simple form, are quite distinctive. From them and from the nucleus cerebelli, efferent fibers stream downward, forward, and backward as cerebello-tegmental fibers, and one large fascicle of these extends farther forward parallel with the sulcus isthmi as brachium conjunctivum.

In my earlier papers a ventricular swelling under the cerebellum was termed "eminentia subcerebellaris tegmenti." This has subsequently been analyzed into several distinctive areas, of which the posterodorsal member was named "eminentia cerebellaris ventralis" ('35a, fig. 1). It is now clear that this is the primordium of the deep nuclei of the cerebellum, and it is here named "nucleus cerebelli" (figs. 2, 10, 32, 33, 91, *nuc.cb.*). This is an ill-defined region of the intermediate zone, not clearly separable from its surroundings. Some fibers of the brachium conjunctivum arise from these cells, though only a small part of them.

The spino-cerebellar tract ascends in company with the spinal lemniscus, some cerebellar fibers being collaterals of lemniscus fibers, as described above. Cerebellar fibers join this tract from the nucleus of the dorsal funiculus and the sensory zone of the medulla oblongata. One confused issue regarding this tract of Amblystoma can now be clarified. A dorsal slip of the spino-cerebellar tract passing through or above the entering fibers of the sensory V root was described by me ('14, p. 7) and by Larsell ('20, p. 277). These observations could not be confirmed by either of us (Herrick, '14a, p. 375; Larsell, '32, p. 414), the supposed tract being then interpreted as the ascending trigeminal root. It now appears that both the earlier and the later accounts are correct. In one of our adult brains, cut in a favorable oblique plane (no. 2245), there is partial Golgi impregnation of the sensory V root fibers and of the spino-cerebellar tract, with no other fibers stained in this vicinity. A separate fascicle of the ventral spino-cerebellar tract turns dorsally at the level of the V nerve, its fibers interdigitate with those of the entering sensory V root, and rostrally of this level the spino-cerebellar fibers and the ascending sensory V fibers are mingled as far as their entrance into the cerebellar commissure. Here a fortunate elective impregnation clarifies the confusion as neatly as could be done by an experimental degeneration (compare Salamandra as described by Kreht, '30, p. 294).

In the literature our tractus spino-cerebellaris is sometimes designated "tractus cerebello-spinalis," implying conduction spinalward. Some descending fibers may be present in this tract of Amblystoma, though our preparations have not revealed them. That the larger part of the tract is ascending is clearly evident.

The tr. tecto-cerebellaris was originally described ('25, p. 483 and figs. 39–41) from two series of sagittal Golgi sections of the adult, and this description has been confirmed by Larsell ('32, p. 425). Additional details can now be added. In several series of Golgi sections of advanced larvae and adults these thin unmyelinated fibers arise from neurons of the nucleus posterior tecti with the dendrites directed forward into the posterior border of the optic tectum. Their axons form a rather compact fascicle close to the dorsal surface of the nucleus posterior, which passes backward though the anterior medullary velum and laterally of it, accompanying thick myelinated fibers of the mesencephalic root of the V nerve which take a similar course. At the anterior border of the cerebellum they turn ventrad and spread out along the anterior side of the body of the cerebellum, some reaching the underlying nucleus cerebelli.

A tr. mamillo-cerebellaris was described by Larsell ('20, p. 279; '32, p. 424), and some evidence of it has been seen in our preparations; but in the absence of elective impregnations no satisfactory description can be given. In view of the large size of this connection in fishes, its presence in urodeles may be expected.

BRACHIUM CONJUNCTIVUM

Large numbers of unmyelinated and lightly myelinated fibers descend from the body of the cerebellum, the auricle, and the nucleus cerebelli. These cerebello-tegmental fibers are in addition to the connections of the auricle with the vestibulo-lateralis sensory field in correlation tracts *a* and *b;* they are directed in dispersed arrangement ventrally, posteriorly, and anteriorly. A large and rather compact fascicle of these fibers, a few of them myelinated, is directed anteroventrally and is clearly a primordial brachium conjunctivum (figs. 10, 21, 71). These axons arise from cells of the body of the cerebellum (fig. 47) and the nucleus cerebelli (14*a*, fig. 52); they assemble anteriorly of the body of the cerebellum and pass forward through the visceral-gustatory nucleus in the isthmus (figs. 33, 34). Here they are joined by fibers of the tertiary visceral tract, which arises in this nucleus (p. 169).

This mixed tract with accessions of fibers from other sources has been termed "fasciculus tegmentalis profundus" ('36, p. 304, figs. 14, 23, *f.teg.p.*); its analysis is possible only with the aid of elective impregnations, and these, fortunately, are available (see the further description in chap. xx). This fasciculus extends forward and ventralward from the cerebellum at the outer border of the gray in the lips and floor of the isthmic sulcus, finally to reach the ventral commissure, where some of its fibers decussate a short distance spinalward of the fovea isthmi, mingled with those of the decussation of the f. retroflexus and spinalward of it. Its course as seen in horizontal sections is shown in figures 29–32 (*f.teg.p.*), and in sagittal sections in figure 104 and previously published figures of the same specimen ('36, figs. 19–21, *tr.cb.teg.*).

Elective impregnations show that the brachium conjunctivum is the largest component of this mixed fasciculus (figs. 71, 72). Many of its fibers spread outward into the alba of the isthmic tegmentum on the side of origin, and these include all the myelinated fibers of the brachium. A residue of the unmyelinated fibers decussates in the ventral commissure. Elective Golgi impregnations show that, at the decussation, fibers of the brachium conjunctivum and tertiary visceral tract are mingled and that both enter a dense superficial neuropil laterally of the crossing. Here the tertiary visceral fibers turn forward in company with those of the secondary visceral tract to reach the area ventrolateralis pedunculi, and the cerebellar fibers turn posterodorsally and spread out in the alba of the isthmic tegmentum. This primordial brachium conjunctivum may activate almost the entire extent of the isthmic tegmentum diffusely on both sides. No evidence has been found of a concentration of cells related with it which could be regarded as a nucleus ruber.

THE CEREBELLAR COMMISSURES

In the body of the cerebellum there are two commissures, which differ in position, connections, and functions. They are specifically related with the two chief subdivisions of the cerebellar complex.

The com. cerebelli (*com.cb.*) is a compact fascicle of well-myelinated fibers crossing more ventrally than the less myelinated com. vestibulo-lateralis cerebelli (*com.cb.l.*), as shown in figures 10 and 91. It crosses between the levels of figures 34 and 35. The com. cerebelli is composed of fibers from two sources, which terminate chiefly in the median body of the cerebellum. These are (1) fibers of the tr. spinocerebellaris (figs. 32, 33, 91) and (2) fibers of the sensory trigeminus system. The latter are, in part, fibers of the ascending sensory V root terminating in the cerebellum and, in part, axons of cells of the superior sensory V nucleus terminating in the cerebellum and the neuropil of the V nucleus of the opposite side (figs. 31, 32, 91). The latter is a true trigeminal commissure.

The com. vestibulo-lateralis cerebelli is a more dispersed collection of unmyelinated and lightly myelinated fibers assembled from the auricles (figs. 31–36, 91). Included among them are root fibers of the vestibular nerve (*com.cb.VIII*) and secondary fibers of the lateral-line system of nerves (*com.cb.l.l.*). These are mingled at their crossing in the cerebellar alba. Whether any primary root fibers of the lateral-line nerves decussate in this commissure is not clear in our material; apparently they do not.

PROPRIOCEPTIVE FUNCTIONS OF THE CEREBELLUM

In the initial differentiation of the cerebellum and in its normal functions in all animals, proprioception has played a dominant part. Nevertheless, it must be recognized that all sensory systems may participate in cerebellar control of muscular movements (Larsell, '45), a control that is applied to the motor adjustors as going concerns at every phase of their operations. The significance of the cerebellum as "the head ganglion of the proprioceptive system" (Sherrington) is discussed in chapter x in connection with a critique of the proprioceptive system as a whole.

CHAPTER XIII

ISTHMUS

THE strategic position of the isthmus as a transitional sector of the brain stem between rhombic brain and cerebrum has been commented upon (pp. 45, 118). In early neural-tube stages of the developing human brain it is a rather large sector, but in the adult it is hardly recognizable as a distinct entity because its tissues are dispersed among other structures of relatively recent phylogenetic origin. In contrast, the isthmus of adult Amblystoma retains its embryonic separateness and, indeed, accentuates it. Here its distinctive features are most clearly seen, and these will be described as fully as available material permits. Its dorsal part is contracted and includes only the anterior medullary velum and some adjoining structures. To the intermediate zone is assigned the superior visceral-gustatory nucleus, and for convenience of description the remainder is included in the motor zone, though its functions are, in large part, of intermediate type. The interpeduncular nucleus is an isthmic structure, to which a separate chapter is devoted.

DEVELOPMENT

The configuration of this sector of the brain exhibits an interesting series of changes in the course of development, due to the rapid and radical dislocations caused by the cerebral flexures and to other inequalities of intrinsic growth. Successive modifications of the isthmic sulcus provide a useful indicator of the course of these changes.

From the coil stage (Harrison's stage 35) to the adult, the isthmic tegmentum is intimately joined with the trigeminal tegmentum, with no constant external or ventricular boundary visible. "Isthmus rhombencephali," therefore, is an appropriate name. But its anterior boundary is sharply marked by the fovea isthmi, the deep ventricular sulcus isthmi, and the external fissura isthmi. Even in early stages these grooves are not exactly parallel, and as development advances they deviate from this relationship more and more widely because of unequal growth of the deep and superficial structures.

Reference was made on page 177 to von Kupffer's sulcus intraencephalicus posterior as the precursor of the sulcus isthmi. In coil ('37, fig. 1), S-reaction ('37, figs. 7, 8), and early swimming stages ('37, fig. 2) the external fissura isthmi lies a little farther spinalward than the internal sulcus. In slightly older early swimmers ('38, figs. 15–18) both the isthmic fissure and the isthmic sulcus are deep, and the sulcus is displaced farther rostrally of the fissure. This means that the gray of the isthmus is proliferating rapidly and pushing the isthmic sulcus farther forward. The sulcus is now a very deep cleft and lies approximately transversely to the long axis of the rhombencephalon. From this stage the enlargement of the isthmic tegmentum continues on the ventricular side, and in the early feeding stage (Harrison's stage 46) the isthmic sulcus extends from the cerebellum almost horizontally forward before dipping downward to the fovea isthmi ('38b, figs. 1, 2). The sulcus and the fissure are still deep, and the sulcus has come to lie far rostrad of the fissure. This is not evident in transverse sections but is shown clearly in horizontal sections of this and subsequent stages ('39b, fig. 16). In later larval stages, internal differentiation, particularly the elaboration of the neuropil and thickening of the auricle, result in smoother contours both internally and externally, and the isthmic sulcus and fissure are shallower in the adult.

At the close of the early swimming stage (about Harrison's stage 38) the cerebellum and auricle are so little developed that the plica rhombo-mesencephalica is marked externally by a wide depression ('38, fig. 17). Between this and the early feeding stage (Harrison's stage 46) the sharp cerebral flexures are straightened, resulting in a backward thrust of the plica rhombo-mesencephalica in the dorsal wall, while the ventral wall of the isthmus remains fixed (compare '38, fig. 18, with '38b, fig. 1). Though there has not yet been any very great enlargement of the isthmic tegmentum, the isthmic sulcus in early feeders lies more nearly horizontally than vertically. In mid-larval stages the deep sulcus isthmi (shown but not named in fig. 2 of '14a) is inclined more than 45° from the vertical, and between mid-larval and adult stages there is enormous enlargement of the gray of the isthmic tegmentum, which thrusts the sulcus isthmi forward and upward toward the horizontal position.

The external fissura isthmi does not follow the sulcus isthmi in these shifts of position, but, as mentioned above, it lies in the adult a considerable distance spinalward of the sulcus. That this involves a mechanical stretching of the wall between the fissure and the sulcus in later stages is evident from the arrangement of the ependymal elements in the lips of the isthmic sulcus. This traction begins early. It was noticed in early swimmers ('38, p. 220) that the ependymal elements of the sulcus isthmi are thickened and sharply bent outward and backward. This was illustrated also in early feeding stages ('38b, figs. 11, 12) and in Golgi impregnations of midlarval stages ('39b, p. 518, and figs. 31, 93). At all stages, including the adult, these thick elements form a dense band of parallel fibers bordering the sulcus on both sides and extending outward and backward to an attachment at the pial surface in the floor of the isthmic fissure. In the adult the sulcus isthmi usually does not extend ventrally so far as the fovea isthmi, but its locus here is indicated by a band of modified ependyma. Two mechanical factors, accordingly, can be recognized in the shifting relations of the isthmic fissure and sulcus: first, the backward thrust of the roof during the straightening of the cerebral flexures between early swimming and early feeding stages and, second, unequal growth of superficial and deep structures, especially in later stages. The rapid enlargement of the gray of the isthmic tegmentum in late larval stages produces an upward and forward thrust ventrally in the direction opposite to the dorsal backward thrust of early stages.

The developmental history and adult position of the sulcus isthmi of Amblystoma are radically different from those of Hynobius as described by Sumi ('26), who illustrates models showing the ventricular sculpturing from closure of the neural tube to the adult. At none of these stages is von Kupffer's sulcus intraencephalicus posterior visible at the plica rhombo-mesencephalica. The sulcus isthmi appears rather late at the fovea isthmi, and it does not at any stage reach the dorsal surface.

SENSORY ZONE

The sensory field of the isthmus is contracted. Its scanty gray substance includes a compact cluster of cells of the mesencephalic V nucleus within and adjacent to the anterior medullary velum. These cells are smaller and more densely crowded than are those of this nucleus in the tectum, and they may have different functional significance (p. 140). Their connections are unknown. Associated with them are other cells at the posterior border of the inferior colliculus which appear to have peripheral connections. From this region a few myelinated, and many unmyelinated, fibers go out to the meninges and chorioid plexus of the fourth ventricle and some with the IV nerve.

These fibers appear to be related with cells bordering the velum, but no clear evidence of their connections has been found ('36, p. 343; '42, pp. 255, 291). Within and adjoining the anterior medullary velum are fibers of passage of several kinds, including the decussation of the IV nerves, mesencephalic root of the V nerve, and tractus tecto-cerebellaris.

INTERMEDIATE ZONE

The intermediate zone as defined here is represented by the small gray area of the superior secondary visceral-gustatory nucleus and associated neuropil. This gray lies dorsally of the isthmic tegmentum and incompletely separable from it. In figures 2 and 13 it is shown wedged between the isthmic tegmentum below and the nucleus cerebelli and dorsal tegmentum above. The cells of the superior visceral nucleus are of medium size, scattered and clumped in an open neuropil, with a few outlying cells in the alba. Their dendrites are directed ventrolaterally through the alba into the posterior isthmic neuropil (described below), where they engage terminals of the ascending visceral tract ('42, pp. 244, 253, and fig. 43) and of fibers from many other sources.

There is an obscure vestige here of the nucleus isthmi, which is well differentiated in the frog ('42, pp. 244, 253). In anurans and reptiles this nucleus is in intimate relations with the lateral lemniscus, and it is probably part of the auditory reflex apparatus, with, perhaps, vestibular connections also. In Amblystoma the auditory system is poorly developed, and the crawling habit maintains the balance of the body without nervous control; the nucleus isthmi, accordingly, is undeveloped.

MOTOR ZONE

This field is so intimately related anatomically and physiologically with the tegmentum of the trigemino-facialis region that these will be described together. The tegmentum isthmi is bounded anteriorly by the isthmic sulcus and fovea isthmi, separating it from the cerebral peduncle and dorsal tegmentum. Posteriorly it is bounded by the nucleus cerebelli and the trigeminal tegmentum. Its gray substance includes the nucleus of the IV cranial nerve (as described in chap. x); a compact central nucleus of the isthmus (fig. 29, *nuc.is.c.*); and, surrounding this, a pars magnocellularis which is continuous spinalward with the large-celled component of the trigeminal tegmentum (fig. 30).

TRIGEMINAL TEGMENTUM

The tegmentum of the trigeminal field forms a low eminence in the floor of the fourth ventricle internally of the V roots, the eminentia trigemini (figs. 2C, 90), which is evident in Necturus also ('30, figs. 11, 12, *em.V*). The gray under this eminence comprises a small-celled dorsal and medial part and a large-celled ventral part (fig. 27). The small-celled part is continuous medially with a band of similar cells, which extends spinalward under the ependyma of the paramedian sulcus (fig. 28), and anterodorsally it is continuous with the ventral border of the central nucleus of the isthmic tegmentum and more intimately with the deep periventricular gray of the isthmus (fig. 29). The large-celled part is continuous anterodorsally with the large-celled component of the isthmic tegmentum (figs. 28, 29, 30). Imbedded among these large tegmental cells are the motor V nuclei. There are two motor V roots (fig. 27), but the two motor V nuclei seen in Necturus ('30, p. 13) are here merged. Figure 40 shows three of these cells in Golgi impregnation. The heavily myelinated fibers of the mesencephalic V root are more dorsal, passing upward toward the tectum at the outer border of the gray (figs. 13, 29–32).

The configuration at the boundary between trigeminal and isthmic tegmentum is variable, and the position of this boundary is debatable. A posterodorsal extension of the isthmic tegmentum is continuous with a low ventricular eminence, the superior visceral nucleus. The latter is separated by a shallow and variable sulcus from the nucleus cerebelli posteriorly and a tegmental area ventrally. Comparison of the two key drawings, figures 2B and 2C, reveals differences due partly to individual variations in relative sizes of parts and partly to lack of any well-defined boundary between the large-celled components of the isthmic and trigeminal tegmentum. In figure 2B the dorsal part of the area marked *teg.is.m.* contains only large cells (fig. 31), and these may be assigned to either the isthmic or the trigeminal tegmentum. More ventrally (figs. 29, 30), both large and small cells occupy a similar ambiguous position.

The large cells of the trigeminal tegmentum take various forms, one of which is shown in figure 62 (others are illustrated in '14a, figs. 21–26). Like those of the nucleus motorius tegmenti elsewhere, they are concerned with motor co-ordination in patterns as yet imperfectly known. Some of these cells in the auricle have dendrites directed forward into the posterior isthmic neuropil, where they engage terminals of the tegmental fascicles (fig. 46). Other similar neurons direct their dendrites laterally among terminals of the ascending sensory V root and related tracts ('39b, figs. 47–49, 53, 62, 63). The axons of some of these elements are directed forward, as shown in figure 43. These short fibers pass from the trigeminal tegmentum to arborize within the gray of the isthmic tegmentum and are doubtless part of the apparatus of intrinsic co-ordination of patterns of action within the motor zone.

Some of the small cells at the anterior end of the trigeminal tegmentum are shown in figures 66 and 91. These resemble those of the adjacent interpeduncular nucleus. Their dendrites extend laterally through the entire thickness of the alba of the ventral part of the auricle, where they are in contact with all kinds of fibers passing this region. The axons of some of them descend to the interpeduncular neuropil (figs. 63, 80, 84; compare similar cells of the isthmic tegmentum, figs. 60, 84).

ISTHMIC TEGMENTUM

Horizontal sections are most instructive in the analysis of this region (figs. 29–33). The nervous elements are arranged in five groups: (1) A narrow and inconstant layer of subependymal cells. (2) These are separated by a thick sheet of neuropil from the lens-shaped area of small and medium cells, here termed the "central nucleus of the isthmus." (3) Externally and spinalward of this area is the pars magnocellularis, containing scattered nerve cells of medium or large size. The extensive neuropil within which these cells are imbedded is broadly continuous with that of the surrounding alba. This large-celled area envelops the lentiform area of small cells laterally, ventrally, and caudally and is continuous spinalward with a similar field of the trigeminal tegmentum. In figure 2B the dotted line separating the pars magnocellularis (*teg.is.m.*) is arbitrarily drawn to indicate a posterior sector which may be assigned to either isthmic or trigeminal tegmentum. (4) Dorsally of all the preceding, the gray of the isthmic sector extends upward into contact with that of the dorsal tegmentum and ventral cerebellar nucleus. This area is the superior secondary visceral-gustatory nucleus (fig. 2B, *nuc.vis.s.*). (5) A probable vestige of the nucleus isthmi.

The first component of the preceding list is the ventral part of the zone of deep subependymal gray, which everywhere surrounds the aqueduct. In the isthmus it is thin and contains few nerve cells, but more posteriorly in the trigeminal tegmentum it expands to a wide zone of subependymal small cells (figs. 29, 30). This and the associated neuropil comprise one of the most primitive and conservative features of this part of the brain and also of the diencephalon and the mesencephalon. In mammals it contributes fibers to the dorsal longitudinal fasciculus of Schütz, as described in the opossum by Thompson ('42).

The second component, the central nucleus, is clearly delimited in its middle part. Anteriorly, it merges with the small cells of the posterior gray of the peduncle (figs. 29, 30, 31) and more dorsally with the ventral border of the dorsal tegmentum (fig. 32). Posteriorly, it is continuous with the small-celled component of the trigeminal tegmentum (fig. 91), though the boundary between these is clearly marked in many preparations. At its ventral border the gray is continuous with that of the interpeduncular nucleus, with difference in form and arrangement of the cells, though many transitional forms are seen ('42, fig. 43). The boundaries here mentioned are in some preparations quite indeterminate; in some reduced silver preparations they are evident; and in Golgi preparations each of the regions mentioned has characteristic structure and connections. All these areas are intimately connected by a web of interstitial neuropil, the structure of which is characteristically different in each of them. All this gray is connected with the interpeduncular nucleus by fibers passing in both directions, and in chapter xiv, devoted to that nucleus, it is suggested that the central nucleus of the isthmus contains, among other constituents, the primordium of the dorsal tegmental nucleus of mammals.

The large-celled component 3 surrounds the central nucleus on all sides, and many of these larger neurons lie deeper, mingled with those of the small-celled component. These cells have massive dendrites, which spread widely in the overlying alba. Among these are cells of the nucleus of the nervus trochlearis (fig. 70). Figure 61 shows three neurons from this region which probably belong to the IV nucleus. They are surrounded by other similar elements of the tegmentum. Most of the axons from the large tegmental neurons descend, crossed or uncrossed, in the tegmento-bulbar tracts, as described in the larva ('39b, p. 590). Some axons from the isthmic tegmentum, as also from the trigeminal tegmentum, are directed forward. So far as observed, these take short courses, though some may extend farther (figs. 43, 46). This may be a precursor of the connection from the substantia nigra to the corpus striatum (Kodama, '29) which has recently been confirmed by several students of mammalian neurology. Component 4, the superior visceral nucleus, has been described above (p. 182).

WHITE SUBSTANCE

In the white substance of the isthmus the most conspicuous features are the numbered tegmental fascicles described in chapter xx. Externally of these fascicles are the massive tecto-bulbar and tecto-spinal tracts and, superficially close to the pial surface, a layer of fine fibers, most of which are unmyelinated. The latter includes tr. thalamo-tegmentalis rectus from dorsal and ventral thalamus (fig. 94; '36, p. 340) and, more dorsally, similar fibers from the dorsal thalamus, which decussate in the postoptic commissure—tr. thalamo-tegmentalis dorsalis cruciatus (p. 299; '39, pp. 95, 116).

Mingled with these fibers and more abundantly at deeper levels are fibers of tr. tecto-peduncularis and tecto-tegmentalis (figs. 18, 22, 24; '42, p. 267 and figs. 14, 30; Necturus, '17, figs. 9–14, 32, 33, *tr.t.ped.p.*). Short fibers from the entire tectum and dorsal tegmentum to the isthmic tegmentum are dispersed in both white and gray substance, and a large number of the longer fibers are assembled in dorsal tegmental fascicles of group (7) accompanying tr. tecto-bulbaris rectus (p. 283).

Figure 21 shows only the more prominent afferent connections; there are many others. In the aggregate these include fibers from the motor zone above and below the isthmus, collaterals of the lemniscus systems in the isthmic neuropil, terminals of the ascending visceral-gustatory tract, cerebello-tegmental fibers, short fibers from the overlying dorsal tegmentum and tectum, longer fibers from the tectum which decussate in the postoptic and ventral commissures, direct fibers from the dorsal and ventral thalamus, strong tracts from the dorsal and ventral thalamus which decussate in the postoptic commissure, fibers from both cerebral hemispheres by way of the lateral forebrain bundle and tr. olfacto-peduncularis, and strong tracts from the hypothalamus, crossed and uncrossed. To this list one might add the fasciculus retroflexus, passing from the habenula to the interpeduncular nucleus and, through the latter, acting upon the isthmic tegmentum.

Most of the efferent fibers descend in the f. tegmentalis profundus (p. 286) and the ventral tegmental fascicles. Some are dispersed in the neuropil accompanying tr. interpedunculo-bulbaris dorsalis, and these are regarded as precursors of the bulbar part of the f. longitudinalis dorsalis of Schütz (p. 208). The relatively short fibers which descend from the large cells of the isthmic and trigeminal tegmentum to the motor zone of the medulla oblongata are especially clearly seen in very young larvae from early swimming to early feeding stages and are interpreted as provision for activation of the mandibular and hyoid musculature involved in feeding. Longer fibers descend into the spinal cord; but these, so far as is shown in our material, are not numerous.

ISTHMIC NEUROPIL

Both the gray and the white substance of the isthmus are permeated with dense neuropil, the axonic component of which is composed mainly of terminals of afferent fibers and collaterals of fibers of passage. The superficial neuropil is composed chiefly of branched terminals of the unmyelinated thalamo-tegmental and tecto-tegmental tracts and dendritic terminals from the underlying gray. The intermediate neuropil is in the form of sheets or plaques insinuated between the tegmental fascicles and composed largely of terminals and collaterals of fascicular fibers. The deep neuropil of the alba and the grisea is a confused entanglement of fibers and dendrites, the analysis of which is quite impossible except where clarified by elective Golgi impregnations. These preparations reveal a number of more or less well-defined tracts imbedded within this interstitial matrix. In the gray substance the cells are arranged in lamellae separated by sheets of neuropil, which is continuous with that of all neighboring parts. A large proportion of these axons trend dorsoventrally from tectum and dorsal tegmentum to isthmic tegmentum and interpeduncular neuropil and posteroventrally, with or without decussation in the ventral commissure. The latter axons accompany thicker fibers of the tr. tegmento-bulbaris arising from large cells of the motor tegmentum.

POSTERIOR ISTHMIC NEUROPIL

The name "posterior isthmic neuropil" is given to a wide zone of neuropil in the alba at the boundary between the isthmus and the bulbar and cerebellar structures. It invades the auricle below and the nucleus posterior tecti above. It is permeated by dendrites from all surrounding parts, including the trigeminal tegmentum (fig. 46; '39b, figs. 47–49, 62, 63), auricle ('39b, fig. 53), cerebellum (fig. 47), superior visceral nucleus ('42, fig. 43), and dorsal and isthmic tegmentum. It receives terminals of peripheral sensory fibers of the V, VII, VIII, and lateral-line nerves, of a bulbo-isthmic tract (figs. 38, 39), of the secondary visceral-gustatory tract; terminals and collaterals of the lemniscus systems; vast numbers of terminals of descending fibers of the tegmental fascicles; and collaterals of tr. tecto-bulbaris rectus.

This evidently is a strategic region in the organization of both ascending and descending systems of conduction. Within this undifferentiated field the primordia of a considerable number of specific nuclei of other animals can be recognized, including specialized structures peculiar to fishes and a number of mammalian nuclei. Among these components (with their probable derivatives in more specialized species) the following can be identified:

1. The chief sensory nucleus of the trigeminus. Some ascending sensory V root fibers terminate here, though most of them more posteriorly, and some pass through to reach the commissura cerebelli and the corpus cerebelli of the same and of the opposite side. Secondary fibers go from this area upward into the cerebellum and forward with or without decussation as the primordial trigeminal lemniscus.

2. Similarly, some of the ascending root fibers of the VIII and lateral-line nerves terminate more laterally in the posterior isthmic neuropil (superior vestibular nucleus), and some of the VIII fibers reach the lateral (auricular) part of the cerebellum of the same and of the opposite side, the latter decussating in the com. vestibulo-lateralis cerebelli.

3. The longitudinal bulbar correlation tracts a and b reach the ventral part of the same neuropil. This field is incorporated within the cerebellum of mammals.

4. Spinal lemniscus fibers, as they recurve dorsalward along the anterior face of the auricle to reach the tectum, are spread through the middle part of this neuropil, with collaterals and branched terminals within it (figs. 42, 43, 44).

5. Associated with the spinal and bulbar lemniscus and laterally of them in the isthmus are fibers of the lateral bulbo-tectal tract (fig. 11, *tr.b.t.l.*), which is the probable primordium of the lateral lemniscus. These ascending fibers are mingled with descending fibers of tr. tecto-bulbaris posterior (fig. 12, *tr.t.b.p.*; '39b, p. 606 and fig. 96). The position of the mixed fascicle in horizontal sections is shown in figures 31–34, here marked *tr.t.b.p.* Terminals and collaterals of the ascending lateral bulbo-tectal fibers spread throughout the posterior isthmic neuropil (figs. 33, 34), and these endings are comparable with those of the lateral lemniscus in the nucleus isthmi of the frog (Larsell, '24). This field also contains precursors of some of the nuclei of the lateral lemniscus of mammals, specifically the dorsal nucleus of the lateral lemniscus (Ariëns Kappers, '29, p. 304; Clark, '33).

6. The nucleus isthmi, which is large in the frog (Larsell, '24), is probably represented in Amblystoma by a few outlying cells in the dorsal alba of the isthmus, which are more numerous in the larva ('42, pp. 244, 254). If present here, it is an insignificant vestigial structure.

7. The neuropil associated with the superior visceral-gustatory nucleus (fig. 33, *nuc.vis.n.*) is an important field of sensory correlation. It is penetrated by dendrites of its nucleus ('42, fig. 43) and by terminals of the ascending secondary visceral tract ('25, fig. 19) and also by fibers from several other sources, including the lemnisci and collaterals of tr. tecto-bulbaris rectus (fig. 37; '42, fig. 67). This complex is very large in fishes (the *Uebergangsganglion* of Mayser) and is present, though small, in the frog (Larsell, '24).

8. The neighboring large cells of the motor tegmentum are undoubted precursors of several mammalian nuclei of this region, including the substantia nigra and the locus coeruleus.

PHYSIOLOGICAL INTERPRETATION

The afferent fibers which terminate in the isthmus come, directly or indirectly, from practically all parts of the brain, and they carry nervous impulses activated from all the major functional systems. The efferent discharge is also widely spread, but the chief distribution seems to be to upper levels of the medulla oblongata which are concerned with movements of the musculature of the head.

It is obvious that the isthmus contains a motor adjustor of prime importance. Its distinctive features can perhaps best be appreciated by comparing it with some of the other important centers of adjustment. The tectum of the midbrain and the dorsal thalamus, which in this animal is ancillary to it, comprise the dominant apparatus of correlation, on the sensory side, of all exteroceptive and proprioceptive systems. In the habenula the olfactory organ as an exteroceptor is tied in with the other exteroceptive systems. In the hypothalamus the olfactory system is similarly related with the visceral and gustatory systems. All fields of sensory correlation discharge into the corpus striatum, ventral thalamus, and peduncle, where they are integrated and co-ordinated primarily in the interest of mass movements of the skeletal musculature, such as those involved in locomotion. Efferent fibers from all the centers just enumerated and from others converge into a common pool in the isthmic tegmentum. Since apparatus adequate for executing the primary mass movements of the trunk and limbs is found elsewhere—primarily in the midbrain, acting through the ventral tegmental fascicles upon the lower bulbar and spinal apparatus—and, since the isthmic tegmentum matures later in development, with notable enlargement preceding metamorphosis, it seems safe to infer that the isthmic apparatus is the chief regulator of the musculature concerned with feeding. The courses of the efferent fibers from the isthmus support this supposition. In Amblystoma the feeding activities are thoroughly integrated movements, where posture of the body, conjugate movements of the eyes, and action of jaws, pharynx, and esophagus are organized as a "total pattern," as Coghill ('36) has demonstrated. For further consideration of the olfactory and hypothalamic components of feeding reactions see pages 210 and 252 and 1934b, page 384.

It is evident that control of feeding reactions is not the only function of the isthmic sector. For instance, it has been shown by Aronson and Noble ('45) that the removal of the entire brain of the male frog in front of this region (including the tectum, cerebellum, and anterior part of the tegmentum) did not interfere with spawning movements but that "lesions in the tegmentum at the level of the motor nucleus of the trochlear nerve markedly disturbed or completely abolished these spawning responses."

In addition to this apparatus of activation of behavior, there are inhibitory functions here also, implying a participation in motor control of wide import. The intricate connections of the isthmic tegmentum described in the next chapter are part of this inhibitory apparatus.

The unitary character of most of the activities of Amblystoma, involving the synergic action of large masses of musculature in invariable orderly sequence, with relatively less capacity for the local autonomous action of the individual members than in higher animals, explains the simplicity and relative homogeneity of the nervous apparatus involved—also why the isthmic tegmentum is so large in Amblystoma. The integrated functional complexes which here are controlled from this major pool are individually differentiated in man, with corresponding specialization and segregation of the apparatus of the several component parts; and, parallel with this differentiation and increase of local autonomy, the central control apparatus has been transferred from the stem to the cerebral cortex. Accordingly, the isthmic tegmentum, which bulks so large in the brain of Amblystoma (and in the early human embryo), is dismembered in the adult man, and its parts are submerged within surrounding structures.

CHAPTER XIV

INTERPEDUNCULAR NUCLEUS

THE interpeduncular nucleus, like the habenula with which it is connected, is one of the most conservative structures in the brains of vertebrates, but in urodeles it has some characteristics which have not been reported elsewhere. The general features are outlined on page 46, and the details are described here, together with some speculations about the probable functions. The chief connections are shown very diagrammatically in figures 19 and 20, the spiral endings of the tractus habenulo-interpeduncularis in figure 50, the composition of the glomeruli in figure 84, and some typical connections of its neurons in figure 83.

COMPARATIVE ANATOMY

This nucleus was discovered by Forel in 1872 and described in the rabbit in 1877, with confirmation by von Gudden in 1881 and Ganser in 1882. It is a constant feature in all vertebrate brains, much larger in some lower groups than in the higher. In urodeles it is a well-defined column of cells, embracing the ventral angle of the ventricle and extending from the fovea isthmi backward to the level of the V nerve roots. It is a ventromedian structure, differentiated *in situ* in the floor plate and the adjoining borders of the basal plates. The fovea isthmi marks the anterior end of the embryonic floor plate as defined by Wilhelm His, and this plate is generally regarded as a nonnervous structure characterized by specially differentiated ependyma (Kingsbury, '30, p. 182); but Coghill ('24, Paper III) has shown that neuroblasts are differentiated here intrinsically. Some of these persist in the adult medulla oblongata as nucleus raphis, and between the fovea isthmi and the V nerve roots this differentiation goes much further, producing the interpeduncular nucleus, as pointed out in my description of this nucleus of Necturus ('34c).

The differentiation is primarily isthmic, as was recognized by van Gehuchten ('00, 2, 199). In Amblystoma it extends spinalward into trigeminal territory, and below the level of the V nerve roots it is continuous with the nucleus raphis of the medulla oblongata. In cyclostomes the fasciculus retroflexus extends spinalward as far as the calamus scriptorius (Johnston, '02, p. 31 and figs. 8–16, 30). A mesencephalic sector of the nucleus, rostrally of the fovea isthmi, has been described in some vertebrates, and there is some evidence of this in Necturus ('34c) and Amblystoma. It has been found in some fishes, reptiles, and mammals; but in urodeles there is no group of differentiated cells in the midbrain comparable with this nucleus in the isthmus, and the secondary connections of mesencephalic terminals of fibers from the habenula are radically different from those of the habenulo-interpeduncular tract. In this account, accordingly, the description relates only to those structures spinalward of the fovea which have habenular connection. This involves a terminological inconsistency, for in my analysis of the brain of Amblystoma the "peduncle" is defined as a strictly mesencephalic structure, and here none of the interpeduncular nucleus lies in the mesencephalon. Since this nucleus, however, is so obviously homologous with the structure so named in mammals, the mammalian name is retained.

The published description of the interpeduncular nucleus of Necturus ('34c) was based on insufficient material and evidently is incomplete. The structure there described is similar to that of Amblystoma but much simpler. Whether this simplicity indicates that Necturus is really more generalized (or more retrograde) or merely that the material available does not adequately show the actual structure cannot be determined. Probably both these suppositions are true. The more abundant Golgi material of Amblystoma reveals a wealth and intricacy of detail that baffle analysis, and much remains obscure. The exceptionally large and elongated nucleus in these animals is favorable for analysis, but the fact that all the connections are unmyelinated and dispersed in very dense neuropil makes accurate description impossible, except where elective silver impregnation isolates particular components. The erratic incidence of these impregnations and the well-recognized hazards of error in their interpretation necessitate great caution and restraint in drawing conclusions. Most of the observations here recorded have been seen in many preparations and can be adequately documented. Some of them, however, rest on slender and ambiguous evidence and require confirmation. The attempt to assemble these fragmentary observations into a coherent unity, such as is shown in figures 19, 83, and 84, is fraught with danger; and it is emphasized that these pictures and the physiological theories expressed are not regarded as final. They present my opinion of the things seen, subject to revision when information is more complete. A few years ago a brief summary of the observations made up to that time was published ('39b, p. 584), and details were later added ('42, figs. 40–43). Here all these data are assembled, some errors in a brief reference in the paper of 1927 (p. 280) are corrected, and further observations and interpretation are recorded.

HISTOLOGICAL STRUCTURE

In Weigert and cytological preparations the cells of this nucleus are densely crowded under and adjacent to the floor of the ventricle. Between the 'evels of the III and IV nuclei this gray column is thin. Ventrally of it there are, successively, the ventral median tegmental fascicles of group (1), the decussating fibers of the ventral commissure, and the superficial interpeduncular neuropil (figs. 60, 80–84, 92; '25, figs. 8, 9; '27, fig. 39; '42, figs. 30, 41–43). At the level of the IV nucleus the median fascicles, $f.m.t.(1)$, turn away from the midplane, and the texture of the ventral medial tissue posteriorly of this level is more open. Some of the cells of the interpeduncular nucleus lie more ventrally among strands of the ventral commissure (fig. 91). This gray is continuous dorsally on each side with that of the isthmic and trigeminal tegmentum, but in cell preparations the boundary is usually evident. Golgi sections show that at this boundary the cells are of transitional form, with some dendrites ramifying in the interpeduncular neuropil and others directed laterally into the alba of the tegmentum (figs. 61, 65, 66).

EPENDYMA

The description of the ependyma of Necturus ('34c, p. 118) applies with little change to Amblystoma. The ependyma at the fovea isthmi is distinctive (fig. 79) and similar to that bordering the sulcus isthmi (p. 181). Posteriorly of this, the ependymal elements are of two forms. In and near the median raphe they are compact fascicles of thick, thorny fibers (fig. 70). More laterally (figs. 63, 64, 81), each element has a much more widely branched arborization of slender, thorny fibers, many of which end at the pial surface in bulbous enlargements. The external limiting membrane is apparently composed of these expanded ependymal terminal bulbs. Both these types of ependyma are seen in midlarval stages ('39b, figs. 89, 94, 95).

The interpeduncular ependyma is an exceptionally dense mat of widely branched and closely interwoven fibers. Many of these branches do not reach the pial surface but arborize throughout the interpeduncular neuropil and especially within the interpeduncular glomeruli (described below), of which they form an integral part. The participation of specially modified ependyma in the formation of these glomeruli raises interesting questions. There is no obvious demand for unusual mechanical support here to account for especially strong ependymal framework, except perhaps in the median raphe, where the ependymal elements are more sturdy. Elsewhere in the interpeduncular field the slender ependymal fibers branch freely, and in the glomeruli they have tufted endings similar to those of dendrites, to be described shortly. This elaboration of the ependymal fabric within specialized synaptic fields suggests that the ependyma has some part to play in metabolism at the synaptic junctions.

INTERPEDUNCULAR NEUROPIL

Ventrally of the interpeduncular gray and the commissure there is a sharply circumscribed area of very dense superficial neuropil, oval in cross-section, which in some of my published figures is marked *nuc.inp*. It is preferable to apply the word "nucleus" to the gray area only. This band of differentiated neuropil is, accordingly, currently called the "specific interpeduncular neuropil" (*inp.n.* or *nuc.inp.n.*), though it should be kept in mind that the entire interpeduncular field is permeated by neuropil, of which this is a specialized part. Its axonic component is a web of interlaced fibers, which is continuous with that of surrounding parts. Imbedded within this fabric are two specialized structures—the spiral terminals of the habenulo-interpeduncular tract and the glomeruli. The whole interpeduncular area is richly vascularized with capillary net and is a synaptic field of great importance. This activity evidently is especially concentrated in the glomeruli and the area of specific superficial neuropil.

The neuropil of the interpeduncular field is continuous dorsally with the deep neuropil of the gray of the isthmic and trigeminal tegmentum and laterally with the intermediate neuropil of the alba of these tegmental areas, as shown very inadequately in figures 60–66. Posteriorly it is continuous with the diffuse neuropil which pervades both the gray and the white substance of the medulla oblongata (figs. 79, 80, 81). In these drawings the structure of the neuropil is shown greatly simplified and schematized, for it is impossible on this scale of magnification to portray the intricacy of its texture. Since all nerve fibers related with the interpeduncular system are unmyelinated and most of the tracts are dispersed in the neuropil, analysis is very difficult and, indeed, impossible except when aided by elective impregnations. The interpretations given in the text and in the diagrams (figs. 19, 83, 84) are based on such evidence. The evidence is incomplete and in many preparations ambiguous, and attention is again called to the fact that the conclusions reached are tentative and subject to revision in detail. It is believed, however, that the main features of the histological analysis are based on adequate evidence and are reliable.

GLOMERULI

Glomerulus-like structures have been known in the mammalian interpeduncular nucleus since 1877 (Forel). In Amblystoma these are small and very numerous elongated areas of very dense neuropil. They are distributed throughout the interpeduncular neuropil, in many places densely crowded; most of them are oriented vertically and extend downward into the ventral area of specific neuropil. Histological analysis of their structure is even more difficult than is that of the rest of the interpeduncular neuropil, for each of the constituent elements may present quite different appearances in Golgi sections, depending on the quality of the impregnation.

These glomeruli resemble in some respects those of the olfactory bulb ('24b), though of much more complicated structure. In addition to the capillary net, there are three constituent elements: (1) a condensation of the ependymal framework as already described; (2) tufted terminals of dendrites; and (3) tufted axonic terminals composed of thin contorted fibers interwoven with the dendritic terminals. Any one of these three constituents may be electively impregnated, or two or three of them may be seen in the same section. In some Golgi preparations the three components are so similar that they cannot be distinguished with certainty except where terminals can be followed to their connections outside the glomerulus, though when two of them are impregnated in the same section the difference between them is usually obvious. In the preparation from which figure 66 was drawn, axons and dendrites are intermingled in each glomerulus. Only the dendritic component is drawn on the left side, and only the axonic component on the right. The visible tufted axons are from interpeduncular neurons.

Most of the glomerular dendrites come from cells of the interpeduncular nucleus, though some of them are branches from the transitional neurons at the ventral border of the tegmental gray. The axonic tufts, as far as observed, come from neurons of the interpeduncular nucleus, cells of the overlying tegmentum, and collaterals of fibers of tr. tegmento-bulbaris. No fibers from other sources seem to participate significantly in the formation of glomeruli. Many axons from the three sources mentioned do not enter glomeruli but arborize in the surrounding neuropil. Glomeruli with dendrites and axons derived chiefly from the interpeduncular nucleus tend to be disposed horizontally, the others vertically. The horizontal glomeruli have been illustrated from the larva ('39b). Figure 58 of the paper cited shows two of the dendritic tufts, and figure 61 shows the axonic component of several of these horizontal glomeruli. Most of the horizontal glomeruli are imbedded in the general neuropil dorsally of the ventral specific neuropil. The vertical glomeruli generally extend downward into the specific neuropil, where they are penetrated by fibers of the interpeduncular spiral.

NEURONS

In transverse Golgi sections the dendrites of the interpeduncular neurons are directed ventrally, passing through the ventral commissure. Within the interpeduncular neuropil they branch widely, and many of the branches have tufted endings in the glomeruli (figs. 19, 65, 66, 83, 84). Longitudinal sections show that the spread of these dendrites is much greater fore and aft than it is transversely (figs. 80, 81, 82; '39b, figs. 57, 58), and the tufted terminals may be oriented either vertically or horizontally. Some of the cells of the isthmic tegmentum at the border of the interpeduncular nucleus may send dendrites into the interpeduncular neuropil, and these, too, may have tufted endings ('39b, figs. 78, 95). Sagittal sections in which both interpeduncular dendrites and ependyma are fully impregnated give spectacular demonstration that the interpeduncular formation ends anteriorly at the fovea isthmi and isthmic sulcus. These interwoven elements directed anteroventrally form a dense mat, which ends abruptly at the locus of the sulcus isthmi. Posteriorly, the interpeduncular formation merges insensibly with the trigeminal tegmentum.

The slender axons of the interpeduncular neurons arise from the cell body or the base of the dendrite (figs. 65, 66, 70, 84; '39b, figs. 61, 66) and immediately branch profusely. These fibers take tortuous courses within the interpeduncular neuropil and can be followed only when electively impregnated. In such preparations the longer branches are seen to be directed spinalward and to form the dorsal and ventral interpedunculo-bulbar tracts, as in Necturus ('30, p. 80; '34c, p 120).

Other branches spread widely in the general interpeduncular neuropil, of which they form an important component. Most of these fibers are dispersed, but some of them form compact longitudinal tufts within the glomeruli (fig. 84; '39b, figs. 41, 61).

AFFERENT CONNECTIONS

Fibers from remarkably diversified sources terminate in the interpeduncular neuropil. The widely branched terminals of some of these fibers spread diffusely; the endings of others are specialized in diverse ways. The dominant member of this complex evidently is tr. habenulo-interpeduncularis, which is the chief component of the f. retroflexus as described in chapter xviii. The interpeduncular connections of this tract are similar in Amblystoma and Necturus ('34c).

Tractus habenulo-interpeduncularis.—The two bundles of unmyelinated fibers converge immediately below the fovea isthmi, and here most of their fibers decussate close to the ventral surface. Their further course is shown in figure 50. Immediately after crossing, each fiber reverses its course, and this is repeated so that a compact spiral is formed, which extends the entire length of the interpeduncular neuropil, diminishing spinalward. This spiral is coextensive with the specific interpeduncular neuropil and is its most characteristic feature. There are many series of sections cut in various planes in which this is clearly shown. In some of these, only one of the habenulo-interpeduncular tracts is impregnated (figs. 51, 52, 56, 57, 58).

Figures 53 and 54 are drawn from a specimen in which tr. habenulo-interpeduncularis is impregnated on both sides at the origin in the habenula. The impregnation fails on the left side from the level of the dorsal thalamus downward. On the right side it is well stained as far as the decussation, below which only a few fibers are blackened, thus revealing clearly the courses of individual fibers, one of which is drawn in figure 54 through several turns of the spiral (cf. fig. 55). These fibers are slightly varicose and thorny. No specialized endings of any sort have been seen, and they have not been observed to branch except for occasional short forked terminals. The spiral penetrates the glomeruli, and its fibers are in synaptic connection with the glomerular dendrites.

Transverse sections show that the spiral is flattened dorsoventrally ('42, figs. 40–43). In the specimen from which these figures were drawn the spiral is fully impregnated, and few other fibers of the interpeduncular neuropil are stained so that the structure of the spiral is clear. No fibers leave the spiral to pass beyond the specialized interpeduncular neuropil. There is a small fascicle of uncrossed fibers at each lateral border of the spiral, and these apparently enter the spiral at successive levels. Another series of transverse sections (no. 2252) from the same lot of young adult specimens shows almost identically the same structure.

Not all the habenulo-interpeduncular fibers enter the initial decussation. As mentioned above, a considerable fraction of them, including some of thicker caliber, descend uncrossed along the lateral borders of the interpeduncular neuropil, which they enter farther spinalward. In some preparations both crossed and uncrossed fibers are impregnated, in some only the spiral fibers (fig. 50), and in one sagittal series only the uncrossed fibers are blackened (fig. 82). In the latter case the f. retroflexus is partially impregnated on each side for its entire length from the habenula to the interpeduncular nucleus. At the decussation none of the crossing fibers or the spiral fibers below it are darkened, but there is a large fascicle of uncrossed fibers, which descends laterally of the decussation and the spiral for almost the entire length of the interpeduncular nucleus. These fibers terminate within the interpeduncular neuropil among those of the unimpregnated spiral.

In 1894 van Gehuchten described the f. retroflexus of the trout. The origin in the habenula is similar to that of Amblystoma, and the course is the same. These fibers decussate in the interpeduncular nucleus and recross with much branching, but they do not form a compact spiral. They are varicose and have no specialized terminals. Simpler spiral terminals of mammals have been many times described (rabbit, von Gudden, '81; mole, Ganser, '82, p. 682; mouse, Ramón y Cajal, '11, 2, 274; Calderon, '28; and many others).

Tractus mamillo-interpeduncularis.—These fibers comprise group (2) of the tegmental fascicles ('36, pp. 303, 338, figs. 3, 8). They arise in the dorsal part of the hypothalamus in company with similar fibers for the peduncle and tegmentum (figs. 19, 71, 79). The mixed bundle decussates partially in the retroinfundibular commissure, which is component 1 of the commissure of the tuberculum posterior (p. 302). The interpeduncular fibers, crossed and uncrossed, pass the fovea isthmi at the ventral surface close to the mid-plane and then descend along the lateral border of the specific interpeduncular neuropil (figs. 27–30, 60; '39b, figs. 22, 41, 42, 57–61; '42, fig. 3), within which they end in open arborizations. They do not join the spiral or participate notably in the formation of glomeruli. Most of them end in the rostral half of the interpeduncular neuropil.

Tractus olfacto-peduncularis.—This important component of the basal forebrain bundles (fig. 101; '39b, p. 534, figs. 1, 57), after contributing fibers to the hypothalamus and peduncle, continues spinalward along the lateral border of the specific interpeduncular neuropil. The endings of these fibers are similar to those of tr. mamillo-interpeduncularis (figs. 18, 19, 21, 25–30, 53, 54, 59, 72, 82). This tract in the interpeduncular region lies laterally and dorsally of tr. mamillo-interpeduncularis, extending spinalward beyond the posterior end of the interpeduncular nucleus. There are some terminals of both this and the preceding tract in the dispersed interpeduncular neuropil.

Nervus terminalis.—In Necturus the longest fascicles of this nerve root probably reach the interpeduncular neuropil ('34c, p. 124). In Amblystoma a small fascicle was observed by McKibben ('11, p. 270) to reach the "dorsolateral part of the hypothalamus and the interpeduncular region." Some of these fibers may connect with the interpeduncular nucleus, but this has not been demonstrated.

Tractus tegmento-interpeduncularis.—This is a very extensive connection, which takes two forms. There is, first, a system of short axons dispersed in the neuropil of the gray and the deep neuropil of the alba which arise from small cells of the dorsal, isthmic, and trigeminal tegmentum and descend directly to the interpeduncular neuropil. Some may come from the tectum. They are especially concentrated in the f. tegmentalis profundus (p. 286). A second type of connection is made by collaterals of the thicker axons of tr. tegmento-bulbaris at their decussation in the ventral commissure.

1. The short tegmento-interpeduncular fibers of the first group are unmyelinated. Their general arrangement is well shown in the sagittal sections, figures 79 and 80. They arborize in the neuropil of the gray of the interpeduncular nucleus and at all depths of the neuropil of the alba, forming an important component of the diffuse neuropil. Transverse sections show that, in addition to the diffuse spread of these fibers in the neuropil, there are also tufted endings which form part of the axonic component of the glomeruli (figs. 61, 62, 63, 66, 84).

2. The tr. tegmento-bulbaris cruciatus as described in the larva ('39b, p. 590) is a series of thick axons of large cells of the isthmic and trigeminal tegmentum, which decussate dispersed in the ventral

commissure and then descend in the ventral and ventrolateral fascicles of the medulla oblongata. Horizontal sections of one of these larvae show slender collaterals of these decussating fibers, which are directed posteroventrally into the interpeduncular neuropil and here have tufted endings in the interpeduncular glomeruli ('39b, p. 600 and figs. 46 and 59). Figures 67 and 68 of the present work show similar collaterals, as seen in transverse sections from the same lot of larvae. Each compact glomerulus-like tuft contains terminals of several of these collaterals. Figure 69 shows similar collaterals in an adult brain. In this preparation only a few of these decussating fibers are impregnated, and these can be followed from section to section. Nothing else is stained here, so there is no opportunity for confusion. Figure 60 is from another adult specimen which shows some of these features more clearly, though the impregnation is less selective. This section is from the series illustrated in 1927 and is a few sections rostrally of figures 39 and 40 of that article. The tufted axonic terminals shown there are similar to the one seen in the present figure 60, and they are now known not to be derived from the mamillo-peduncular tract as there suggested ('27, p. 280).

Tertiary visceral-gustatory tract.—These fibers, as elsewhere described (p. 169), pass from the secondary visceral nucleus in the isthmus to the area ventrolateralis pedunculi. They lie close to the gray in the posterior lip of the sulcus isthmi and form one component of the complex f. tegmentalis profundus. Some fibers of this fasciculus decussate in the ventral commissure dorsally and spinalward of the decussation of the f. retroflexus, and here in some preparations there is evidence that fibers of the tertiary visceral tract turn spinalward to enter the interpeduncular neuropil. Several other systems of similar fibers are mingled here, and, in the absence of elective impregnation of this connection, its presence remains doubtful. Whether or not this direct connection exists, the visceral-gustatory system is related less directly with the interpeduncular nucleus by way of the hypothalamus. The secondary visceral nucleus is connected with the dorsal (mamillary) part of the hypothalamus and from here the large tr. mamillo-interpeduncularis may transmit visceral sensory and gustatory influences to this nucleus.

Other afferent fibers to the interpeduncular nucleus have been described in various animals. Gillilan ('41) reports that in bats and rodents the basal optic tract sends some fibers directly to this nucleus.

This connection may be present in Amblystoma also, though we have no demonstration of it.

EFFERENT CONNECTIONS

The widely branched axons of the interpeduncular neurons apparently ramify throughout the entire interpeduncular field, and some of them pass beyond this field. The latter go out in three directions and are designated as three tracts named after their terminal connections—tr. interpedunculo-bulbaris dorsalis and ventralis and tr. interpedunculo-tegmentalis (figs. 19, 83, 84). All these efferent fibers go out in a web of mixed neuropil, within which the two bulbar groups are more or less clearly fasciculated in elective impregnations. The tegmental group of efferents are nowhere fasciculated, and we have no Golgi preparations in which they are electively impregnated.

Sagittal and transverse sections show that the interpeduncular neuropil extends dorsally into the isthmic and trigeminal tegmentum. Most of these fibers trend dorsoventrally, these comprising part of the mixed system, termed "fasciculus tegmentalis profundus." These fibers are marked $f.teg.p.$ in the following figures: 29, 30, 31, 79; '42, figs. 30, 41, 42, 43. Elective impregnations have revealed some of the components of this fasciculus, including the brachium conjunctivum, tertiary visceral tract, and tr. tegmento-interpeduncularis. The latter is shown in figures 61, 62, 63, 65, and 66. These sections show also (though not clearly drawn in the figures) that from the interpeduncular neuropil a large number of fibers swing dorsolaterally into the alba of the tegmentum, where they engage dendrites of tegmental neurons. These fibers are interpreted as an interpedunculo-tegmental connection. In the light of the mammalian connections of the chief efferent tract from the interpeduncular nucleus to the dorsal tegmental nucleus discussed below, it seems probable that in Amblystoma the small-celled central nucleus of the isthmic tegmentum contains the precursor of the dorsal tegmental nucleus, though this evidently is only one of its relationships.

The interpedunculo-bulbar connections are clearly seen in our material. These fibers emerge from the diffuse interpeduncular neuropil in dorsal and ventral strands, which are connected with each other and with the surrounding neuropil, as shown in figures 79 and 81. The tr. interpedunculo-bulbaris ventralis descends for an undetermined distance at the lateral border of the interpeduncular neuropil and farther spinalward in the lip of the ventral fissure. The dorsal tract descends dorsally of the f. longitudinalis medialis (fig. 92). As pointed out below, this position of the dorsal tract agrees with that of the dorsal longitudinal fasciculus of Schütz, and its fibers take a similar course, turning laterally to spread in the trigemino-facial tegmentum and as far back as the level of the motor nucleus of the IX nerve. In fact, the dorsal bulbar tract is really an extension posteriorly of the tegmental tract, for these comprise a continuous series of efferents to the tegmentum. The dorsal longitudinal fasciculus of Amblystoma is not a well-formed tract as in mammals, but comparable fibers can be recognized in the deep neuropil of the brain stem. In both lower and higher animals this complex seems to be concerned with general central excitatory state and disposition of the individual rather than with specific sensori-motor responses (pp. 208, 217). The interpedunculo-tegmental and dorsal interpedunculo-bulbar system of fibers is evidently the primordium of an important component of the "caudal division of the dorsal longitudinal fasciculus" as described in the opossum by Thompson ('42). This morphological conclusion rests, I think, on sufficient evidence, whether or not the following physiological speculations can be confirmed experimentally.

INTERPRETATION

Though much remains obscure, the main features of the histological structure and connections of the urodele interpeduncular nucleus are now clear, and these are assembled schematically in figures 19, 83, and 84. The most striking feature is the ventral plaque of specific neuropil containing spiral terminals of the habenulo-interpeduncular tract and glomeruli. Other afferent fibers enter both the specific and the diffuse interpeduncular neuropil and are so arranged as to permit their activation from almost all parts of the cerebrum. All these fibers come from areas of intermediate-zone type, not from primary sensory centers.

These afferents are of four general classes, depending on the source of their activation: (1) olfactory from the hemisphere, ending in the diffuse neuropil; (2) olfacto-somatic from the habenula, ending in the specific (spiral) neuropil; (3) olfacto-visceral from the hypothalamus, ending in diffuse neuropil; and (4) tegmento-interpeduncular from the overlying tegmentum and ending in both diffuse neuropil and glomeruli. Those of the last class come from somatic sensory fields of correlation with no appreciable olfactory component. The glomeruli provide apparatus for nonspecific summation and reinforcement.

Each one of the other classes of afferents may have its own specific type of trans-synaptic transmission, depending on the histological and chemical structure and the strength and timing of volleys delivered.

Provision is thus made for discriminative responses to a considerable range of different kinds of activity that may be in process elsewhere in the brain—olfactory, visceral, somesthetic, and so on. The number of interpeduncular neurons activated and the temporal rhythm of their discharge may be determined not only by the algebraic sum of afferent impulses received but also by their timing. It is believed that this sort of analysis is characteristic of the nervous adjustors.

The outflow from the interpeduncular nucleus is distributed in part to remote regions and in part within the interpeduncular field, the latter providing apparatus for summation and intensification of the discharge. The remote effects of this outflow will depend on the location, structure, and connections of the motor pools entered by the efferent fibers.

The entire interpeduncular field is a well-circumscribed motor pool, in Sherrington's sense, in which every intrinsic nervous element is subject to activation by a wide variety of nervous impulses coming from diverse sources, each of which has a specific form of synaptic junction. The different structural patterns of these synapses presumably involve physiological differences in the type of activation. The dominant member of this complex of afferents evidently is the spiral of habenulo-interpeduncular fibers; every fiber of this tract may activate (or inhibit) every neuron of the nucleus, and, indeed, in successive turns of the spiral it may act upon several dendrites of the same neuron. There is no provision for separate localization of specific function here. The nucleus inevitably acts as a whole. But there is provision for summation of the effect in a large way. The diffuse neuropil also may act upon these neurons. Volleys may be discharged into it from many sources, each of which may reinforce (or inhibit) the activity instigated by the spiral.

Though temporal summation at an individual synapse is regarded as practically inoperative (Fulton, '43, p. 58), there is ample provision in this pool for spatial summation. The glomeruli present a series of difficult problems. Each of these small nodes of denser neuropil contains tufted dendritic terminals and tufted axonal terminals, and the latter come from two sources, intrinsic and extrinsic. The horizontal glomeruli seem to be especially adapted for summation of the discharge from the nucleus. If one of these neurons is experiencing subliminal activation, a collateral discharge from an active neuron may impinge upon it in a horizontal glomerulus, and the sum of the two excitations may be sufficient to fire the element. In a similar way, terminals of intrinsic fibers in the vertical glomeruli may reinforce the action of the extrinsic fibers. The glomerular activities may reinforce (or inhibit) those of the spiral, or conceivably they may act independently if the habenular component is inactive.

The observed structure seems to favor the supposition that, whatever may be the functions performed by this complex, they are of generalized type. The apparatus which patterns the specific reflexes, both somatic and visceral, is located elsewhere in the brain stem. The intrinsic structure of the interpeduncular field seems to preclude any well-defined local representation of different functions within it. The nucleus apparently acts as a unit, and any functional specificity that may exist would be in terms of differential excitation at the various forms of synaptic junction or of the functions of the motor fields into which its efferent fibers discharge. The ventral bulbar tract descends in the primary motor column of Coghill, and this suggests that it is in some way concerned with mass movements of the musculature of the trunk and limbs. The tegmental and dorsal bulbar efferent tracts form an anatomical unit, and they probably are similarly related physiologically. These fibers discharge into a motor field within which the reflexes of the head are organized, notably those concerned with feeding. The interpeduncular complex seems to be so organized as to act as a whole without intrinsic localization of function. What part, then, does this complex play in patterned behavior?

In the first place, it is a noteworthy feature of the interpeduncular system that it is connected "in parallel," as the electrician would say, with the major systems of fore-and-aft conduction of the brain stem. The olfactory, optic, and lemniscus systems on the afferent side and the strio-pedunculo-bulbar systems on the efferent side, with their related centers of adjustment, seem to provide the apparatus which patterns overt behavior. The interpedunculo-habenular system is in addition to this apparatus and in some way ancillary to it.

In noncorticated vertebrates the activators of the motor zone receive their nervous excitations mainly from three regions which are physiologically distinctive: (1) the somatic sensory field of the tectum and dorsal thalamus, (2) the visceral field of the hypothalamus, and (3) the olfactory field of the cerebral hemispheres. The parts of this primary activator system of the cerebrum are interconnected, and all are represented in the higher adjustors of the hemispheres.

The interpeduncular complex is separate from this great system of activators except at its upper and lower ends. It is activated from the same three physiologically distinctive regions, and its efferent impulses are discharged into the same lower motor fields. These relationships are in some respects similar to those of the cerebellum with the activators of the skeletal musculature. The cerebellum does not pattern behavior, but it acts upon the motor systems as going concerns, facilitating the activity in process by reinforcement and inhibition, appropriately placed and timed. The structure and connections of the interpeduncular nucleus suggest that it is similarly related with the great descending (extra-pyramidal) systems of the cerebrum, and specifically with those concerned in the feeding reactions.

In the light of Coghill's definition of the amphibian reflex "as a total behavior pattern which consists of two components, one overt or excitatory, the other covert or inhibitory," and making application to a specific problem of the general discussion of reflex and inhibition in chapter vi, an attractive hypothesis might be framed along the following lines:

The brain stem may be conceived as a labile, equilibrated, dynamic system, within which the excitatory and inhibitory components are balanced against each other in reciprocal interaction at every successive phase of motor activity. The efferent activating systems of patterned behavior discharge through the basal forebrain bundles and the tegmental fascicles (fig. 6). Above and below this central core of activating fibers and parallel with them, there are inhibitory systems of fibers which have the same origins and terminations as do the activating systems but which pursue different courses with different connections. Dorsally there is an olfacto-somatic inhibitory system centered in the habenula, and ventrally there is an olfacto-visceral inhibitory system centered in the mamillary region of the hypothalamus. Both these inhibitory systems converge into the interpeduncular nucleus (figs. 19 and 112). Here the inhibitory influences are integrated and distributed to lower motor centers as going concerns, in accordance with momentarily changing physiological requirements. Presumably not all components of the action system are under this sort of regulatory control by the interpeduncular complex, for, as pointed out below, there are other local fields with inhibitory functions.

The unusually large size of the interpeduncular nucleus of urodeles may be correlated with the fact that in the normal behavior of these animals total inhibition is a conspicuous feature (p. 77). Adopting Sherrington's terms, there is usually a long period of inhibition between the anticipatory and the consummatory phases of the reaction, notably in the feeding reflexes.

This hypothesis can have only academic interest unless and until it is confirmed by physiological experiment or clinical evidence, but it may have heuristic value as an indication of profitable points of attack in further exploration. From such scanty evidence as is now available, it seems to the writer to rest on an insecure foundation because it probably is an oversimplification of the problem. Our knowledge of the mechanism of central inhibition is so inadequate that inferences about physiological action from anatomical evidence alone are unsafe in this field, though such inferences are justifiable in many other fields where adequate physiological controls are available and have been made. Keeping these admonitions in mind, let us inquire further into the possible role that inhibition may play in the interpeduncular complex.

Though we do not know exactly how inhibition is effected in the central nervous system, we do know that in some parts of the brain the dominant function is activation, and in other parts it is inhibition. Current analysis of the primate cerebral cortex has revealed specific activating fields in the motor cortex and also specific inhibitory fields, the so-called "suppressor bands" (p. 79). In the stem the head of the caudate nucleus and the reticular formation of the upper medulla oblongata are also known to have inhibitory functions. That the interpeduncular nucleus is a similar field in which inhibition is dominant is a legitimate working hypothesis that can be tested experimentally. This, of course, does not imply that inhibition is the only function here localized, for inhibitory action is balanced against excitatory action in normal behavior. It follows that experimental disturbance of this normal balance may come to expression in so great a variety of ways depending on so great a variety of factors that it is difficult to devise crucial experiments.

This difficulty is illustrated by the published report of experiments by Aronson and Noble ('45) designed to localize in the brain the mechanisms which mediate specific components of the mating behavior of male frogs. In hundreds of experiments involving local ablations or transections, there were a few which gave clear-cut evidence of precise localization of function. In this connection Dr. Aronson writes me: "I believe that, while some activities, e.g., warning croak, ejaculatory response, and release of the female by the male at the termination of oviposition, are quite definitely localized, the mechanism controlling other phases of mating behavior are rather diffuse. In the latter cases more precise experiments would, I think, demonstrate diffuseness rather than localization."

The authors' very conservative analysis of their protocols does reveal a few areas with local specificity of function. Among these is the clasp reflex, which is one of the most characteristic features of this mating behavior. Normally, the mating clasp is relaxed or inhibited immediately after oviposition, and this presumably involves inhibitory action. There is evidence that the primary center of this reflex is in the spinal cord and that the inhibitory phase is under some measure of cerebral control. Relaxation or inhibition of the clasp is quite consistently abolished by destruction of the preoptic nucleus or dorsal part of the hypothalamus, but this result does not consistently follow injury to the habenulae or interpeduncular nucleus. Though the clasp involves the use of skeletal muscles, the reflex as a whole is a visceral reaction related with oviposition and ejaculation of sperm. Cerebral control of the release of the clasp in the hypothalamus may be transmitted to the upper levels of the spinal cord by a direct pathway via tr. mamillo-tegmentalis and ventral tegmental fascicles of group (3), as described on page 278, without involvement of the interpeduncular complex. But Dr. Aronson would not limit the inhibitory influence resulting in release to this pathway. Again in personal correspondence he says: "While the spinal clasp center might be quite limited, the mechanism inhibiting or modifying the spinal clasp reflex is very diffuse, especially in the midbrain and diencephalon. . . . . The interpeduncular nucleus might well be involved here."

An interesting series of experiments on the Japanese toad reported by Kato ('34) reveals an inhibitory center for the muscles of the contralateral limbs in a region described as "on the anterior end of the lamina terminalis," but not accurately defined. The diagrammatic figures seem to place it in the region of the anterior commissure between the septum and the preoptic nucleus. The efferent pathway from this center decussates in the vicinity of the interpeduncular nucleus. Further information about the connections and physiological significance of this center is desirable.

These experiments yield no crucial evidence of participation of the interpeduncular nucleus in the components of mating behavior under investigation or of movements of the limbs in general. This nucleus may play a more specific part in some other types of behavior; and some experiments indicate that this is true.

The only direct experimental evidence about the functions of the interpeduncular nucleus of mammals known to me is a recent note by Bailey and Davis ('42a) in which they describe a "syndrome of obstinate progression" in cats following destruction of this nucleus. This is a locomotor impulsion, which continues without remission until the death of the animal. Mettler and Mettler ('41) have produced similar symptoms by lesions involving the head of the caudate nucleus, and more recently Mettler ('45, p. 180) describes in primates a direct connection from the globus pallidus to the interpeduncular nucleus. The caudate is supposed to be one link in a chain of conductors with inhibitory functions (Fulton, '43, p. 456) and the syndrome following the destruction of the interpeduncular nucleus suggests to me that this nucleus also may be an inhibitor. The "obstinate progression" may then be interpreted as a release phenomenon.

In mammals efferent fibers from the interpeduncular nucleus have been described with distribution to neighboring parts of the brain stem, of which by far the most important is the large pedunculotegmental tract to the dorsal tegmental nucleus and the related dorsal longitudinal fasciculus of Schütz. In the recent study of this fasciculus of the opossum by Thompson ('42), she describes this interpeduncular connection as an important component of it. In this marsupial the dorsal longitudinal fasciculus is a complicated system of fibers, chiefly descending, connecting the deeper areas of gray between the diencephalon and the spinal cord.

The structure and connections of the dorsal longitudinal fasciculus suggest general functions of some sort rather than control of local reflexes. Another series of experiments by Bailey and Davis ('42) shows that this fasciculus has some activating function which is essential for maintenance of normal motor behavior. They succeeded in destroying by electrocautery the periaqueductal gray of cats without injury to surrounding parts. This lesion must have involved the dorsal longitudinal fasciculus and the dorsal tegmental nucleus. If the injury is slight, the cats on awakening from the anesthetic are very wild and active but after a few days recover normal behavior. This seems to be an irritative excitation. If the lesion is extensive, the cats "lie inert, silent and flaccid as a wet rag." They never again show any spontaneous activity, though after a few days placing reactions are elicited normally and, if stimulated, they may walk slowly.

If now it is assumed that the dorsal longitudinal fasciculus contains fibers which normally transmit some sort of continuous nonspecific activating or facilitating influence upon the entire motor field of the lower brain stem—an influence which is essential for effective co-ordination and integration of these local systems of synergic muscular activity—then the destruction of this pathway would leave the animal helpless. This is not a flaccid paralysis, because the animal ultimately regains control of righting and placing reactions and even co-ordinated locomotor movements. All initiative and spontaneous activity are permanently lost because the animal is deprived of some essential component of the integrating apparatus. The "obstinate progression" that follows destruction of the interpeduncular nucleus may result from loss of an essential inhibitory influence normally acting upon the dorsal tegmental nucleus and the f. longitudinalis dorsalis. Loss of this influence allows the remaining apparatus of facilitation in the f. longitudinalis dorsalis to "race" like a steam engine deprived of its governor, and in this particular setup of conditions the result is uncontrolled progression.

In Kabat's ('36) study of alterations in respiration resulting from electrical stimulation of cats, it was reported that stimulation of the periaqueductal gray and of diencephalic regions known to contribute fibers to the dorsal longitudinal fasciculus was followed by increase in rate and amplitude of respiration. Shallower and unusually slower breathing followed stimulation of the habenula and habenulo-interpeduncular tract.

Detwiler has shown (p. 62) that in Amblystoma a nonspecific mesencephalic influence is essential for the maintenance of motor efficiency. This is in addition to the action of the apparatus which activates patterned behavior, the latter maturing earlier and quite independently of cerebral influence. This mesencephalic influence in the amphibian brain seems to be comparable, in a general way, with the activating function of the dorsal longitudinal fasciculus of mammals.

These fragmentary observations cannot justify any final conclusions, but they support the working hypothesis that the habenulo-interpeduncular complex is balanced against the strio-pedunculo-bulbar and f. longitudinalis dorsalis systems, the latter being the activating members and the former the inhibiting member of a dynamic system adapted to insure co-ordinated action of specific synergic systems of muscles.

It is evident that in macrosmatic animals the olfactory system is dominant in the interpeduncular complex. It is the chief functional component of the stria medullaris thalami, habenula, and f. retroflexus and probably also of the f. longitudinalis dorsalis, though all these structures survive its absence in anosmic animals. In Ichthyopsida the olfactory system is dominant in the hypothalamus. Attention has been called (p. 99) to the double role played by the olfactory system in the cortical activities of macrosmatic mammals; that is, in addition to the specific olfactory functions, there is a nonspecific facilitation by activation or inhibition of other cortical activities. This conception of the nonspecific facilitating action of the olfactory cortex may be extended into the subcortical field also, and in the interpeduncular complex the dominant part played by olfaction is probably of this nonspecific nature.

CONCLUSION

The hypothesis here tentatively suggested divides the adjusting apparatus of the brain into two reciprocally interacting components, the one a system of activators, the other a system of inhibitors. In the normal patterning of behavior these systems are balanced one against the other. In the brains of primitive vertebrates the dominant highest center of correlation of the exteroceptive sensory systems lies in the tecto-thalamic sector and of the interoceptive systems in the hypothalamus. From both these regions the activating fibers converge into the ventral thalamus and motor zone of the midbrain and isthmus, where they are joined by olfactory fibers and others from the supra-sensory centers of the hemispheres. Above and below this central core of activating fibers are two systems of inhibitors—the olfacto-somatic system in the habenula and the olfacto-visceral system in the hypothalamus—both of which discharge into the interpeduncular nucleus. The efferents from this nucleus reach the same lower motor centers as do the efferents of the activating systems.

In all macrosmatic species, and especially in those of more primitive groups, the olfactory system is dominant in the forebrain, and in all groups it plays an important part in nonspecific activation and inhibition. In lower groups the deep neuropil of the gray substance is a diffuse activator, and this sytem survives in mammals as the dispersed periventricular fibers which converge to descend in the f. longitudinalis dorsalis of Schütz. Associated with this fasciculus is the dorsal tegmental nucleus, which receives inhibitory fibers from the interpeduncular nucleus. This complex discharges nonspecific activating and inhibiting impulses into the bulbar tegmentum, and these play an essential part in the maintenance of the appropriate balance between activation and inhibition in all muscular activity. The destruction of any major component of this equilibrated dynamic system may result in pathological behavior which has no counterpart in the normal animal but which may furnish clues pointing the way toward successful analysis.

Though, as suggested above, this schematic outline is doubtless oversimplified, the hypothesis or some variant of it may suggest profitable lines of experiment. For this purpose the urodeles, with great enlargement and elongation of the interpeduncular nucleus, are favorable subjects. The larger species, like Necturus and Cryptobranchus, can be operated upon more conveniently than can the salamanders and frogs, though their more sluggish behavior may make interpretation more difficult.

CHAPTER XV

MIDBRAIN

IN THE preceding chapters the structure of the rhombencephalon and the functions intrinsic to it were summarized. We now pass to the cerebrum, with a radically different type of organization, dominated by the optic and olfactory systems. In lower vertebrates the major optic terminals in the midbrain evidently have determined the course of its differentiation; but, since optic fibers terminate also in the diencephalon, a separate chapter is devoted to the visual system as a whole.

DEVELOPMENT

Because the mesencephalon is the site of a remarkable series of changes in the shape of the brain in early developmental stages, it is appropriate to review here some features of this growth. Attention has been called (p. 117) to the significance of von Kupffer's anterior and posterior intraencephalic sulci as stable landmarks in the development of all vertebrate brains. These mark the loci of two strong cerebral flexures, which are extremely variable in different species and among individuals of the same species. These flexures so modify the topographic arrangement of parts as to make morphological analysis difficult. The posterior sulcus of von Kupffer (the development of which is described in chap. xiii) marks the posterior limit of the great mesencephalic flexure, which bends the neural tube ventralward and backward so that, at the close of the early swimming stage of Amblystoma tigrinum (Harrison's stages 36–38), the hypothalamus is closely appressed against the ventral surface of the peduncle and isthmus ('38, figs. 1–4, 15–18). The tuberculum posterius marks the ventral fulcrum around which this bend is made.

Beginning somewhat later than this ventral bending, there is a flexure in the reverse direction more anteriorly at the site of von Kupffer's anterior sulcus, which bends the cerebral hemispheres forward and upward. In the early feeding stage (Harrison's stages 44, 45) this movement is far advanced ('38*b*, figs. 1, 2); and from this stage onward to the adult stage these flexures appear to be somewhat straightened and masked by intussusception of newer tissues.

Baker and Graves ('32) described the external form of the brain in six stages of A. jeffersonianum (3–17 mm.). Their figures 1 and 2 show that in their 3-mm. embryo, which is 7 days and 16 hours old and in a premotile stage, the mesencephalic flexure has begun. In Harrison's stage 30 of A. punctatum, about 24 hours younger than Coghill's nonmotile stage, this flexure is far advanced ('38, fig. 1), and the reverse flexure leading toward the evagination of the cerebral hemisphere is incipient.

Correlated with these changes in external form, cells are proliferating internally, and the fibrous connections are established. Coghill's charts show that differentiation is precocious in the peduncle, the chiasma ridge, and the olfactory area of the hemisphere. Other local areas of differentiation appear successively, and, before the early swimming stage is reached, confluence of these areas has begun, and fibers are descending from the peduncle and ventral thalamus in the primary motor tract, which is the precursor of the fasciculus longitudinalis medialis. This development is well advanced before connection is made between the sensory field of the tectum and the motor field of the peduncle (compare '37, figs. 1 and 2; '38, figs. 5 and 20). Fibers from the retina enter the brain in early swimming stages, but they do not reach the tectum until the close of this period. These optic fibers enter the tectum at its anterior end in the eminence of the posterior commissure, and the first connection between optic tectum and peduncle is by way of this commissure, in which fibers first appear in the coil stage (Harrison's stage 35), long before optic fibers reach the tectum.

The commissura tecti in the remainder of the midbrain roof appears much later, beginning as an extension spinalward from the posterior commissure. As late as the midlarval period (A. tigrinum of 38 mm.) this commissure has matured only in the anterior half of the tectum ('39a, figs. 1, 11, 12; '39b, fig. 1). Full details of the development of the optic nerve and its connections have been published ('41), as well as some features of the development of the tectum ('42, p. 243).

In development the nucleus posterior tecti is retarded as compared with the tectum opticum, more so in A. tigrinum than in A. punctatum ('38b, p. 413; '39a, p. 265). The ventricle is here dilated as the recessus posterior mesencephali, and in larval stages the deep sulcus isthmi ends dorsally in this recess ('38, fig. 18; '14a, fig. 2). The roof and side walls of the posterior recess remain relatively thin until late larval stages, and they are expanded so as to form a noticeable external eminence, which disappears in the adult except in the midplane.

SENSORY ZONE

TECTUM OPTICUM: SUPERIOR COLLICULUS

Nothing need be added here to the statements in chapters iv and xvi and the description of the mesencephalic V nucleus in chapter x. For details my paper of 1942 may be consulted.

NUCLEUS POSTERIOR TECTI: INFERIOR COLLICULUS

The small nucleus posterior tecti is an undifferentiated primordium of the inferior colliculus, with perhaps some additions in higher animals from the posteroventral part of the optic tectum.

The most characteristic cells of the posterior nucleus are small elements with thick bushy dendrites directed forward into the tectum opticum. Slender axons arise from these dendrites and are directed forward to the thalamus as an important component of the brachium of the inferior colliculus ('42, p. 265 and figs. 51 and 79). Other efferent fibers go out in large numbers in tractus tecto-thalamicus et hypothalamicus cruciatus posterior (p. 297; '42, p. 221), of which the strongest component decussates in the postoptic commissure and distributes to the tegmentum in tegmental fascicles (8) and (6). Other shorter efferents reach all surrounding parts. The chief connections of this region are shown diagrammatically in figures 10–18 and 21.

In Amblystoma there is no sharp boundary between the optic tectum and the nucleus posterior. The distinguishing features of the latter are the absence of optic terminals and the presence of two special tracts—the primordial lateral lemniscus (*tr.b.t.l.*) and the efferent tr. tecto-bulbaris posterior (*tr.t.b.p.*). The spinal lemniscus and the general bulbar lemniscus terminate chiefly in the optic tectum, though the spinal lemniscus has a larger proportion of terminals in the nucleus posterior. The tecto-cerebellar tract arises chiefly in the nucleus posterior. These connections suggest proprioceptive functions (chap. xii). This suggestion is strengthened by the presence of many cells of the mesencephalic V nucleus here and by the undeveloped condition of the auditory apparatus. The dominance of the cochlear connection is a later acquisition in land animals, with resulting transformation of the nucleus posterior into the colliculus inferior and differentiation of the medial geniculate body. This transformation is far advanced in the frog, for Aronson and Noble ('45) report that the warning croak uttered during mating is abolished by extensive injury of the inferior colliculi, though it is not lost after complete ablation of the hemispheres, diencephalon, superior colliculus, cerebellum, and anterior parts of the tegmentum.

In Anura the optic tectum is greatly enlarged so as to cover the dorsal convexity of the midbrain. The nucleus posterior is also enlarged under the influence of the well-developed lateral lemniscus, and it is folded into the aqueduct as the so-called "corpus posticum" or "torus semicircularis." In some reptiles and in all mammals the definitive inferior colliculus reappears on the dorsal surface.

INTERMEDIATE ZONE

The distinctive characteristics of the intermediate zone of Amblystoma are more clearly shown in the midbrain than elsewhere. This band of subtectal tissue is well differentiated from the overlying tectum and the underlying peduncle and isthmic tegmentum and is given a distinctive name.

TEGMENTUM DORSALE

The dorsal tegmentum is separated from the isthmus by the sulcus isthmi (*s.is.*) and from the peduncle by the limiting sulcus (*s.*) of the nucleus of the tuberculum posterius (fig. 2). On the external and ventricular surfaces there is no visible boundary between this zone and the overlying tectum, but the internal structure is sharply contrasted. The gray of the tectum is irregularly laminated, with plaques of cells separated by thin sheets of neuropil; that of the subtectal area is more homogeneous (figs. 34, 36, 93, 94). The texture of the white substance is even more distinctive.

The neurons are rather small and are imbedded in dense neuropil, which is continuous with that of adjoining areas. Typical illustrations are shown in figures 24, 48, and 49 (see also '42, figs. 29, 30, 44, 45, 48). Their dendrites ramify among dorsal and ventral tegmental fascicles and into the tectum, peduncle, and isthmic tegmentum. The axons arise from the dendritic tree, and most of them are short, arborizing locally and in adjoining areas. Many divide with ascending and descending branches. Axons of the larger elements (fig. 49) descend to the isthmic tegmentum in the dorsal tegmental fascicles (fig. 21, *tr.teg.is.*), these fibers being the largest component of fascicles of group (7). All these neurons seem to be concerned with local correlation.

This is typical correlational tissue interpolated between the sensory zone above and the motor zone below, with both of which it is connected by numberless short fibers, by long dendrites, and by the intrinsic neuropil (fig. 24). In addition to these local connections, it receives longer fibers from the cerebral hemispheres, dorsal and ventral thalamus, hypothalamus, and tectum, many of these fibers decussating in the anterior and postoptic commissures. There are long efferent fibers which descend in the dorsal tegmental fascicles to the isthmic and trigeminal tegmentum.

This field evidently contains important components of the apparatus of sensori-motor adjustment, but how this mechanism operates is unknown. This relatively homogeneous area probably loses its identity in higher animals, in which its parts are specialized and dispersed in the brain stem or perhaps supplanted by the more efficient adjustors of the thalamus and hemispheres.

MOTOR ZONE

PEDUNCLE

At the sharp ventral cerebral flexure of the brain stem, a ventricular eminence marks the position of the nucleus of the tuberculum posterius, which is called the "peduncle" in a restricted sense (p. 21). This peduncular area, which is bounded by a shallow and variable sulcus (fig. 2, s.), is an arbitrarily defined field at the ventral surface of the midbrain, bounded dorsally by the dorsal tegmentum and posteriorly by the isthmic tegmentum. It arises from the anterior end of the embryonic basal plate of the neural tube, and nerve fibers appear within it very early in embryogenesis. In the coil stage, fibers are added to this tract from the ventral thalamus and dorsal tegmentum ('37, fig. 1), and, before the early swimming stage is reached, there are additions also from the tectum and other neighboring parts ('37, fig. 2). This basal sector of the midbrain is the nucleus of origin of the first efferent fibers of this region for lower motor centers, and, as early as the S-reaction stage, fibers go out from it to the periphery in the oculomotor nerve. This is before any afferent fibers are connected with this region. The motor mechanism develops autonomously.

In Necturus the f. longitudinalis medialis is definitely organized in the peduncle and is a well-defined anatomical structure from here caudad ('36, p. 349); but in Amblystoma the fibers which compose this bundle are mingled with others in tegmental fascicles numbered (1), (4), (5), and (6) in the peduncle and isthmus, and below the isthmus a residue of fibers from all these fascicles is assembled to continue spinalward in the f. longitudinalis medialis ('36, figs. 9–16).

The gray layer is relatively thin, and in the mid-plane most of its cells are displaced by the massive decussations of the commissure of the tuberculum posterius (figs. 2C, 29, 30, 31, 94). More laterally, large and small cells are mingled in the gray, with little evidence of local segregation except for the nucleus of the oculomotor nerve at the posteroventral border.

The largest cells include dorsally the primordium of the nucleus of Darkschewitsch, which is related primarily with the posterior commissure (figs. 6, 18, 22, 24; shown but not named in fig. 32), more ventrally Cajal's interstitial nucleus of the f. longitudinalis medialis (figs. 18, 22, 104), and posteriorly the III nucleus (figs. 18, 22, 24, 93, 104). Larval forms of these cells have been illustrated ('39b, figs. 42, 53, 72, 73). They resemble those of Necturus ('17, figs. 23, 24, 30, 33).

These cells evidently are collectors of a wide variety of nervous impulses brought into the peduncle by the numberless fibers which converge to end here and by fibers of passage with collaterals in this area. The axons of these cells descend in the ventral tegmental fascicles and seem to be primarily concerned with activation of mass movements, particularly of locomotion. Most of these thick fibers descend in ventral tegmental fascicles of group (5) and are uncrossed, but some of them decussate in the ventral commissure, mingled with the tecto-bulbar fibers of the ventral median fascicles of group (1) ('39b, figs. 23, 24).

Small cells in the anteroventral part of the peduncle are intimately connected with the underlying hypothalamus and are doubtless concerned primarily with olfacto-visceral adjustments (figs. 18, 21, 53). There are numberless terminals here of the nervus terminalis, tr. hypothalamo-peduncularis from the ventral hypothalamus ('42, p. 226 and figs. 3, 22, 23, 68, 69), tr. mamillo-peduncularis from the dorsal hypothalamus ('39b, fig. 22; '42, fig. 39), and tr. olfacto-peduncularis from the anterior olfactory nucleus and primordial head of the caudate nucleus ('39b, fig. 1). This area also receives terminals of the secondary and tertiary visceral-gustatory tracts, and there is a strong pedunculo-hypothalamic connection (fig. 8; '42, p. 227). Ventral tegmental fascicles (3) and (4) from the hypothalamus ('42, fig. 3) are related with this area ('42, figs. 39, 40, 44), and these fascicles, together with the related small cells, probably contain primordia of those components of the mammalian dorsal longitudinal fasciculus of Schütz which are related with the peduncle and hypothalamus (Thompson, '42, p. 249).

Cells of medium size at the posterior end of the peduncular gray in the vicinity of the nucleus of the III nerve (fig. 59) have the chief dendrites directed rostrad, some of them longer and more sharply bent forward than those shown in figure 59. These dendrites engage fibers from the basal optic tract, visceral-gustatory system, hypothalamus, tr. olfacto-peduncularis, and mesencephalic terminals of the f. retroflexus, among other systems. Their axons also are directed forward; how far they extend within the basal forebrain bundles is not evident, probably for a long distance because they are thick, smooth fibers similar to those of the long descending and ascending tracts. It is not improbable that they reach the primordial corpus striatum, where similar fibers end in wide terminal arborizations ('39b, figs. 39, 41, 50, 72, 73; '42, fig. 69). These fibers may be comparable with those described in mammals as ascending from the entopeduncular nucleus and substantia nigra to the corpus striatum.

The neuropil of the alba resembles that of the ventral thalamus, but that of the gray substance is more dense and of different composition. In the white substance there are mingled terminals of axons derived from a great variety of sources—tectum, basal optic tract, posterior commissure and its nucleus, dorsal and isthmic tegmentum, the entire diencephalon, and the basal forebrain bundles (figs. 12, 14, 15, 17, 18, 20, 21, 22). There is little evidence of regional localization of these terminals except in two places.

Posteriorly, in the vicinity of the nucleus of the III nerve, the deep neuropil of the gray is very dense, with strands of fibers descending from the tectum and the dorsal and isthmic tegmentum (fig. 24). These clearly envelop the cell bodies of the III nucleus, and many of them decussate here in the ventral commissure. A second and more sharply localized area of neuropil is the area ventrolateralis pedunculi, extending superficially from the III root forward (fig. 23), as described in chapter iii.

The commissure of the tuberculum posterius includes all fibers of the ventral series of commissures which cross in the midbrain; for details of its connections see chapter xxi.

Summary.—The peduncle of Amblystoma as here defined receives fibers from practically all parts of the brain above the isthmus, and its primary function is control of mass movements of the trunk and limbs and conjugate movements of the eyes. Control of local reflexes is effected elsewhere. It is intimately connected with the hypothalamus, and various visceral motor adjustments are made here, though of these little is known.

CHAPTER XVI

OPTIC AND VISUAL-MOTOR SYSTEMS

THE eyes in Amblystoma are much better developed than they are in Necturus, where they are degenerate, though functional. The retina of Triturus is more highly differentiated, and vision is evidently more efficient. In frogs this advance is carried much farther, with corresponding structural elaboration of the visual centers of the tectum opticum and thalamus.

OPTIC NERVE AND TRACTS

The optic tracts and their connections of Necturus have been described ('41a). An account of the development of the optic nerves and tracts of Amblystoma was published in the same year ('41), and subsequently ('42) a detailed analysis of the visual centers and their connections, together with commentary on related physiological problems. Some further theoretical considerations were published later ('44a). These details need not be repeated. The salient features were outlined in chapter iv, and some topics are amplified here.

In the coil stage (Harrison's stage 34), optic nerve fibers begin to grow from the ganglion cells of the retina through the optic stalk before the other retinal layers are histologically differentiated ('41, p. 477; Coghill, '16, Paper II, p. 274). In the S-reaction stage (Harrison's stages 35, 36) these fibers reach the brain; in early swimming stages (Harrison's stages 36, 37) they decussate in the chiasma ('41, p. 480); and shortly thereafter they reach the tectum (Harrison's stage 38) and the primordial basal nucleus in the peduncle. Up to this stage, rods and cones and other retinal layers are incompletely differentiated. In stages immediately preceding effective feeding reactions (about stage 44) the retina is functional, and the larva can orient itself with reference to an object moving in the field of view; and shortly thereafter well co-ordinated lurching and snapping movements efficiently capture moving prey. The first visual responses are probably activated by fibers from the tectum through the posterior commissure to the nucleus of Darkschewitsch and the fasciculus longitudinalis medialis and perhaps also by the basal optic tract, though the time of the appearance of these fibers has not been recorded. Tecto-bulbar and spinal fibers mature very soon after those of the posterior commissure, the uncrossed fibers preceding the crossed fibers. These and the tecto-peduncular fibers described below are apparently the pathways employed in the activation of the skeletal musculature of the trunk and head when they are employed in the capture of food. Early feeders (stage 46) have attained full larval status, and in subsequent stages the visual system is not radically changed except for an increase in the number of thin retinal fibers and an enormous increment of these at metamorphosis, the total number of fibers in adult Amblystoma tigrinum probably reaching about 8,000 ('41, p. 506).

Correlation of this developmental history with experimental work by others has led me to the hypothesis that in Amblystoma the thick retinal fibers, which appear first in ontogeny, activate generalized total behavior patterns and that the thin fibers are concerned with more refined analysis of visual experience. (It does not follow that this is true in mammals, in which the visual responses are under thalamo-cortical control.) What kind of mechanism is employed in this analysis is still obscure. My opinion ('41, p. 528) that endings of thin and thick fibers have separate localization in the tectum proved on further study to be unfounded ('42, pp. 284, 293), and there is little anatomical evidence of any other kind of visual localization here, though experimental evidence cited below indicates that retinal loci are projected locally upon the tectum.

The more important known connections of the visual system are shown in the simplified diagrams (figs. 6, 11–16, 18, 20–24, 93, 101; '36, figs. 2, 6, 7; '42, figs. 4, 5, 79). Fibers of the optic nerve are distributed to five quite separate fields in the brain stem: (1) tectum, (2) pretectal nucleus, (3) cerebral peduncle, (4) thalamus, and (5) hypothalamus, as shown schematically in figure 14. This wide distribution is in marked contrast with the tendency of the other functional systems of peripheral nerves to converge into a single primary central field. This dispersal appears to have been determined by the types of motor response to be evoked rather than by the sensory qualities of the visual excitations. These five receptive fields will now be examined.

1. The tectum opticum evidently is the most extensive of these fields and physiologically the dominant member (see below).

2. The pretectal nucleus may participate in the regulation of the

intrinsic musculature of the eyeball and have other functions still unknown.

3. The basal optic tract is present in all classes of vertebrates from cyclostomes to man. It is evidently of major importance, though little attention has been given to it in current studies of the physiology of vision. Two possible functions have been suggested for it in Amblystoma. The more primitive of these is to act as a general activator of the ocular and somatic muscles in response to visual stimuli, manifested in such behavior as the "regarding reaction" described by Coghill (pp. 38, 78). In the second place, the structure and connections of the area ventrolateralis pedunculi, within which the basal optic fibers terminate, suggest that this neuropil may be part of the apparatus of conditioning of visual reflexes (p. 38).

4. Optic terminals are widely spread in the thalamus and are especially concentrated in its posterior part. This ill-defined area of neuropil receives the brachia of the superior and inferior colliculi and is regarded as the primordium of the lateral and medial geniculate bodies (p. 239; '42, p. 280). Large numbers of efferent fibers go out from this field to the peduncle and the dorsal and isthmic tegmentum. The fibers to the tegmentum take four courses. Two are superficial, the uncrossed tractus thalamo-tegmentalis rectus and a crossed path ('42, p. 224, *tr.th.teg.d.c.A.*) which decussates in the postoptic commissure. Two other tracts take deep courses close to the gray, one of which decussates ('42, p. 224, *tr.th.teg.d.c.B.*) and the other descends uncrossed in dorsal tegmental fascicles of group (7). For a detailed description see the analysis of the postoptic commissures in chapter xxi. These four tracts end chiefly in the dorsal and isthmic tegmentum, two of them in the superficial neuropil and two at deeper levels. They provide pathways for visual control of local reflexes, particularly those employed in capturing food. There is no separate optic projection tract from the thalamus to the cerebral hemisphere, though visual influence may reach the hemisphere through the thalamo-frontal tract, which is a common pathway of sensory projection from the dorsal thalamus (pp. 95, 238, and fig. 19, *tr.th.f.*).

5. There is a small number of optic terminals and collaterals in the hypothalamus near the chiasma ('42, pp. 219, 233, 237). They connect with the large cells of the preoptic nucleus, axons of which form the hypophysial tract (p. 245). This is doubtless the pathway for visual influence upon hypophysial endocrine activity. This connection is

probably smaller in Amblystoma than in some other vertebrates, and different opinions have been expressed about it. Eugen Frey's description ('38) of a large hypothalamic optic root in some amphibians was evidently based on inadequate material and faulty observation, as I have pointed out ('39b, pp. 558–76; '41a, p. 498; '42, pp. 218, 233). The same is probably true of Geiringer's ('38) description of a passage of fibers from the optic nerve directly into the hypophysial tract. I have preparations that give this appearance, but more critical examination disproves it. There is some evidence that in fishes and amphibians efferent fibers go out from the preoptic nucleus through the optic nerve to the retina ('33b, p. 254); but I have not been able to confirm this or the claim of some authors that efferent fibers go out to the retina from the tectum. Both these connections may exist.

In the chiasma the coarser optic fibers are segregated from the finer, and most of them take deeper courses in the optic tracts—the axial bundles. As the optic tracts traverse the thalamus, they separate into medial (dorsal) and lateral (ventral) tracts, each of which contains both thick and thin fibers. In the tectum most of these fibers end by wide arborizations in a common pool of intermediate neuropil (fig. 93, layer 2).

TECTUM OPTICUM

In embryological development the first afferent fibers to reach the tectum are those of the optic tract. These terminate in its anterior part near the posterior commissure, and tectal structure matures in subsequent stages from this region posteriorly. In the adult the common pool of neuropil, to which reference was made above, also receives similar terminals of the spinal and bulbar lemnisci, the brachium of the superior colliculus from the thalamus, the strio-tectal and habenulo-tectal tracts, and some other fibers. This pool of neuropil is spread throughout the entire tectum, with little evidence of localization of function within it.

The optic tectum is structurally nearly homogeneous, with two exceptions: (1) a dorsal thickening on each side, which is related with the dorsal tectal commissure and probably corresponds (in some respects only) with the torus longitudinalis of fishes ('42, pp. 250, 287), and (2) the eminence of the posterior commissure at the anterior border of the tectum, likewise chiefly commissural in function. The stratification of the tectum, which is so conspicuous in the frog and many other vertebrates, is here notably absent. The only layers that are clearly evident are the deep gray and the superficial white, though there are obscure indications of the incipience of further stratification, more evident in the alba than in the grisea. For descriptive purposes I have divided the optic tectum, somewhat arbitrarily, into eight concentric layers, as shown in figures 36 and 93; for their characteristics see page 244 of 1942.

Two features of tectal structure merit special emphasis. The first is that the convergence of most of the afferent fibers into a common pool of neuropil of almost homogeneous structure suggests a totalizing or integrative function for the tectum as a whole. The second point of major interest is that the efferent fibers from the tectum are extremely diversified in structure and distribution. There are all gradations from diffusely spread fibers in the deep neuropil to well-defined long myelinated tracts like the tecto-bulbar and tecto-spinal systems. This suggests that such localization of function as exists in this tectum is determined more by what is going on in the efferent side of the arc than in the afferent side, a point which has been emphasized by Crosby and Woodburne ('38).

Five strong systems of myelinated fibers leave the tectum. These take widely divergent courses. They are as follows:

1. *Commissura posterior.*—These are the first to appear in the embryo. The crossed fibers are joined by uncrossed fibers from the eminence underlying the commissure, and they spread widely in the peduncle and adjoining parts of the thalamus and dorsal tegmentum. The primary connection is with the nucleus of Darkschewitsch (p. 217 and figs. 6, 18, 22) and thence to the f. longitudinalis medialis, providing innervation of the trunk musculature in response to visual stimulation.

2. *Tractus tecto-peduncularis rectus et cruciatus* (p. 303 and figs. 18, 22, *tr.t.p.c.*; '42, p. 267).—These thick fibers arise from the anterior part of the tectum, descend parallel with fibers from the posterior commissure, decussate (in part) in the commissure of the tuberculum posterius, and arborize in the alba of the peduncle of both sides more ventrally than those of the posterior commissure. Their course as seen in horizontal sections is shown in figures 30–36, *tr.t.p.c.1.* (see also '42, p. 267, and the figures there cited). They activate the interstitial nucleus of the f. longitudinalis medialis and probably also the oculomotor nucleus and appear to be adapted to regulate mass movements of trunk, limbs, and eyes in the orientation and movements of the body with reference to objects in the visual field. These

fibers are joined by similar thick fibers from the pretectal nucleus and dorsal thalamus (*tr.th.p.c.* of the figures).

Superficially of the connection just described, there is a large collection of uncrossed fibers passing between the pretectal nucleus (pars intercalaris diencephali) and the peduncle. These were described in Necturus ('17, p. 264; '33*b*, p. 110 and fig. 57) under the name "tr. thalamo-peduncularis dorsalis superficialis." These fibers are spread close to the pial surface in the wide di-mesencephalic fissure and end in large numbers in the neuropil of the area ventrolateralis pedunculi. Some of them also reach the hypothalamus. The condition in Amblystoma is similar, thus bringing the pretectal nucleus into intimate relation with terminals of the basal optic tract and the oculomotor nucleus.

3. *Tractus tecto-thalamicus et hypothalamicus cruciatus* (fig. 12, *tr.t.th.h.c.a.* and *p.*; see also the section on com. postoptica in chapter xxi).—The anterior and posterior divisions of this complex were described in 1942 (p. 221). The anterior tract passes from the dorsal part of the superior colliculus to the hypothalamus, decussating in the postoptic commissure. It is here termed "tr. tecto-hypothalamicus anterior" (figs. 25–36, *tr.t.hy.a.*; see further on p. 296). It has connections with the thalamus and hypothalamus of both sides, and some of its crossed fibers may reach the peduncle and tegmentum.

The posterior division arises chiefly from the inferior colliculus, with accessions from the ventral border of the superior colliculus. It lies parallel with the anterior division and, more posteriorly, partially decussates in the postoptic commissure and has terminals in the thalamus and hypothalamus of both sides (figs. 25–35, *tr.t.th.h.c.p.*). Its thicker and more myelinated fibers continue as tr. tecto-tegmentalis cruciatus into tegmental fascicles of groups (6) and (8), to end in the peduncle and tegmentum as far back as the VII nerve roots. The distribution of these fibers suggests that they are primarily concerned with mass movements and local reflexes of the musculature of the head and probably also with conditioning of these movements. What seems to be the equivalent of this tract is described as connected with the nucleus isthmi in all animals in which that nucleus is well developed; but in urodeles this tract is very large, though the nucleus isthmi is vestigial.

4. *Tecto-bulbar and tecto-spinal tracts.*—These tracts have been described and illustrated in the papers of 1936 (p. 340) and 1942

(p. 268). They are the largest of the efferent tracts, and most of their fibers are well myelinated. There are two crossed tracts (figs. 12, 27–36, *tr.t.b.c.1.* and *tr.t.b.c.2.*) and an uncrossed tract (*tr.t.b.c.r.*) from the tectum opticum. The posterior crossed tract (*tr.t.b.c.2.*) decussates transversely in the vicinity of the nucleus of the IV nerve. The anterior crossed tract (*tr.t.b.c.1.*) takes a peculiar course, its fibers entering the ventral medial tegmental fascicles, *f.v.t.(1)*, within which they decussate obliquely. There are about a dozen of these anastomosing fascicles extending from the commissure of the tuberculum posterius backward to the level of the IV nucleus, where these crossed fibers join those of *tr.t.b.c.2.* and descend to the medulla oblongata and spinal cord (figs. 6, 12).

The uncrossed tecto-bulbar fibers—tr. tecto-bulbaris rectus—were seen in my earlier studies to arise only from the posterior part of the tectum, as drawn here in figure 12 (*tr.t.b.r.*); but later ('42, p. 269) other fibers of this system were found to arise more anteriorly and to pass backward by several pathways to join the posterior group. The most anterior of these fibers enter dorsal tegmental fascicles of group (7), from which they separate to join the posterior tract. Others pass spinalward in layers 3 and 4 of the tectum, at the posterior end of which they join the other fascicles. Here they send collateral branches into the isthmic visceral-gustatory nucleus, as shown in figure 23. Below the junction of these three divisions, the tr. tecto-bulbaris rectus descends superficially in the isthmus, then turns spinalward, under the lateral recess of the ventricle, passing through the ventral border of the auricle into the medulla oblongata, where it joins the crossed tecto-bulbar tract.

5. *Brachia of superior and inferior colliculi.*—These are large collections of fibers passing in both directions between the tectum and the thalamus. They are widely dispersed, and most of them discharge from tectum to dorsal thalamus. Those from the superior colliculus are spread throughout the alba and form a massive tract (figs. 34, 35, 36, *tr.t.th.r.*). These are joined by some fibers from the inferior colliculus, and from the latter area other strands of unmyelinated fibers take separate courses, some passing forward in the dorsal part of the tectum and some at its ventral margin; most of these fibers pass from inferior to superior colliculus, but some of them continue into the thalamus, particularly ventrally, to enter the geniculate neuropil (figs. 14, 15, 16, *br.col.*; '42, p. 264).

CONCLUSION

The internal structure of the optic tectum of Amblystoma and the wide spread of its fibrous connections, both afferent and efferent, suggest that in this animal the primary function of the tectal system is visual control over movements of the body as a whole and, in particular, the orientation of the body and conjugate movements of the eyeballs with reference to objects in the visual field. Such other local visual reflexes as the animal possesses are probably organized elsewhere.

TECTO-OCULOMOTOR CONNECTIONS

The visual location of food and enemies involves the innervation of the muscles of the eyeballs, and our knowledge of the apparatus employed in the conjugate movements of the eyes is scanty. The anterior and posterior divisions of the crossed tecto-bulbar tract decussate, respectively, in the vicinity of the nuclei of the III and IV nerves (fig. 12). The anterior division (*tr.t.b.c.1.*) crosses in the ventral medial tegmental fascicles which span the rather long distance between the III and the IV nuclei. When these ventral fascicles—group (1) of my analysis in 1936—were first identified as tecto-bulbar fibers, it was supposed that this anomalous arrangement was a provision for activation of the eye-muscle nuclei by collaterals from these fascicles. No evidence of this has been found, but the myelinated fibers of both tr. tecto-bulbaris cruciatus 1 and 2 are accompanied by very many shorter, unmyelinated fibers, which pass from the tectum to the III and IV nuclei of both sides. This tr. tecto-peduncularis is shown diagrammatically in figures 18, 22, and 24 (and in more detail in '42, figs. 14, *tr.t.p.2.*, and 45, *tr.t.p.c.2.*). In addition to the tecto-peduncular fibers in the alba, there is a deep series (tr. tecto-peduncularis profundus, '42, p. 267 and fig. 14, *tr.t.p.3.*), which takes tortuous courses in the neuropil of the grisea. From the anterior part of the tectum and pretectal nucleus the well-defined tr. tecto-peduncularis cruciatus and the uncrossed tr. thalamo-peduncularis dorsalis superficialis may participate in the regulation of conjugate movements of the eyes.

If pupillary constriction is controlled from the pretectal nucleus, as in mammals, there are two available pathways to the III nucleus, first, by way of the pretectal components of the two tracts last mentioned and, second, by tr. pretecto-hypothalamicus (p. 296 and fig. 15, *tr.pt.hy.*) and thence to the III nucleus by tr. hypothalamo-peduncularis (figs. 18, 23, *tr.hy.ped.*).

VISUAL FUNCTIONS

Most of our experimental knowledge of visual functions of amphibians has been derived from Amblystoma, Triturus, and frogs. In these animals there is no visual cortex, and in urodeles there is no specific optic projection tract to the cerebral hemisphere. Here the intrinsic features of the stem portion of the visual apparatus can be studied, simplified by absence of the cortical connections; and yet the complications of internal structure as just summarized present baffling physiological problems. Triturus has more efficient eyes than Amblystoma, and in frogs this differentiation is much further advanced, so that the Anura offer the most promising approach for experimental analysis, and for this we need more information about the histological structure of the anuran brain than is now available.

The electrical activity of the optic tectum of the frog has been investigated by Beritoff ('43, pp. 215–40) in co-operation with Tzkipuridze, who present oscillographic records of its spontaneous activity and of the effects of various kinds of stimuli. A distinction is drawn between the quick oscillations due to excitation of the ganglion cells of the retina and slow oscillations ascribed to activity of the neuropil. Similar studies with the aid of microelectrodes may profitably be made of the properties of the layers of the tectum and other central stations of visual activity.

Attention has been called ('41, p. 521) to evidence that the retina of Triturus is more highly differentiated histologically than that of Amblystoma. That it is functionally superior is confirmed by the experiments of Stone and Ellison ('45), who exchanged eyes of adult A. punctatum and T. viridescens. The Triturus grafts degenerated, but the Amblystoma eyes on Triturus hosts followed a course of recovery as they do in homoplastic transplantations, retinal degeneration being followed by complete regeneration. Visual acuity appeared to be higher than in normal Amblystoma but lower than in normal Triturus.

In the feeding reactions of larval Amblystoma, the location and seizure of prey can be successfully done, in the absence of eyes, with the aid of the lateral-line organs (Scharrer, '32; Detwiler, '45). The lemniscus systems may discharge lateral-line impulses into the tectum, in company with those of other sensory systems; the mesencephalic V system may participate in the reaction; and the habenulo-tectal tract may transmit an olfactory influence. In addition to these, there is a large thalamo-tectal connection and a small strio-tectal tract. But there is no experimental evidence about the specific functions of any of these connections.

The tectum and thalamus are connected by fibers passing in both directions (figs. 11, 12). These are precursors of the mammalian brachia of the superior and inferior colliculus. It is evident that the tectum and the dorsal thalamus may act as an integrated unit in regulating the major activities of the body in adjustment to external situations. With the emergence of the cerebral cortex in higher animals, this pattern is radically changed, and for the understanding of these changes it is essential that the pre-existing organization be adequately known. The history of the transformation of this generalized amphibian structure to those of reptiles and mammals has been sketched by Huber and Crosby ('43).

Evidence has been published that in some vertebrates the several quadrants of the retina have local representation in the optic tectum. In a series of vertebrates, including several species of amphibians, Ströer ('39, '39a, '40) described continuous fascicles of fibers from the retinal quadrants to the tectum, where they end locally in an arrangement which projects the retinal quadrants upon the tectum in reversed orientation. I have made diligent search for evidence of such an arrangement in Necturus and in young and adult Amblystoma without success ('41, '41a, '42). In Amblystoma thick and thin fibers from the different retinal quadrants are inextricably intertwined in the optic nerve. At the chiasma the thick fibers are segregated from the thin fibers, and in the tectum the terminals of both kinds of fibers mingle and seem to be distributed nearly uniformly to all parts. There is physiological evidence that Amblystoma can visually localize objects in the field of view; and, by analogy with other animals, the presumption is strong that retinal loci are projected upon the tectum in more or less definitely circumscribed areas.

In Triturus and Amblystoma good localization of objects in the visual field is restored after transplantation of the eyes (Stone, '44; Stone and Zaur, '40; Stone and Ellison, '45), and it is preserved after rotation of the eyeball through 180°, though in the latter case motor reactions to objects in the visual field are reversed from the normal (Sperry, '43). Similar physiological results follow operations which eliminate the chiasma and connect the retina with the tectum of the same side, and also after transplanting an eye to the opposite side of the head (Sperry, '45). These observations have been repeated upon seven species of anurans, all of which have much more highly differentiated visual apparatus, with similar results. After the cutting of the optic nerve and its regeneration in the adult animal, the same results were obtained. Vision was normal in those animals whose retinas had been kept in normal position, but it was reversed about the optic axis in animals whose retinas had been rotated through 180° prior to nerve section (Sperry, '44, '45). In frogs, destruction of quadrants of the tectum of both normal and operated animals resulted in scotomas of local quadrants of the visual field in a pattern which is in agreement with the anatomical observations of Ströer ('39a); and, in cases in which the eyeball was rotated, the scotomas were in arrangement reversed from that found when the eyes were in the normal position.

These experiments increase the probability that, in Amblystoma, retinal loci are in some way projected locally upon the tectum; but the mechanism employed in localization of visual functions can be clarified only by further experimentation. Sperry's experiments demonstrate "the high degree to which the complex and precisely patterned neural mechanisms subserving adaptive visuomotor coordination are dependent upon inherently predetermined rather than upon functionally acquired neural adjustments."

CHAPTER XVII

DIENCEPHALON

GENERAL FEATURES

IN THE following analysis the structures at the di-telencephalic junction are omitted, since they are fully described in the next chapter. The retina and the optic nerve are integral parts of the diencephalon. The retina itself is more than a simple receptor, for it contains elaborate apparatus for analysis and synthesis of visual excitations (Polyak, '41).

The diencephalon combines characteristics of the sensory, intermediate, and motor zones, and the tissues involved are so closely interwoven that topographic subdivisions into zones is arbitrary and misleading. The largest number of afferent fibers enter by the optic nerve and are distributed to the thalamus, pretectal nucleus, and hypothalamus, in addition to their mesencephalic connections. The nervus terminalis has widely spread terminals in the preoptic nucleus and the hypothalamus. The parietal nerve contains a few fibers of uncertain connections and functions. The hypophysial nerve carries many fibers to the pars nervosa of the hypophysis and a smaller number to the pars distalis.

Between the central stations of these peripheral nerves and interpenetrating them is correlating tissue of intermediate-zone type. As repeatedly emphasized in several preceding contexts, the diencephalon as a whole (except for the retina) is a transitional sector of the brain, interpolated between the olfactory field anteriorly and all other sensori-motor fields posteriorly. This apparatus of correlation is supplementary to the more direct paths of through traffic which organize the basic patterns of the stable action system characteristic of the species. In urodeles the most direct descending olfactory connections are by way of the tractus olfacto-peduncularis for somatic responses and to the hypothalamus for visceral responses. Additional diencephalic olfactory connections are ancillary to these. The apparatus of other stereotyped patterns of sensori-motor adjustment is similarly organized in the stem below the diencephalic level. These systems all have a collateral discharge into the thalamus, which is probably concerned with conditioning and other types of "inflection" of the stable components of behavior. In the most primitive vertebrates this primordial thalamic apparatus of higher synthesis and conditioning is small and at a low grade of differentiation. In Amblystoma it is notably larger, but still relatively unspecialized and probably concerned more with regulation of central excitatory state and the general physiological disposition and diathesis than with refined analysis. In mammals this tissue is elaborately specialized, with provision for precise localization of function correlated with the simultaneous differentiation of the neopallial cortex. But even in man the intrinsic functions of the thalamus are not supplanted by cortical development. The thalamo-cortical connections carry two-way traffic, and all cortical activity may be profoundly affected by diencephalic influence.

The primary subdivisions of the adult diencephalon are more clearly evident in the gross preparation of the urodele brain than in most other vertebrates, as illustrated in figures 1B and 2. I emphasized this in the paper of 1910, and the analysis there proposed has been widely accepted as applicable in all vertebrates. The epithalamus includes the habenular nuclei, the pars intercalaris, the epiphysis, and the membranous dorsal sac and paraphysis. The dorsal thalamus is an undifferentiated nucleus sensitivus, the ventral thalamus a field of motor co-ordination, and the hypothalamus an olfacto-visceral adjustor, including the pars nervosa of the hypophysis. These subdivisions and their more important connections have been described in previous publications (see the summary, '42, p. 204; also, '10, '21a, '27, '35a, '36, '39, '39b, '41, '42; Necturus, '17, '33b, '34b, '34c, '41a; for other urodeles see the works cited on p. 11).

DEVELOPMENT

The larger part of the wall of the diencephalic sector of the early neural tube is evaginated to form the retina and the optic nerve. The details of this development, which begins at the anterior end of the medullary plate, have been recorded by Adelmann ('29, '36). I have described later stages of the development of these structures ('41). During these stages most of the dorsal median wall of this part of the neural tube remains membranous and is elaborately enlarged and folded to form the epiphysial vesicle, paraphysis, and diencephalic chorioid plexuses, the development of which has recently been described in several species of amphibians by Rudebeck ('45). The adult structure of these epithelial organs and the related meninges and blood vessels I described in 1935. The ventral medial wall of the adult (fig. 2) between the anterior commissure ridge and the tuberculum posterius is thin but nervous, except for the dorsal wall of the infundibulum. It has a massive thickening—the chiasma ridge—midway of its length. The arrangement of the chiasma ridge, the anterior commissure ridge, and the adjoining preoptic recess and precommissural recess is peculiar in the amphibian brain (fig. 2 and p. 291).

The massive side wall of the adult diencephalon is composed of four columns of gray substance, which project into the third ventricle as longitudinal ridges separated by deep sulci. As shown here in figures 1 and 2, these are epithalamus, dorsal thalamus, ventral thalamus, and hypothalamus. This adult configuration is achieved by a series of local pulses of proliferation and differentiation of cells, which shift in location and rate of growth from stage to stage, as indicated by changes in the sculpturing of the ventricular surface.

Rudebeck ('45) illustrates embryonic stages of Necturus, Triturus, and the anuran Pelobates which are younger than any of my models of Amblystoma. His pictures show that, in each of these species in stages preceding hatching of the eggs, the first ventricular markings to appear in the rostral part of the neural tube take the form of a series of transverse sulci. The most anterior of these, his "sulcus subpallialis," separates the subpallial from the pallial parts of the hemisphere. Next follows von Kupffer's sulcus intraencephalicus anterior (p. 117), which separates the hemisphere and preoptic nucleus from the diencephalon. Posteriorly of this sulcus is a high transverse ridge, the di-telencephalic ridge, the adult derivatives of which apparently are the habenula and those structures ventrally of it which are listed in our next chapter as bed-nuclei of the di-telencephalic junction—nucleus of Bellonci, eminentia thalami, nucleus of tr. olfactohabenularis (p. 239), and part of the preoptic nucleus and hypothalamus. This ridge is bounded spinalward by his "sulcus diencephalicus ventralis," the dorsal part of which persists as the sulcus posthabenularis and the ventral part as the sulcus ventralis thalami plus the sulcus hypothalamicus. Posteriorly of the di-telencephalic ridge is an undeveloped area, giving rise to the adult pars intercalaris diencephali and the major parts of the dorsal thalamus, ventral thalamus, and hypothalamus.

After swimming is well established (about Harrison's stage 37), differentiation is much further advanced, though there is no radical change in the topographic arrangement of the areas involved ('38, figs. 15–18). But, by the time the larvae begin to feed (Harrison's stage 46), the parts have shifted their positions toward the adult condition ('38b, figs. 1, 2), which is fully attained in midlarval stages (38 mm. in length) of Amblystoma tigrinum ('39a, fig. 1).

These changes are brought about in two ways. There is, in the first place, a gradual straightening of the two great flexures of the neural tube, with a resulting rearrangement of the several areas in relation to the long axis of the brain and to one another. In the second place and accompanying these changes in the general shape of the forebrain, the regions of most rapid proliferation and differentiation of cells shift their positions. There results a change in the pattern of the ventricular eminences and the intervening sulci. The original transverse ridges and sulci are broken up, and their parts are rearranged in the longitudinal series which we see in the adult brain.

The details of these changes have been described during recent years by a number of workers in Professor Holmgren's Zootomical Institute at Stockholm. Rudebeck ('45), in his admirable paper on the development of the forebrain of lungfishes, shows that the sequence of these changes is essentially the same in Dipnoi, Urodela, and Anura. In the youngest stages, proliferation of cells is most active along the lines of the primitive sulci. These active zones enlarge, and this results in obliteration of the primary sulci or a shift in their positions. The primary grooves, accordingly, do not mark the boundaries of the primordia of the different diencephalic centers, but, on the contrary, they are zones of more active growth from which these centers are developed. In some regions the original proliferation grooves disappear, in others they are gradually displaced and so become limiting sulci at the boundaries of the differentiated areas of the adult.

My interpretation of these changes as outlined above differs somewhat from that of Rudebek as summarized on page 63 of his monograph. Certainly, the details of these developmental processes vary from species to species, and examination of still more species of vertebrates along the lines of the critical studies made in the Stockholm Institute will doubtless clarify the questions which are still in controversy. One is impressed by the close resemblance of this development in lungfishes and amphibians, which suggests that these groups of animals are more intimately related phylogenetically than has hitherto been generally admitted.

EPITHALAMUS

The epithalamus, like all other parts of the sensory zone, combines reception from the periphery with functions of correlation. It has two radically different parts that are separated by the pineal recess and the overlying pineal vesicle. The habenular nuclei comprise the anterior division. The posterior division is the pars intercalaris diencephali, which is relatively larger in urodeles than in higher animals. Because of this enlargement and the small size of the dorsal thalamus, the habenula lies farther forward than usual. The two habenulae are connected by the habenular or superior commissure and the pars intercalaris by the commissura tecti diencephali.

The pars intercalaris is sharply defined in front, but posteroventrally it merges with the posterior sector of the dorsal thalamus. There is a narrow subependymal layer of small nerve cells imbedded in dense periventricular neuropil, and externally of this lies a lentiform mass of crowded cells, mostly small, with some of larger size, which I term the "pretectal nucleus" (figs. 2B, 14, 15, 35, 36, *nuc.pt.*) without commitment about its homologies in other animals. It is probable that the pulvinar of mammals has been derived from the undifferentiated gray of the posterior part of this field.

As described in 1942 (pp. 205, 259, 279), the pretectal nucleus is permeated and covered with dense neuropil, which is continuous with that of all surrounding parts—tectum, habenula, and dorsal thalamus—with fibers passing in both directions. It receives terminals and collaterals from all chiasmatic bundles of the optic tracts (figs. 14, 22), tectum (figs. 12, 14, *tr.t.pt.*), posterior commissure, dorsal thalamus, habenula, and tr. strio-tectalis (figs. 14, 101, *tr.st.pt.*). Efferent fibers leave this nucleus for the tectum (figs. 11, 15, *tr.pt.tec.*), thalamus (fig. 15), hypothalamus (figs. 15, 16, 25–36, *tr.pt.hy.*), and peduncle (tr. thalamo-peduncularis cruciatus and tr. thalamo-peduncularis dorsalis superficialis). Some fibers accompany the fasciculus retroflexus to reach the area ventrolateralis pedunculi (fig. 23).

In my previous publications the tr. pretecto-hypothalamicus was not identified. Here its course is shown in figures 25–36. Many unmyelinated fibers, accompanied by a few with myelin sheaths, leave the pretectal nucleus (figs. 35, 36) and descend near the surface rostrally and internally of the marginal optic tract. Many of these fibers end in the thalamus (tr. pretecto-thalamicus, *tr.pt.th.*), and the remainder descend to decussate in the postoptic commissure and spread in the hypothalamus. These fibers and those here termed tr.

tecto-hypothalamicus anterior (*tr.t.hy.a.* of the figures) comprise the mixed system which in my previous papers was called "tr. tecto-thalamicus et hypothalamicus cruciatus anterior" (p. 296). The thickest myelinated fibers of this complex arise in the tectum and are clearly followed through the postoptic commissure into the hypothalamus.

The dominant afferent connections of the pretectal nucleus apparently are optic. The efferent connections to the hypothalamus and peduncle provide pathways for optic control of the intrinsic muscles of the eyeball, as in mammals, and also for possible conditioning or other influence upon the primary motor paths of the brain stem.

The pineal body appears in the premotile embryo as a low evagination from the roof of the diencephalon ('38, fig. 2). In early swimmers it is a hollow vesicle with a lumen, communicating with the third ventricle by a patent recessus pinealis ('38, figs. 3, 4, 18). Within three days after the early swimming stage this communication is closed ('38a, p. 18). The adult epiphysis is a small, flattened, simply lobulated epithelial vesicle, entirely detached from the brain except for a few fibers of the parietal nerve. No nervous elements other than these fibers have been found in it, and their connections are still unknown.

In Necturus there are more of these fibers, and their courses are more easily followed ('17, p. 236). These myelinated fibers are spread between the cells of the epithelial wall of the pineal vesicle, from which they pass into the brain in several small fascicles associated with the habenular commissure and posteriorly of it. They do not cross the mid-plane and they appear to have no functional connection with the epithalamus but to pass through it and join the f. retroflexus. They can be followed ventralward nearly to the cerebral peduncle, where they mingle with other myelinated fibers. The parietal nerve of Amblystoma has a smaller number of myelinated fibers than in Necturus (not more than 10), but their arrangement is similar ('42, p. 205). These certainly are not aberrant fibers of either the habenular commissure or the com. tecti, but their central connections and functions are uncertain. They resemble those associated with the IV nerve, which are distributed peripherally to the meninges and chorioid plexus (p. 181). It is not improbable that all these fibers have sensory functions, but for this there is no evidence. All chorioid plexuses are abundantly supplied with unmyelinated nerve fibers, probably of vasomotor function, but in the search for the sources of these fibers in Amblystoma I have been no more successful than I was with Necturus ('33b, p. 15).

DORSAL THALAMUS

The dorsal thalamus is a well-defined part of the diencephalon and is intimately related by a web of connecting fibers with all contiguous parts and with many remote parts of the brain. It evidently is an important center of sensory correlation, but it is also the primary terminal station of a considerable fraction of the fibers of the optic nerve which come directly from the retina. The thalamic connection of optic fibers is not a late acquisition incidental to the differentiation of cerebral cortex, for it is present in all vertebrates, even the most primitive types. In corticated animals the larger part of the thalamus is clearly a cortical dependency, the neothalamus; but there is also a paleothalamus, which is a primordial part of the vertebrate brain, being present in all vertebrates, from the lowest to the highest. This paleothalamic component is not lost in mammals, as is shown by the surviving thalamic residue, which does not degenerate after total decortication.

The amphibian dorsal thalamus shows only vague outlines of an incipient subdivision into the local nuclei so characteristic of mammals. This is more evident in anurans than in urodeles, and in Amblystoma than in Necturus. In the urodeles, as in fishes, the undifferentiated dorsal thalamus may be regarded as a single "nucleus sensitivus," which acts, in the main, as a whole, without well-defined functional localization. Nevertheless, an early stage of local differentiation can be recognized. In Amblystoma there are three areas of gray, separated by shallow and variable ventricular sulci, namely:

1. Anteriorly the nucleus of Bellonci (fig. 2, *nuc.B*) produces a slight ventricular eminence wedged between the middle area, the ventral habenular nucleus, the eminentia thalami, and the ventral thalamus. Superficially of this gray is an area of very dense and complicated neuropil, which is intimately connected with that of all the surrounding areas. It receives terminals of the optic tract, and this justifies its inclusion within the sensory zone, though its other connections are those of correlating tissue of intermediate-zone type. The nucleus of Bellonci is primarily a bed-nucleus of the neighboring tracts—stria medullaris, optic and postoptic systems, and others (chap. xviii).

2. The middle part of the dorsal thalamus is marked by a low ven-

tricular eminence inclosed by the dorsal and middle thalamic sulci (fig. 2, *s.d.* and *s.m.*). It is the precursor of several of the sensory nuclei of the mammalian thalamus, though there is no visible local differentiation of its gray substance corresponding to these nuclei. This homology is clearly established by the location of these cells and their fibrous connections, which conform, as far as they are present, with the mammalian pattern (figs. 14, 15). This area receives terminals and collaterals of the optic tract, terminals of the spinal and bulbar lemniscus, strong tecto-thalamic tracts (precursors of the brachia of the superior and inferior colliculi), some fibers from the habenula, and probably also fibers from the hypothalamus by way of the postoptic commissure systems and from the cerebral hemisphere accompanying the tr. strio-pretectalis. These fibers all enter a common pool of neuropil, which is most dense in the intermediate alba but which also pervades the entire structure.

From this thalamic pool, efferent fibers go out to all surrounding parts of the brain, some in the diffuse neuropil and some as well-fasciculated tracts. An extensive and complicated system of efferents discharges into the underlying motor zone of the same and the opposite side, including direct and crossed tracts to the ventral thalamus, hypothalamus, cerebral peduncle, and isthmic and bulbar tegmentum. The more important of these are shown in figure 15. A large tract containing some thick myelinated fibers (*tr.th.p.*) joins similar fibers from the pretectal nucleus and tectum (fig. 18, *tr.t.p.c.*), partially decussates in the commissure of the tuberculum posterius, and spreads through the alba of the peduncle on both sides, as illustrated in figures 30–36 (*tr.th.p.c.* and *tr.t.p.c.1.*). Parallel with these deeper fibers there is a very large superficial uncrossed tract from the dorsal thalamus and pretectal nucleus to the peduncle (not shown on the figures) similar to that described in Necturus as tr. thalamo-peduncularis dorsalis superficialis ('17, p. 264). Other similar fibers are dispersed in the neuropil, some of which go directly to the ventral part of the thalamus—tr. dorsoventralis thalami (shown but not labeled in figs. 15 and 16; see '17, p. 266). A small number of fibers go to the habenula (*tr.th.hab.*), and a larger number to the tectum in the brachia of the superior and inferior colliculus (*br.col.*). The large tr. thalamo-tegmentalis rectus includes deep and superficial fibers to the dorsal and isthmic tegmentum (figs. 32–34, 94, *tr.th.teg.r.*); for those which decussate in the postoptic commissure see chapter xxi and '42, p. 223.

The fibers which ascend to the hemisphere are in small, compact fascicles (tr. thalamo-frontalis; see figs. 15, 19, 30–34, 71, 72, 75, 95, 101, 102, 103, *tr.th.f.*), which enter the lateral forebrain bundle. Within this bundle the fibers pass to the strio-amygdaloid field in the ventrolateral wall of the hemisphere. Here they spread out and, so far as is now known, they end here, thus constituting a thalamo-striatal system of projection fibers, which persists in mammals with the addition of the thalamo-cortical projection system. The thalamo-frontal fibers arise as axons of cells of the middle part of the dorsal thalamus only, so that this "nucleus sensitivus thalami" may be regarded as the primordium of most of the mammalian thalamic nuclei which have cortical connections. This is not a new connection in the Amphibia, for it is present in various groups of fishes ('22a).

3. The posterior part of the dorsal thalamus is less differentiated and less clearly delimited. Its tissue is confluent with that of the pars intercalaris of the epithalamus, the eminence of the posterior commissure, the dorsal tegmentum, the peduncle, and the ventral thalamus. This suggests that it is functionally related with all these parts. All the connections of the middle part are more or less evident, except the thalamo-frontal tract. What structures of higher brains have been derived from this area is not clear. Some of this tissue should probably be assigned to the ventral thalamus.

This posterior sector develops precociously in company with the ventral thalamus (Coghill, '28, Paper VIII). Uncrossed impregnated fibers pass from it to the ventral thalamus and peduncle in the S-reaction stage—Harrison's stages 35, 36 ('37, fig. 2). In early swimmers these are more numerous, and similar fibers, which decussate in the postoptic commissure, make their appearance ('38, fig. 5; Coghill, '30, Paper IX, fig. 4). Differentiation of the anterior parts of the dorsal thalamus is relatively retarded.

The dense and intricate neuropil of the posterior part of the dorsal thalamus spreads to surrounding parts with no well-defined boundaries, being intimately joined with that of the middle part and the ventral thalamus (fig. 73). This field (figs. 14, 16, *np.gen.*) receives two systems of fibers which seem to have special phylogenetic significance, viz., abundant terminals of the optic tracts and the brachium of the inferior colliculus. The arrangement of these connections in Amblystoma is different from that seen in both Necturus and the frog ('42, p. 278). Optic terminals are spread through the alba of the entire dorsal thalamus and most abundantly in this posterior neuropil. I have called this common pool of undifferentiated tissue of the dorsal thalamus and dorsal part of the ventral thalamus the "geniculate neuropil," on the assumption that out of it the medial and lateral geniculate bodies have emerged (p. 221). In the course of phylogeny the more sharp segregation of the optic terminals was accompanied by the differentiation of the lateral geniculate body, a process which is well advanced in the frog ('25), and the differentiation of the cochlear apparatus and lateral lemniscus was accompanied by the emergence of the medial geniculate body from the same common pool. This hypothesis is held subject to revision, pending further study of the related species.

VENTRAL THALAMUS

On the ventricular surface (figs. 1, 2, 13, 95) the ventral thalamus is sharply delimited by the sulcus medius thalami (*s.m.*) above and the sulcus ventralis (*s.v.*) below, and it is separated into anterior and posterior parts by a depressed area containing in some specimens a shallow sulcus. The two parts differ in embryological origin and connections, yet in the adult brain their most fundamental features are similar ('42, p. 207).

The posterior part of this field extends forward from the peduncle, and in early functional stages it has similar structure and connections, though it is clearly in diencephalic territory. In these stages the primordium of the anterior part (area 7a of my analysis, '37, p. 393) lies at the di-telencephalic junction ('38, p. 213 and fig. 18), and in prefunctional stages it is joined with area 7, which becomes the corpus striatum. This anterior part of the ventral thalamus may be genetically telencephalic, depending on how the arbitrary di-telencephalic boundary is defined.

In the adult ventral thalamus none of the nuclei of more highly differentiated brains are well defined, though some local differentiation is evident. Anteriorly, the eminentia thalami (figs. 2, 16, *em.th.*) belongs in a series of bed-nuclei related with important tracts of fibers at the di-telencephalic junction (chap. xviii). Both this area and the underlying nucleus of the olfacto-habenular tracts discharge fibers backward into the unspecialized gray of the anterior part of the ventral thalamus (fig. 17). Between the anterior part of the ventral thalamus and the dorsal (mamillary) part of the hypothalamus there are fibers passing in both directions which seem to be precursors of the mammalian tr. mamillo-thalamicus ('39b, p. 554 and figs. 22, 35).

The posterior part has a thicker gray layer, which produces a rather high ventricular eminence, the structure of which is similar to that of the peduncle. As illustrated by previously published figures ('39b, p. 544 and figs. 23, 28), the large neurons of the ventral thalamus have widely spread dendrites and are evidently collectors of impulses from many sources (figs. 16, 17). Spinalward of the bednuclei at the anterior end, the two parts receive the strong tr. striothalamicus, which descends from the strio-amygdaloid field by way of the lateral forebrain bundle (*f.lat.t.*). Short fibers descend into both parts from the dorsal thalamus (tr. dorsoventralis thalami) and from the pretectal nucleus by tr. pretecto-thalamicus et hypothalamicus. There are much larger connections from the tectum, some uncrossed in the brachia of the superior and inferior colliculi (*br.col.*) and some decussating in the postoptic commissure (*tr.t.th.h.c.p.*).

Fibers stream backward from the ventral thalamus into the motor zone of all lower parts of the brain stem and into the hypothalamus, some uncrossed and some decussating in the postoptic commissure and in the ventral commissure of the isthmus. These have been seen especially clearly in larval brains ('39, fig. 2; '39b, p. 546 and fig. 23). Most of these fibers end in the cerebral peduncle and tegmentum of the midbrain and isthmus; many of them extend into the trigemino-facialis region; and relatively few which enter the f. longitudinalis medialis may go into the spinal cord.

The connections just described indicate that the ventral thalamus collects fibers from all parts of the cerebrum, and chiefly those concerned with somatic responses to exteroceptive stimuli. Its efferent fibers descend to those motor fields which supply the skeletal musculature. These fibers take a surprising variety of courses, the details of which need not be recounted here ('39b, p. 544), but all the more important pathways are links in the chain of conductors which activate the skeletal muscles. For this reason this field is regarded as part of the motor zone, though it has no direct connections with the periphery.

This ventral thalamus corresponds in all essential respects with the subthalamus of mammalian neurology. In contrast with the mammals, it is here far larger than the dorsal thalamus, as is evident upon inspection of figure 2. This difference is probably due not to shrinkage of the ventral thalamus in mammals but to the great enlargement of the dorsal thalamus, i.e., to the addition of the neothalamus to the primordial paleothalamus as the latter is seen in Amblystoma.

HYPOTHALAMUS

The hypothalamus as here defined includes the ventral part of the brain stem between the anterior commissure ridge and the tuberculum posterius. At its anterodorsal border the nucleus of the olfacto-habenular tract might be included or assigned to the ventral thalamus. Its fibrous connections, like those of the preoptic nucleus, are mainly of hypothalamic type. Posteriorly of the chiasma ridge the deep sulcus hypothalamicus separates the dorsal from the ventral part of the hypothalamus, and each of these parts is further subdivided. A shallow and variable sulcus hypothalamicus dorsalis separates the dorsal hypothalamus again into dorsal and ventral lobes, both of which are confluent with the ventral thalamus and peduncle (fig. 2). These parts contain the primordium of the mamillary body, but this structure is not recognizably differentiated. The ventral hypothalamus is obscurely separable into posteroventral and anterodorsal parts. All these subdivisions are more clearly seen in Necturus than in Amblystoma ('35a, p. 253).

The sulcus preopticus separates the preoptic nucleus into anterior and posterior lobes. The lateral preoptic recess at the anteroventral border of the chiasma ridge is a remnant of the lumen of the hollow epithelial optic stalk of the embryo ('41), which persists in the endocranial part of the optic nerve of adult Necturus ('41a, p. 494).

The sequence of differentiation of the nervous connections of the hypothalamus has been summarized elsewhere, together with a description of these connections in midlarval stages ('39b, p. 550). The adult structure and connections of the hypothalamus of Necturus have been fully described ('33b, '34b, '41a), and those of Amblystoma are essentially similar, though with considerable advance in differentiation throughout ('42, pp. 211 ff.). None of the hypothalamic nuclei described in mammals are here clearly segregated, though some of them are recognizable in more dispersed arrangement. The preoptic nucleus is an exception to this. It is very large, and its boundaries are well defined except anterodorsally, where it merges with the nucleus of the olfacto-habenular tract, and posterodorsally, where it merges with the ventral part of the hypothalamus.

AFFERENT CONNECTIONS

The nervus terminalis, as described by McKibben ('11), has endings distributed throughout the hypothalamus both before and behind the chiasma ridge. The only other peripheral fibers which reach the hypothalamus are a few which separate from the optic tracts and ramify in the vicinity of the chiasma (p. 221).

By far the larger part of the white substance of the hypothalamus is occupied by the great medial forebrain bundles (*f.med.t.* of the figures); and, since these bundles are composed chiefly of fibers descending from the olfactory field of the hemispheres, the implication is that olfactory functions are dominant here. The olfacto-peduncular tract arises chiefly from the head of the caudate nucleus and distributes some of its fibers to the dorsal part of the hypothalamus. Fibers descend in the dorsal fascicles of the medial forebrain bundle from the septum and primordium hippocampi and distribute chiefly to the preoptic nucleus and dorsal hypothalamus. Included here are the precommissural fornix and part of the stria terminalis system. The ventral fascicles have descending fibers from the olfactory bulb and from the ventral and medial sectors of the anterior olfactory nucleus, which spread throughout the preoptic nucleus and the ventral part of the hypothalamus. Strong collaterals of both dorsal and ventral fascicles enter the stria medullaris thalami. Many fibers descend from the preoptic nucleus in diffuse formation, to spread throughout the hypothalamus posteriorly of the chiasma ridge. One component of this preoptico-hypothalamic connection is the compact tr. preopticus, as described below.

The dorsal olfactory projection tract (*ol.p.tr.*) is a compact fascicle of unmyelinated fibers, which descend from the dorsal striatal nucleus and in larger number from the amygdala. This fascicle accompanies the lateral forebrain bundle, and above the chiasma it turns ventrad to connect with its nucleus at the posterior border of the chiasma ridge. This tract also contains ascending fibers. It was first described in Amblystoma and the frog ('21a) and subsequently in Necturus ('33b, p. 158); its course was fully illustrated in 1921 and is shown in many other figures (figs. 19, 25, 26, 27, 95, 96, 97, 101, 102, 103; '27, figs. 11–20, 24–32; '36, fig. 5).

Afferent fibers of the visceral-gustatory system enter the dorsal part of the hypothalamus (fig. 8). The ventral part receives fibers from the pretectal nucleus by tr. pretecto-hypothalamicus (figs. 15, 16), and much larger numbers by the accompanying tr. thalamo-hypothalamicus, and also by tr. thalamo-hypothalamicus cruciatus (fig. 16). The entire tectum is connected with the hypothalamus by two large tracts—(1) the anterior division of tr. tecto-thalamicus et hypothalamicus cruciatus (tr. tecto-hypothalamicus anterior) from the superior colliculus and (2) the posterior division of this tract from both inferior and superior colliculi (fig. 12).

Through these manifold afferent connections nervous impulses are discharged into the hypothalamus from almost all parts of the brain, directly or indirectly. All sensory systems are represented here, but the olfactory connections evidently are preponderant. The significance of this convergence will appear after comparison with similar pools in the habenula and several other places (p. 252).

INTRINSIC HYPOTHALAMIC CONNECTIONS

All parts of the hypothalamus are interconnected by dense and intricately woven neuropil in both gray and white substance. Though all activities of the hypothalamus may thus be integrated, the structure is diversified, indicating the inception of specialization of the local nuclei as these are seen in higher animals. The most complicated and interesting of these local fields of neuropil surrounds and permeates the chiasma ridge, of which I have given a detailed description ('42, p. 214). Neuroblasts of this area are differentiated very early in embryogenesis ('37, '38), and some of their axons form tr. hypothalamo-peduncularis (figs. 18, 23) and tr. hypothalamo-tegmentalis (fig. 21), as elsewhere described ('42, p. 226). This area seems to be the primary focus into which converge all tracts of the ventral part of the hypothalamus for discharge through the two tracts just mentioned. Among these afferents are the intrinsic fibers of tr. preoptico-hypothalamicus from in front and of tr. infundibularis ascendens from the posteroventral lobe.

The last-mentioned ascending tract divides. One moiety is directed dorsally, providing a broad connection from the ventral part of the hypothalamus to the dorsal, and a larger moiety is directed forward into the medial forebrain bundle. Most of these ascending fibers apparently end in the postchiasmatic neuropil and preoptic nucleus, but some may go farther into the hemisphere ('33b, p. 250; '34b). This tract is quite independent of the ascending fibers of the olfactory projection tract previously mentioned. There are probably other fibers which ascend from the hypothalamus in the basal forebrain bundles, but our preparations have not revealed them.

From the anterior part of the ventral thalamus, fibers stream backward into the dorsal part of the hypothalamus. These thalamo-mamillary fibers are accompanied by fibers passing in the reverse direction—tr. mamillo-thalamicus—the combined tract being the probable precursor of the mammalian mamillo-thalamic bundle of Vicq d'Azyr.

EFFERENT CONNECTIONS

From the whole extent of the preoptic nucleus fibers pass dorsad to enter the stria medullaris thalami. Others probably ascend to the hemisphere in the basal bundles. The dorsal (mamillary) part of the hypothalamus is connected with the anterior part of the thalamus, as just mentioned. There is probably also a mamillo-cerebellar connection (p. 170). The other efferents from the mamillary region are the very extensive and complicated systems of mamillo-tegmental, peduncular, and interpeduncular fibers illustrated crudely in figures 18 and 21 and mentioned on page 278, where some references to literature are given (for fuller description see '39b, p. 551).

The efferents from the ventral part of the hypothalamus include the ascending fibers and the hypothalamo-peduncular and tegmental systems already mentioned. Fibers may go out with some components of the postoptic commissure to the thalamus and tectum, but these have not been recognized.

The pars magnocellularis of the preoptic nucleus gives rise to the large hypophysial tract, which is one of the major features of the ventral hypothalamus. In most lower vertebrates, including some amphibians, these large cells are aggregated as a well-defined nucleus, homologous with the supraoptic nucleus of mammals. But in Amblystoma, as in Necturus, they are dispersed, being most numerous above and anteriorly of the chiasma ridge. These widely scattered large cells in the aggregate are here termed the nucleus magnocellularis. Their long dendrites are widely branched and may be activated by practically all nerve fibers of the preoptic and epichiasmatic areas. In this mixed collection of fibers there are two systems which seem to be specifically related with these large cells.

The first of these systems is the tr. preopticus (fig. 2C, 96, 97, *tr.po.*). These fibers are axons of cells at the posteroventral border of the anterior commissure ridge and the anterior part of the preoptic nucleus, which descend in the thin floor of the long preoptic recess, then recurve dorsally along the anterior face of the chiasma ridge, where they spread in the alba among dendrites of the cells of the nucleus magnocellularis. In the catfish, Ameiurus, the course is similar, and here the connection of these fibers specifically with the compact nucleus magnocellularis is unmistakable ('41b). This, accordingly, is one source of activation of the hypophysial tract, the course of which in Ameiurus has been described by Palay ('45). His observations also indicate that in these fishes the nucleus magnocellularis itself is a source of endocrine secretion, as is clear also from the work of Scharrer and Scharrer ('40).

A second specific source of excitation of the nucleus magnocellularis is the small number of fibers which separate from the optic tracts near the chiasma and arborize in the area where these cells are most abundant (p. 221), thus providing visual control of endocrine secretion. This, however, is minimal in Amblystoma, and the very extensive hypophysial innervation must be concerned in the main with other functions.

The very numerous unmyelinated axons of the cells of the nucleus magnocellularis are difficult to follow, for they descend in dispersed arrangement, penetrate the chiasma ridge, and then converge into the compact tr. hypophysius in the thin floor of the infundibulum. Their further course is clear. We have many brilliant Golgi impregnations of their distribution, some of which have been illustrated ('42, figs. 55–65). In the pars nervosa of the hypophysis they form a very dense neuropil, are less abundant in the pars intermedia, and a few of them penetrate the pars distalis. They terminate in tiny end-bulbs resting upon the epithelial cells. The pars nervosa is a rather thin sheet of convoluted epithelium which forms the posterodorsal wall of the wide infundibulum. These lobules take the form of irregular villi, in the axis of each of which is a capillary loop.

CONCLUSION

In most lower vertebrates the hypothalamus is relatively much larger than in the higher groups, and this seems to be correlated with the dominance of the olfactory system in the organization of the forebrain of the lower forms. All other functional systems have extensive representation here, and in this respect the hypothalamus resembles the habenular complex, but with the difference that in the latter the afferents converge into a very small compact area with one major efferent path, the f. retroflexus, while the much larger hypothalamus is diversified in structure and has a wider range of distribution of its efferent fibers. Though both these complexes are evidently concerned with the correlation of olfactory with a wide variety of other types of sensory experience, it is equally evident that the type of adjustment made is radically different. To this topic we shall return (p. 252).

The intricate connections described above are very different from those of the specialized fishes. How they are related with those of mammals remains uncertain, pending further study of intermediate species. The lack of differentiation of a localized mamillary body here is probably explained by the failure of the fibers of the primordial postcommissural fornix to reach the hypothalamus (p. 255); and this, in turn, is correlated with the primitive structure of the hippocampal formation. The analysis of the intricate system of post-optic commissures (chap. xxi) probably provides an instructive point of departure for further study of these connections, leading up to a solution of still controversial problems about the mammalian supra-optic commissures.

The nervous connections of the hypophysis in urodeles are very large, and they are so arranged as to be easily accessible for experimental study. It is hoped that advantage will soon be taken of this favorable material for investigation of some problems of endocrinology.

CHAPTER XVIII

THE HABENULA AND ITS CONNECTIONS

IN ALL vertebrates with fully evaginated cerebral hemispheres the deep transverse stem-hemisphere fissure separates the dorsal parts of the hemisphere from those of the diencephalon. Ventrally of the floor of this fissure there are two great systems of fore-and-aft conduction—the basal forebrain bundles and the stria medullaris thalami. These form a superficial ventrolateral eminence, which is more conspicuous in the frog, where it was termed by Gaupp the "prominentia fascicularis." Attention has been called to the fundamental difference between these two great systems of fibers. The basal bundles are concerned primarily with the patterning of behavior, and, accordingly, their composition and connections vary widely from species to species in conformity with diverse modes of life. The composition of the stria medullaris, on the other hand, is remarkably constant throughout the vertebrate series, and its functional role is obscure. This chapter is devoted to these fibers and their widely dispersed connections, together with a few other systems of fibers which are associated with them.

THE DI-TELENCEPHALIC JUNCTION

All afferent fibers to the habenula, with the exception of the tecto-habenular tract, enter the stria medullaris thalami for longer or shorter parts of their courses. These fibers are assembled from all parts of the cerebral hemisphere and preoptic nucleus and, in smaller number, from the thalamus. They form a massive fasciculus, which ascends vertically in the posterior lip of the stem-hemisphere fissure; and associated with them is a series of bed-nuclei at the di-telencephalic junction, which will next be described. These nuclei have diverse and complicated connections, and all have this in common, that they are in functional relations with the stria medullaris. The following are listed in this series; for their arrangement see figure 2B.

1. *Preoptic nucleus.*—Among the connections of this nucleus, as described in the preceding chapter, are terminals and collaterals of descending fibers of the nervus terminalis, medial forebrain bundle,

precommissural fornix, stria terminalis, and ascending fibers of the medial bundle. Fibers ascend from all parts of this gray to the habenula in the stria medullaris, some externally of the lateral forebrain bundle (tractus olfacto-habenularis lateralis) and some medially of it (tr. olfacto-habenularis medialis). Among these fibers are strong collaterals from both dorsal and ventral fascicles of the medial forebrain bundle ('39b, p. 538).

2. *Bed-nuclei of the anterior commissure* (figs. 25, 26, 27, 97).—These are cells sparsely scattered in the anterior commissure ridge and massed laterally of it. They are related with tracts of the basal forebrain bundles decussating here, the stria terminalis, and the tr. septo-habenularis.

3. *Nucleus of tr. olfacto-habenularis* (figs. 27, 28, 29, nuc.tr.ol.h.; '35a, p. 250; '39b, p. 538).—This gray is interpolated between the anterior commissure ridge, the anterodorsal part of the nucleus preopticus, the anterior end of the ventral thalamus, and the eminentia thalami. It is penetrated by fibers of the tr. olfacto-habenularis medialis, with which it is related by terminals and collaterals. It is also connected with the amygdala by dispersed thick fibers, some of which are myelinated, termed "tr. amygdalo-thalamicus," though the direction of conduction is unknown (shown but not named on figs. 19 and 30; also shown on fig. 96, where the nucleus of the tr. olfacto-habenularis medialis is marked, *p.v.th.*). This tract is probably a part of the complicated stria terminalis system. It was described in Necturus ('33b, p. 218) as connected with the nucleus of Bellonci, but its main connection is with the nucleus of the olfacto-habenular tract, as shown in figure 40 of the paper cited. Impregnated neurons of the posterior part of this nucleus are shown in figure 24 of the paper of 1942. Their unmyelinated axons take widely divergent courses—to the stria medullaris, the amygdala, medial and lateral forebrain bundles, and the ventral thalamus.

In describing the early development of both dipnoan and amphibian brains, Rudebeck ('45) recognizes two parts of the nucleus preopticus: (1) an inferior "nucleus preopticus proper," most of which is developed rostrally of the sulcus intraencephalicus anterior, and (2) a pars superior, developed posteriorly of this sulcus in the ventral part of the di-telencephalic ridge. The superior part he regards as the primordium of the amygdala of the adult amphibian. This part is larger in dipnoan embryos than in amphibians and may be the source of a more extensive area of the adult brain. In Amblystoma the posterior lobe of the inferior or "proper" nucleus preopticus develops posteriorly of the sulcus intraencephalicus anterior; and, more dorsally, Rudebeck's pars superior of this nucleus is clearly the primordium of my nucleus of the olfacto-habenular tract, including, perhaps, also the posterior end of the amygdala. In any case the two nuclei last mentioned are very intimately related.

4. *Eminentia thalami.*—This name was given ('10, p. 419) to a prominent ventricular eminence lying behind the interventricular foramen between the anterior commissure ridge and the habenula, which has been variously interpreted by authors. It is appropriately named "nucleus commissurae hippocampi" by Addens ('46). I have regarded it as the anterior end of the ventral thalamus, differentiated as bed-nucleus of several large tracts which converge here. It is well defined in early swimming stages (area 7a), when the first fibers of the stria medullaris are visible ('38, p. 213 and figs. 18, 19); and in early feeding stages, with good development of stria medullaris and hippocampal commissure, it has acquired essentially adult form ('38b, p. 401, figs. 1, 2, 6). It is derived from the middle of the di-telencephalic ridge of Rudebeck's ('45) pictures of the embryonic brain.

This eminence receives collaterals from the stria medullaris and in larger numbers from the fibers of the hippocampal commissure (figs. 31, 32, 71, 76; '27, p. 294; '35a, p. 250). As the latter fibers swing downward behind the interventricular foramen toward their crossing in the anterior commissure ridge, they are accompanied by similar fibers which pass from the primordium hippocampi to the ventral thalamus—tr. cortico-thalamicus medialis (figs. 31, 32, 33, 72, 75). Like the fibers of the hippocampal commissure, these are mostly unmyelinated, with a few thick, well-myelinated fibers scattered among them. Many of them end in the eminentia thalami, and others continue into the anterior part of the ventral thalamus. This tract is probably the precursor of the mammalian column of the fornix, though here none of its fibers have been seen to reach the hypothalamus.

Numberless short, thin, unmyelinated axons descend from the small cells of this eminence to spread in the adjoining ventral thalamus ('21a, figs. 26–28; '27, p. 269), and thicker axons enter the complex crossed and uncrossed system of tr. thalamo-tegmentalis ventralis ('39, p. 120 and fig. 2; '39b, p. 546; '42, p. 225).

5. *Nucleus of Bellonci.*—This name is applied to a group of cells which form a low ventricular eminence between the anterior parts of

the dorsal and the ventral thalamus. I have regarded it as part of the dorsal thalamus, though its morphological status is uncertain. This nucleus was not recognized by Bellonci ('88), but he did describe in the frog a peculiar field of very dense neuropil related with it, which I term the "neuropil of Bellonci" in preference to his name, "nucleus anterior superior corporis geniculati thalami." This nucleus and associated neuropil I first described in Necturus ('17, p. 243) under the name of "pars optica thalami," but this term also is inappropriate and has been discarded. In Necturus and Amblystoma this peculiar neuropil receives terminals and collaterals from the optic tract, stria medullaris, tr. cortico-thalamicus medialis, the anterior tectal fasciculus (p. 297), and other sources. Axons of the cells of the nucleus go to the habenula and to the ventral thalamus and peduncle. This neuropil has received much study ('25, p. 454; '27, p. 295; '33b, p. 216; '34, p. 107), but, as recently remarked ('42, p. 279), many details of its structure and connections are still obscure. The probable equivalent of this nucleus has been found in all major groups of vertebrates by Addens ('46 and papers there cited).

6. *Habenula.*—The general structure and chief connections of the amphibian habenular nuclei are well known and have been summarized by Ariëns Kappers, Huber, and Crosby ('36). There are two habenular nuclei on each side, different in position from that of mammals but with similar fibrous connections. They are arranged one above the other at the anterodorsal border of the diencephalon, dorsally of the eminentia thalami. The taenia thalami, or line of attachment of the membranous dorsal sac, extends from the eminentia thalami along the anterior border of both nuclei and backward along the dorsolateral border of the dorsal nucleus as far as the habenular commissure. Posteriorly of this commissure the roof is again membranous as far as the recessus pinealis. The dorsal and ventral nuclei are separated on the ventricular surface by a shallow sulcus intrahabenularis and bounded posteriorly by the sulcus posthabenularis, which is the surviving dorsal end of the sulcus diencephalicus ventralis of Rudebeck ('45). Below the ventral nucleus is the deep sulcus subhabenularis. This is essentially the embryonic arrangement, in contrast with that of adult man, where the great enlargement of the dorsal thalamus and reduction of the membranous dorsal sac and paraphysis result in great changes in the relative positions of parts.

The ventral and dorsal habenular nuclei of Amblystoma are probably the equivalents, respectively, of the lateral and medial nuclei of higher forms, but the details of their connections are not exactly comparable. Both nuclei receive terminals of the stria medullaris thalami (apparently all components of it), and the chief efferent path of both nuclei is the tr. habenulo-interpeduncularis, which is the largest component of the fasciculus retroflexus (of Meynert). The dorsal nucleus is connected with the tectum and pretectal nucleus by fibers running in both directions, and the ventral nucleus is similarly connected with the dorsal thalamus, the ventral border of the tectum, and the dorsal tegmentum.

In the ventral habenular nucleus the nerve cells are relatively few in the usual arrangement at the ventricular side. The alba is almost completely filled by the massive stria medullaris, internally of which is neuropil, which invades the gray substance. Most of the cell bodies are widely separated within this dense neuropil. The larger dorsal nucleus has very numerous small cells, densely crowded on all sides except dorsally of the commissure. This layer of cell bodies incloses a central core of dense neuropil, which connects with the stria medullaris below and the habenular commissure above. The thick contorted dendrites ramify in the central neuropil (fig. 73; '42, figs. 77, 79).

Primitively, these nuclei had direct connection with the parietal eye, as is the case in some still living species. It is possible that in some early ancestor of the vertebrates, now extinct, the dorsal parietal eye was better developed than the lateral eyes and that for this reason the primordial olfacto-visual correlation was made in the epithalamus rather than in the tectum of the midbrain.

The habenular system is one of the most conservative parts of the vertebrate brain. Edinger's statement ('11, p. 370) that this is perhaps the only part of the brain the organization and connections of which show no alterations during the whole course of vertebrate phylogeny requires some qualification; yet, as we pass from cyclostomes to man, with revolutionary changes in all surrounding parts, the chief habenular connections show a surprising uniformity.

In urodeles afferent fibers enter the habenula from all parts of the cerebral hemisphere, except perhaps the olfactory bulb (some authorities would not make this exception) and also from the preoptic nucleus, thalamus, and tectum. The largest of these tracts come from areas which are under the strongest olfactory influence, and the habenular complex is generally regarded as primarily concerned with olfactory adjustments. That this is not its only function is evident from the fact that in anosmic animals, like some birds and cetaceans, the habenular system, though reduced in size, retains most of its components. The associational connections of the habenula with the tectum, dorsal thalamus, and adjacent regions justify the conclusion that the complex, viewed as a whole, is adapted to insure the correlation and integration of the activities of all parts of the olfactory field with those of all other exteroceptive systems.

The sense of smell is both interoceptive and exteroceptive. Olfactory nervous impulses of exteroceptive type are somehow sorted out from the entire olfactory field and converged into the epithalamus, and those of interoceptive type are similarly converged into the hypothalamus. In the former of these areas they may be correlated with all relevant somatic sensory experience, and in the latter with the sum total of visceral experience. The habenula discharges into that portion of the motor field known to control the activities of the skeletal musculature and chiefly into the interpeduncular nucleus, which is imbedded within this motor field and articulated with it in very complex patterns (chap. xiv). The olfacto-visceral correlations of the hypothalamus may come to expression in overt action of the skeletal muscles or in less obvious visceral changes. The chief nervous pathways involved in the overt responses can be identified, but those of visceral activities are still obscure. There is a strong connection from the hypothalamus to the interpeduncular nucleus.

Since both the hypothalamic and the epithalamic olfactory systems are large in all vertebrates, it was suggested by Edinger ('11, p. 371) that these regulate the feeding activities, or muzzle reflexes (*Oralsinnapparat*)—dorsally the exteroceptive components and ventrally the interoceptive. This attractive hypothesis seems to explain the different courses and connections of the medial forebrain bundle and stria medullaris, but it leaves out of account the equally large lateral forebrain system. Moreover, the chief connections of the medial and lateral forebrain bundles are localized in the hemisphere, respectively medially and laterally of the ventricle; but the habenular connections are drawn equally from both sides of the hemisphere. The habenular connections seem to be of a different kind, something added to an existing adequate provision for both visceral and somatic adjustments.

The chief efferent discharge from the habenula is to the interpeduncular nucleus, and there is a large efferent tract from the hypothalamus to this nucleus (figs. 18, 21). It is suggested in chapter xiv that the habenulo-interpeduncular system is the inhibitory component of an equilibrated dynamic system, of which the basal forebrain bundles comprise the activator component. On this hypothesis the activation of all olfacto-motor systems goes out through the basal forebrain bundles, and this activation is accompanied by an inhibition of all conflicting activities, the inhibitory component of the reaction being centered in the interpeduncular nucleus. This implies that inhibitory influences derived from exteroceptive fields are transmitted to the interpeduncular nucleus by the f. retroflexus and from interoceptive fields by the mamillo-interpeduncular tract. This hypothesis seems to be consistent with the known structure, but it lacks experimental proof. If there is factual basis for it, the activating and inhibitory systems must not be regarded as independent units of structure; they are everywhere interconnected, and all their activities are balanced one against the other in an integral dynamic system.

The literature contains many fragmentary accounts of the habenular connections of Amphibia, with some conflict of observation and confusion of nomenclature. When these descriptions are assembled, together with additional observations here reported, the salient features of the habenular connections may be grouped into three classes. These are:

1. There are two groups of commissural connections between the two cerebral hemispheres. Group 1 comprises the commissura superior telencephali, with fibers from the anterior olfactory nucleus, piriform area, and amygdala (components 3, 4, 5, and 6 of the list given on pp. 257–60). This is the larger part of the habenular commissure. It was seen by van Gehuchten in Salamandra ('97a), though its true origins were not recognized and his belief that none of these fibers have connection with the habenular nuclei needs confirmation. Group 2 is the com. pallii posterior (component 8 of the list). These fibers come from the primordium hippocampi and are probably comparable with the com. aberrans of some reptiles.

2. The habenula is an important way-station for through traffic from all parts of the cerebral hemisphere and preoptic nucleus to the brain stem below this level, particularly to the peduncle and interpeduncular nucleus. All these conduction pathways except the commissural fibers are interrupted by synaptic junctions in the habenula, and the habenular synaptic field is connected with neighboring regions by fibers passing in both directions.

3. The through traffic is influenced by these local connections with areas of the sensory zone concerned primarily with exteroceptive

functions. The habenula, accordingly, may be regarded as a bed-nucleus interpolated in one of the main pathways from higher centers of correlation, where olfaction as an exteroceptive function is integrated with other functions of this type and transmitted to lower centers in the motor zone, where patterns of response are organized. The hypothalamus is a similar way-station for through traffic from the same higher centers to lower motor fields, with olfacto-visceral functions dominant.

Most of the afferent connections of the habenula are in the stria medullaris, and most of the efferent connections in the f. retroflexus. Before describing this chief thoroughfare of through traffic, mention should be made of two systems of fibers which are in intimate functional relation with the stria medullaris, though not component parts of it. These are the fornix and stria terminalis.

FORNIX

In mammals this name is given to a complicated system of fibers which descends medially from the hippocampal formation to the underlying brain stem in two groups separated by the anterior commissure. Both groups descend within or adjacent to the lamina terminalis dorsally and rostrally of the interventricular foramen. The postcommissural fornix, commonly called columna fornicis, passes downward between the foramen and the anterior commissure and thence across the thalamus to end chiefly in the mamillary body. The precommissural fornix is a complex system of more loosely arranged fibers descending in front of the anterior commissure to the gray of the septal nuclei and neighboring parts and continuing spinalward in the medial forebrain bundles to preoptic nucleus, hypothalamus, and (according to some descriptions) as far as the cerebral peduncle.

Primordia of the two main divisions of the mammalian fornix are obviously present in the amphibian brain, but the topography of the region of the lamina terminalis is here so different that these tracts take peculiar courses. The anterior and hippocampal commissures cross, not in the lamina terminalis, but in a commissural ridge behind the interventricular foramen (fig. 2; '27, p. 235; '35, p. 299). To reach this crossing, the fibers of the massive hippocampal commissure swing downward behind the foramen (figs. 72, 76), and here they are accompanied by the fibers of tr. cortico-habenularis medialis and tr. cortico-thalamicus medialis (fig. 20, nos. 8 and 9); which, as pointed out in the preceding description of the eminentia thalami, marks the beginning of the differentiation of the mammalian columna fornicis. Some of these fibers are collaterals of those of the com. hippocampi and tr. cortico-habenularis medialis. In Amblystoma none of these fibers have been followed as far as the dorsal (mamillary) part of the hypothalamus, so that the homology with the mammalian postcommissural fornix is incomplete. In other urodeles it has been described as reaching the hypothalamus (Röthig, '24, pp. 10, 15; Salamandra, Kreht, '30, p. 252). The precommissural fornix is an ancient system, being well developed in many fishes. In Amblystoma it is large and of typical pattern. These fibers descend from the hippocampal formation rostrally of the foramen to the medial forebrain bundle (figs. 98, 99), within which they descend for an undetermined distance.

The transverse sections described in 1927 revealed clearly the relations of the precommissural fornix ('27, p. 311), but the postcommissural connections were obscure. In Necturus these connections are perfectly clear, and the morphological relations of precommissural and postcommissural fornix fibers have been discussed ('33b, p. 188). Amblystoma resembles Necturus. Accompanying the unmyelinated fibers of the hippocampal commissure there are similar fibers of tr. cortico-habenularis medialis and tr. cortico-thalamicus medialis, and among these are a few myelinated fibers. In the horizontal Cajal sections (figs. 28–32) the very numerous thin fibers of this complex are not impregnated, but the thick axons are deeply stained. Few of the thick fibers decussate in the hippocampal commissure (figs. 28, 29, 30). Many of them ascend in the stria medullaris (figs. 32, 33, 34, *tr.c.h.m.*), and most of the others pass backward and downward through the eminentia thalami into the ventral thalamus and (probably) the preoptic nucleus (figs. 31, 32, *tr.c.th.m.*, 72, 75). These thick fibers and the much more numerous thin fibers by which they are accompanied I have termed "tr. cortico-thalamicus medialis." An extension of this tract into the hypothalamus would give the connection described by Kreht ('30) in Salamandra and by Loo ('31, p. 84) as tr. cortico-hypothalamicus in the opossum. The mammalian tr. cortico-mamillaris is a further addition to this system.

STRIA TERMINALIS

In mammals the ventromedial septal field is connected with the ventrolateral amygdala by fibers, passing probably in both directions, some ventrally of the basal forebrain bundles (diagonal band of Broca) and some dorsally of them (stria terminalis, stria semicircularis, or stria cornea). Here the enormous enlargement of the basal forebrain bundles, particularly the internal capsule system of fibers, displaces the dorsal stria so that its course between medial and lateral ends is a wide arc, in some species almost a complete circle. At both of these terminal areas the stria makes complicated connections with the preoptic nucleus and hypothalamus, some of which were termed by Cajal the "olfactory projection tract." In Amblystoma, with smaller basal forebrain bundles, these fibers take more direct courses, and they are more dispersed so that there is no compact stria terminalis as in mammals. For description and illustrations of these tracts the reader is referred to my paper of 1927 (p. 302). The following connections have been described:

1. The com. amygdalarum, crossing in the anterior commissure (fig. 97).

2. The stria terminalis *sensu stricto* (fig. 96), a massive connection passing downward from the amygdala medially of the basal forebrain bundles. Some of its fibers are directed rostrad toward the septal area, but most of them turn posteriorly in the dorsal fascicles of the medial forebrain bundle to reach the preoptic nucleus and dorsal part of the hypothalamus. The rostrally directed fibers of this group and the commissural fibers are comparable with the fascicles of the stria which in mammals follow the taenia semicircularis and tail of the caudate nucleus, arching over the internal capsule and connecting the amygdala with the septal field. None of the fibers of this tract are drawn in the figures of horizontal sections. At the levels of figures 27–30 the descending fibers accompany the tr. olfacto-habenularis medialis and then turn spinalward in the dorsal fascicles of the medial forebrain bundle. Accompanying these are shorter myelinated fibers which pass between the amygdala and the nucleus of the olfacto-habenular tract—the tr. amygdalo-thalamicus (p. 248).

3. The ventral olfactory projection tract, descending from the amygdala externally of the basal forebrain bundles. These fibers are shown but not named in figure 16 of 1927.

4. The dorsal olfactory projection tract (p. 242, figs. 19, 25, 26, 27, *ol.p.tr.*).

STRIA MEDULLARIS THALAMI

In cyclostomes, with incomplete evagination of the cerebral hemisphere, the connections between the dorsal parts of the telencephalon and of the diencephalon are massive and direct ('22a, fig. 8), and the habenular tracts are not fasciculated to form a stria medullaris. With further evagination of the hemisphere in amphibians and the appearance of a deep stem-hemisphere fissure, the habenular connections of the pallial parts of the hemisphere must turn downward to pass under the floor of this fissure. Here they are joined by other habenular tracts to form the stria medullaris thalami. These tracts of Amblystoma were described in 1910 (and further details, '27, p. 284 and figs. 15–18; '39b, p. 538), and now some corrections and additions can be contributed. Compare also my description of the habenular connections of Necturus ('33b, pp. 204–14).

The components of the stria which have been identified in Amblystoma are shown in figure 20, where they are projected upon the median plane. The courses of four of the more lateral components are shown as projected upon the lateral aspect of the brain in figure 85. Their arrangement as seen in horizontal sections is shown in figures 25–36 and in sagittal sections in figures 74–78. Some components are electively impregnated in our Golgi preparations, and these specimens have been especially useful in demonstration of the truly commissural connections of several of the tracts listed below. Components 3, 4, 5, and 6 are known to decussate in com. superior telencephali and component 8 in com. pallii posterior. On the figures the components of the stria are numbered as in the following list:

1. Tractus olfacto-habenularis medialis.
2. Tractus olfacto-habenularis lateralis.

Most of the fibers of these two tracts are axons of cells of the preoptic nucleus, ascending, respectively, medially and laterally of the basal forebrain bundles (figs. 25–30, 74–77; '27, fig. 16; '42, figs. 18, 24), but some of them are collateral branches of axons of the medial forebrain bundle ('39b, p. 538). This is doubtless true in Necturus also, though I was not able to demonstrate it ('33b, p. 208). In the passage last cited, references are given to descriptions of this connection in other urodeles. The fibers of these tracts are assembled from all parts of the preoptic nucleus, ascending on both sides of the sulcus preopticus. The fibers of the medial tract pass through the gray of the nucleus of this tract, with many terminals and collaterals, and here the tract receives accessions from the nucleus. More dorsally both tracts have collateral connections with the eminentia thalami and neuropil of Bellonci. They form the most posterior component of the stria medullaris, and within the habenular nuclei they spread out in the neuropil. Whether any of them cross in the habenular commissure has not been determined. A few of these fibers are myelinated.

3. Tractus olfacto-habenularis anterior, ventral division.
4. Tractus olfacto-habenularis anterior, dorsal division.

As shown in figure 85, these fibers arise chiefly in the medial sector of the nucleus olfactorius anterior, from which they diverge ventrally and dorsally in a medial fascicle of fibers, which I have called the "f. postolfactorius" (fig. 100; '27, p. 283, figs. 2–4). The fibers of this system which are directed ventrally (no. 3) arise also in the ventral sector of the anterior nucleus and the septum. They take a posterolateral course along the ventral surface of the hemisphere, turning dorsally at the level of the anterior commissure to enter the stria medullaris under the posterior pole of the hemisphere ('27, p. 284 and figs. 5–17, *tr.ol.hab.ant.*). In the stria they lie anteriorly of the tr. olfacto-habenularis (1 and 2) and posteriorly of the tr. cortico-habenularis lateralis (5), as shown in sagittal sections (figs. 75, 76, 77, *tr.ol.h.a.v.*). In the horizontal sections (figs. 25–34) this tract is marked *tr.ol.h.a.* Some of these fibers terminate in the habenular neuropil, and some cross at the posterior end of the habenular commissure in the com. superior telencephali.

The dorsal fibers of this system (no. 4) cross from the medial to the lateral side of the hemisphere in the f. postolfactorius between the olfactory bulb and the anterior olfactory nucleus (fig. 5) and then turn posteroventrally across the lateral aspect of the hemisphere in company with the large tr. olfactorius dorsolateralis ('27, p. 284). Over the piriform area these fibers are joined by those of tr. cortico-habenularis lateralis (no. 5), and the mixed fascicle enters the stria medullaris under the posterior pole of the hemisphere. In the stria the mixed fascicle 4 + 5 ascends anteriorly of no. 3, and at the boundary between ventral and dorsal habenular nuclei it splits into lateral and medial bundles. The lateral fibers are dispersed in the habenular neuropil, and the medial fibers decussate in the com. superior telencephali (figs. 74–77, *tr.c.h.l.*). This component is shown in horizontal sections in figures 30–34, and its entire course is seen in transverse sections in figures 6–17 of 1927 (*tr.c.hab.l.*).

The anterior olfacto-habenular tract is the first component of the stria medullaris to mature in ontogeny, its fibers appearing in early swimming stages ('38, pp. 222, 238, figs. 10, 19); and its relations in cyclostomes indicate that it is very ancient phylogenetically ('33*b*,

p. 212). In Necturus the arrangement of these tracts is somewhat different ('33b, pp. 210–13).

5. Tractus cortico-habenularis lateralis. These unmyelinated fibers, as just described, pass from the piriform and dorsal pallial areas in company with those of no. 4 (fig. 20) to the habenula, where the majority of them decussate in the com. superior telencephali. Although there is no differentiated cortex in the amphibian brain, this tract is so evidently homologous with the one so named in mammals that this name is preferable to Ariëns Kappers' term, "tr. olfacto-habenularis lateralis," because the latter term is commonly applied to a different tract, no. 2 of the present list. In my paper of 1910 (p. 428) the tr. olfacto-habenularis anterior was not recognized as a separate entity but was regarded as a forward extension of the tr. olfacto-habenularis from the preoptic nucleus. This usage has some justification in cyclostomes and other most primitive vertebrates and is still employed by some authors, but the distinction between the fibers which arise from the preoptic nucleus and those from the hemisphere is significant and should be recognized in the nomenclature. In most species tr. cortico-habenularis lateralis is well defined. To avoid confusion, the terms "tr. olfacto-habenularis lateralis et medialis" should be restricted to fibers which enter the stria medullaris between the anterior commissure and the optic chiasma, arising either in the preoptic nucleus or as collaterals from the medial forebrain bundle.

6. Tractus amygdalo-habenularis. These are thick fibers, some of which are myelinated, passing from the gray of the amygdala to the habenula. Some of them may connect farther forward with the primordial corpus striatum—tr. strio-habenularis. These fibers ascend in the stria between nos. 5 and 8, and many of them end in the habenular neuropil (figs. 31–34, 74–78; '27, p. 302 and figs. 13, 14). As they approach the habenular commissure, these thick fibers mingle with those of the medial cortico-habenular tract, and these components can be separated only in electively impregnated specimens. In one such preparation of adult Amblystoma (no. 2257) the sections are inclined about 45° to the sagittal plane and the tr. amygdalo-habenularis is heavily and electively impregnated on both sides. The only other component of the stria stained is a small number of fibers of tr. cortico-habenularis medialis on one side only. The large tract from the amygdala is clearly followed through the commissure to the amygdala of the opposite side.

7. Tractus septo-habenularis. These fibers arise from the septal area and the bed-nuclei of the anterior commissure and ascend to the stria medullaris between this commissure and the interventricular foramen (fig. 75). They spread in the ventral habenular neuropil and have not been followed farther. In some preparations there is evidence that these fibers are accompanied by others from the anteroventral part of the corpus striatum complex (primordial nucleus caudatus)—a strio-habenular connection.

8. Tractus cortico-habenularis medialis (figs. 32, 33, 34, 71, 76; '27, figs. 11–17, here marked, *str.med.*). This is a large component of unmyelinated fibers and a few thick fibers with myelin sheaths. They arise from all parts of the primordium hippocampi, leaving its posteroventral border in company with similar fibers of the hippocampal commissure and tr. cortico-thalamicus medialis. Under the stem-hemisphere fissure they turn sharply dorsad, to ascend as the most anterior component of the stria medullaris. They spread widely in the habenular neuropil, and some of them decussate at the anterior end of the habenular commissure, thus forming the com. pallii posterior.

9. Tractus cortico-thalamicus medialis (figs. 31, 32, 72, 75). These fibers are regarded as primordia of the columna fornicis. They accompany the tr. cortico-habenularis medialis, and some of their fibers are collaterals from that tract and the hippocampal commissure. They cross the stria medullaris obliquely and are not integral parts of it except for the collateral connection mentioned.

10. Tractus olfacto-thalamicus. This name was given to a collateral connection from the stria medullaris to the neuropil of Bellonci in Necturus ('33b, p. 205). It is present also in Amblystoma.

11. Tractus thalamo-habenularis. These fibers arising from the dorsal thalamus and regions posteriorly of it enter the stria medullaris within the ventral habenular nucleus (fig. 76). They are accompanied by fibers passing in the reverse direction, the tr. habenulo-thalamicus ('39b, p. 539; '42, p. 261 and fig. 77). This connection has been described in mammals by several authors, recently by Marburg ('44, p. 220) in man, where its fibers arise from the anterior nucleus and pulvinar.

The tr. strio-habenularis is probably present in Amblystoma but has not been clearly seen. If so, its fibers may accompany those of the tr. amygdalo-habenularis and tr. septo-habenularis, as mentioned above. The former of these probable striatal connections is comparable with the pallido-habenular connection described in man by Marburg ('44).

Number 12 of figure 20 is an afferent tract to the habenula from the tectum and nucleus pretectalis, which has no connection with the stria medullaris—tr. tecto-habenularis. These fibers are accompanied by others which pass in the reverse direction—tr. habenulo-tectalis.

FASCICULUS RETROFLEXUS

Efferent fibers leave the habenula, so far as is known, in three tracts: (1) from the dorsal nucleus to the tectum by the tr. habenulo-tectalis, (2) from the ventral nucleus to the dorsal thalamus by the tr. habenulo-thalamicus accompanying the tr. thalamo-habenularis, and (3) between these from both nuclei in the much larger f. retroflexus of Meynert. There may be other efferent fibers, e.g., to the cerebral hemispheres accompanying those of com. superior telencephali and com. pallii posterior, but these have not been observed. One habenula of the newborn rabbit was destroyed by von Gudden ('81), and subsequently Meynert's bundle of the same side was found to be atrophied.

The chief component of the f. retroflexus is the habenulo-interpeduncular tract, the fibers of which arise from both dorsal and ventral habenular nuclei (figs. 20, 71, 73, 77, 103). This tract is composed chiefly of thin unmyelinated fibers, which form the central core of Meynert's bundle. Surrounding this core, other thicker fibers are loosely arranged, some of these being myelinated. Weigert sections show that a few myelinated fibers arise from both dorsal and ventral habenular nuclei and that these are joined by a few others from the dorsal thalamus, pretectal nucleus, and eminence of the posterior commissure. This fasciculus is also accompanied for part of its course by the few myelinated fibers of the parietal nerve (p. 235). No fibers have been seen to enter it from the tectum. The myelinated fibers leave the fasciculus and scatter in the alba of the nucleus of the tuberculum posterius; none of them enter the interpeduncular nucleus. In addition to these myelinated and other thick axons which terminate in the alba of the peduncle, there are many fine fibers from the axial core which take similar courses; some of these are probably of habenular origin.

The fasciculus descends across the thalamus at the outer border of the gray and is partly imbedded within it. Its course as seen in horizontal sections is shown in figures 30–35 (for the course in transverse sections see '25, figs. 2–7, and '27, figs. 28–33, *tr.hab.ped.*). As it enters the peduncle, it turns outward, passing ventrolaterally through the alba to reach the surface near the superficial origin of the III nerve root. Passing ventrally of these emerging root fibers, it turns medially at the fovea isthmi, immediately spinalward of which it enters its decussation in the ventral commissure (figs. 50, 51, 53). The spiral endings of these fibers are described in chapter xiv. In the alba of the peduncle it passes internally of the area ventrolateralis pedunculi and well separated from this superficial neuropil (fig. 94). Here many thin axons separate from the fasciculus to enter the superficial neuropil. A little farther spinalward, at the level of the III nerve root, a larger number of fine fibers leave the fasciculus and turn forward into the posterior end of the neuropil, as shown in a favorable Golgi section ('42, fig. 40, fibers from *f.retr.* to *a.l.t.*).

This is the usual arrangement. That the connection of the f. retroflexus with the area ventrolateralis pedunculi is important physiologically is evident from atypical courses of the fasciculus which have been observed in a number of specimens, always on one side only. One such anomaly is well illustrated in figures 56, 57, and 58. In this specimen (horizontal Golgi sections) the right f. retroflexus is unstained but can be seen to take the usual course; the left is abundantly impregnated from the habenula to its decussation. As it enters the alba of the peduncle it divides into two bundles. The medial bundle, containing about one-third of the fibers, takes the typical course; the larger lateral bundle passes along the inner border of the area ventrolateralis pedunculi with numberless terminals or collaterals entering this neuropil and spreading in the surrounding alba. Both bundles pass spinalward under the emerging III root fibers and converge to the decussation below the fovea isthmi. Many of these fibers spread in the alba rostrally and caudally of the fovea without decussation.

Another adult specimen (no. 2217) from the same lot and similarly prepared is like that just described except that the lateral bundle is smaller than the medial. This smaller bundle enters the area ventrolateralis and here breaks up into several slender fasciculi. Posteriorly of this area of neuropil, these fasciculi, somewhat diminished in volume, rejoin the medial bundle at the level of the III nerve root and both bundles then enter the decussation. This arrangement is found, on one side only, in the series of sagittal Golgi sections from which figures 72 and 73 were drawn. A somewhat different atypical feature is seen in the series of sagittal Cajal sections from which figures 74–78 were drawn. The fasciculus is well impregnated on both sides, and on one side it takes the typical course. At the decussation the thinnest fibers cross in the usual way, and some fibers of thicker caliber descend uncrossed at the lateral border of the interpeduncular neuropil. Whether they ultimately decussate and enter the interpeduncular spiral is unknown, because the impregnation fails below the decussation. On the opposite side the f. retroflexus divides as it enters the peduncle. The larger medial bundle takes the typical course. A small compact fascicle separates from it and passes more laterally to reach the area ventrolateralis pedunculi, where its fibers spread in the neuropil and disappear. Modifications of the arrangements just described have been seen in several other series of Golgi sections of late larvae and adults. The lateral bundle may be larger or smaller than the medial bundle, and in the latter instances it may end in the peduncle or rejoin the medial bundle at the decussation.

From these observations it is concluded that in Amblystoma the f. retroflexus is a mixture of fibers of different physiological characteristics. The main axial bundle of fine unmyelinated fibers is the habenulo-interpeduncular tract. Most of these fibers decussate below the fovea isthmi and enter a peculiar elongated spiral in the interpeduncular neuropil within which they end. A smaller number, including some thicker axons, descend uncrossed along the lateral border of the interpeduncular neuropil. These endings are in isthmic territory. Rostrally of the fovea many thin unmyelinated fibers, evidently of habenular origin, end uncrossed in the posteroventral border of the cerebral peduncle near the level of the nucleus of the III nerve. This area may be regarded as a mesencephalic sector of the interpeduncular nucleus, though there is no cellular differentiation here, like that of this nucleus below the fovea. Other similar fibers, doubtless also from the habenula, leave the f. retroflexus farther forward and terminate in the superficial neuropil termed "area ventrolateralis pedunculi." In some specimens these are very numerous. The few myelinated fibers of f. retroflexus derived from the habenula end in the cerebral peduncle. This fasciculus receives accessions of fibers, a few of which are myelinated, from the dorsal thalamus, pretectal nucleus, and eminence of the posterior commissure. All these fibers probably end in the cerebral peduncle; there is no evidence that any of them reach the interpeduncular nucleus. In a series of sagittal Golgi sections (no. 2215), in which the tr. habenulo-interpeduncularis is impregnated, a few thick axons are seen to arise in the gray of the dorsal thalamus and join the f. retroflexus, which they accompany almost to the ventral surface. Here some of them end by wide arborization in the area ventrolateralis pedunculi. This area of neuropil, accordingly, receives fibers by way of the f. retroflexus from both the habenula and the dorsal thalamus.

The connections just described are a simplified version of the complicated structure of the f. retroflexus of Ceratodus described by Holmgren and van der Horst ('25, p. 105). The fact that the atypical arrangements occur on only one side in Amblystoma probably is explained by the asymmetry of the habenular system in Ceratodus and other primitive species.

CHAPTER XIX

THE CEREBRAL HEMISPHERES

SUBDIVISIONS OF THE HEMISPHERE

LITTLE need be added here to the general descriptions given in chapters iv and vii. For details the reader is referred to the paper of 1927. That description was based mainly on a survey of a small number of well-preserved specimens cut in the transverse plane. There are in our collection many more instructive Golgi sections cut in longitudinal planes which have not been critically studied, though preliminary surveys have been made. It is deemed unprofitable at this time to continue the study of these sections because their interpretation should be based on physiological experiments correlated with the anatomical analysis.

At the present time our knowledge of the details of the internal structure of the cerebral hemispheres of Necturus ('33b) is more complete than of any other amphibian. This brain is not only larger than most of the others, but it is less compact and its great elongation is favorable for accurate localization of experimental studies by a wide variety of methods. This generalized brain provides a norm or basic pattern for the vertebrate phylum as a whole. The other urodeles and the anurans present a series of progressively more differentiated brains, and the sequence of stages of this process of specialization can readily be followed. That such a program of correlated histological and experimental work is practicable was demonstrated by Coghill in a restricted field of embryological research. With the refined experimental methods now at our disposal and with some reorientation in the fields of developmental mechanics, localized experimental extirpations, and electrical excitations, supplemented by oscillographic records, the steps in progressive phylogenetic differentiation of structure can be correlated with changes in overt behavior. For the completion of such a program frogs will probably prove to be more serviceable animals than the more sluggish salamanders (p. 98). These data will enable the comparative psychologists to identify and interpret prodromal stages of some of the basic patterns of human mentation.

Comparison of the amphibian cerebral hemisphere with the human shows a common plan of organization, and in the amphibian brains we find evidence of the beginning of differentiation of some mammalian structures at the earliest stages of their emergence from an undifferentiated matrix. The formative agencies which are operating to produce this local specialization are open to inspection and experimental investigation.

On the basis of position, internal structure, and connections the following mammalian organs have been identified in the brain of Amblystoma. First, the pallial field is distinguishable from the stem, and within this field primordia of hippocampal and piriform cortical areas are unmistakable. Some connections are suggestive of influences which may be precursors of neopallial differentiation, but these are vague and uncertain. Most of the mid-dorsal pallial area is probably represented in higher brains at the margins of archipallial and paleopallial cortex—such transitional cortex as the subicular and perirhinal areas.

In the subpallial part the lateral and medial walls of the hemisphere are organized essentially as in mammals. Laterally, the strio-amygdaloid complex is well defined, though its subdivisions are not clearly separable. Of these, the amygdala is definitely organized, with connections very similar to those of mammals. In the corpus striatum the nucleus accumbens septi is present as in lower mammals, and associated with it is an area which probably corresponds with the head of the caudate nucleus. The remainder of the corpus striatum is an undifferentiated lentiform nucleus, within which large and small cells are mingled. The connections of these cells suggest that the dorsal part of this area becomes the putamen and the ventral part the globus pallidus (p. 96).

On the medial side of the hemisphere the structure and connections of the septum conform with the mammalian arrangement, and below this is an undifferentiated area which gives rise in some of the fishes and in mammals to the tuberculum olfactorium.

THE OLFACTORY SYSTEM

As outlined in chapter vii, the olfactory nerve and its connections have played the dominant role in the morphogenesis of the cerebral hemispheres of lower vertebrates. The brief summary of the structure at the end of chapter iv is here supplemented by further description of the distribution of the olfactory tracts.

Nervus terminalis.—These unmyelinated fibers enter the brain in small compact fascicles mingled with those of the olfactory nerve. Their peripheral and central courses can be accurately followed only in elective Golgi impregnations, which, fortunately, are frequently obtained. All fibers of the olfactory nerve end in the olfactory bulb, but none of the terminalis fibers do so. The latter enter the brain at the ventral border of the olfactory bulb and course backward in several small fascicles, which terminate in the septum, preoptic nucleus, and hypothalamus. In Necturus some of them reach the interpeduncular nucleus, and this may be true in Amblystoma also. This nerve is present in vertebrates generally, from fishes to man, but our knowledge is incomplete about its terminal connections and functions (McKibben, '11; for Necturus see my '33*b*, p. 120, and '34*b*, '34*c*; for the frog, '09). It is regarded here as a sensory nerve, but even this is a debatable question.

Olfactory bulb.—In the olfactory bulb, as in the retina, the peripheral receptors discharge into a field which receives few afferent fibers of other functional systems. In the other primary sensory centers there is a common pool of neuropil within which terminals of peripheral fibers of diverse sensory modality are mingled, and to these there are added terminals of other correlating fibers of central origin. The bulbar formation of Amblystoma receives an enormous number of fibers of the olfactory nerve and no others from the periphery. There are also terminals of fibers ascending from other parts of the brain: (1) many of these are collaterals from the secondary olfactory tracts; (2) some are axons of cells of the anterior olfactory nucleus; (3) some may be commissural fibers by way of the anterior commissure; (4) some may come from more remote parts of the hemisphere. By far the larger number of these ascending fibers belong to the first two classes, in which olfactory influence is clearly dominant. From this it follows that the impulses conducted by the secondary olfactory tracts are influenced relatively little by other functional systems. In this respect they differ from the lemniscus systems of the lower brain stem; and the fact that these almost purely olfactory tracts reach all parts of the cerebral hemisphere is probably the reason why this hemisphere remains at a low level of structural differentiation and physiological specificity.

Necturus and Amblystoma exhibit two well-defined stages in the histological differentiation of the olfactory bulb, but the mammalian type of structure has not been attained ('24*b*, '31). Throughout the bulbar formation, except for the accessory bulb, the structure is nearly homogeneous, with little evidence of localization of function. The sense of smell lacks any provision for localizing in external space the source of odorous excitations. In the retina there is very complicated mechanism for analysis of the components of visual excitation (Polyak, '41). The analysis of olfactory sensibility for discrimination of odors is evidently a much simpler process. Judging by analogy with Polyak's description of the retina, there is little provision for this in the olfactory bulb. It is possible that the periglomerular cells may perform this function, but the structural organization of the bulbar formation gives clear evidence that the dominant activity here is not analysis but summation and intensification. The correlation of olfaction with other sensory systems is effected throughout the cerebral hemisphere, hypothalamus, and epithalamus, beginning in the nucleus olfactorius anterior.

Anterior olfactory nucleus.—This nucleus was first defined in Amblystoma ('10, p. 497) as undifferentiated olfactory tissue of the second order, closely associated with the olfactory bulb and extending backward a longer or shorter distance between the bulbar formation and the more specialized parts of the hemisphere. It is of large extent in the amphibian brain, and to it considerable attention has been given ('24b; '31; '27, p. 288; '33b, p. 133; '34, p. 99). In higher brains it shrinks in size as progressively more of this generalized tissue is specialized. Its comparative anatomy was discussed in connection with a detailed description of it in the Virginia opossum ('24d). In the amphibian brain it is a broad ring of gray bordering the bulbar formation on all sides. This cylinder is divided topographically into ventral, medial, dorsal, and lateral sectors, each of which has its own distinctive connections with other parts of the hemisphere. The ventral sector and the lower part of the medial contain the primordium of the tuberculum olfactorium. The arrangement of these sectors in Necturus is shown in figures 111 and 112, and their structure and connections have been described in detail ('33b, p. 133). Transverse sections through this region of Amblystoma are in the paper of 1927 (p. 288 and figures 2–5). The neurons and neuropil of this nucleus are illustrated in figures 105, 108, and 109 and in 1934, figures 1 and 2.

Typical neurons of the anterior nucleus have widely spread thorny dendrites, and axons which enter the olfactory tracts. Many other forms of cells are seen, some of which are transitional to those of the olfactory bulb. The axons of some of its cells are directed peripherally, to end with wide arborizations in the granular layer of the bulb. Most of the smaller cells have short, much branched axons, which participate in the formation of the dense axonic neuropil of this region.

Olfactory tracts.—Strictly defined, a tractus olfactorius includes axons of olfactory neurons of the second order only, that is, axons of mitral cells; but practically all these fibers are mingled with those of higher order from the anterior olfactory nucleus and other parts of the hemisphere, so that the tracts so designated on the figures are all mixed bundles. These axons of mitral cells stream backward from all margins of the olfactory bulb. Only the shorter fibers to the anterior nucleus are drawn in figures 111 and 112. As shown in figure 6, these are accompanied by longer fibers, which join the tracts descending from the anterior nucleus. Olfactory tracts from the lateral and ventral borders of the bulb take direct courses backward in three series. The more dorsal fibers enter tr. olfactorius dorsolateralis for distribution to the dorsolateral olfactory nucleus, which is primordium piriforme. This is the largest of the olfactory tracts and is comparable with the lateral olfactory stria of mammals. Other lateral fibers pass to the corpus striatum and amygdala. Some of these fibers join tr. olfacto-peduncularis, most of the fibers of which arise in the anterior nucleus and primordial caudate nucleus. Fibers from the ventral border of the bulb enter tr. olfactorius ventralis and descend for an undetermined distance in the medial forebrain bundle.

As shown by figure 4, the lateral ventricle extends forward almost to the anterior end of the olfactory bulb. Many of the longer fibers from the bulb take tortuous courses to reach their terminal stations. They accumulate in the medial sector of the anterior olfactory nucleus and primordium hippocampi, where they form a very large compact sheet of fibers termed "fasciculus postolfactorius" (fig. 100, *f.po.*; '27, figs. 2, 3). These fibers run vertically around the tip of the lateral ventricle, some directed ventrally to enter tr. olfactorius ventralis and some dorsally to enter tr. olfactorius dorsolateralis (fig. 5).

Olfactory tracts of the third and higher orders, i.e., those separated by two or more synapses from the periphery, are generally designated by hyphenated compound words, as tr. olfacto-peduncularis; but, as mentioned above, many of these tracts are mixtures containing some axons of mitral cells. For further details of these connections see the summaries ('33*b*, p. 124; '27, p. 282).

CHAPTER XX

THE SYSTEMS OF FIBERS

THE principles of classification and nomenclature of nerve fibers here employed are explained on page 9. A systematic account of them is very difficult because of their dispersed arrangement, their many deviations from the familiar mammalian pattern, and a cumbersome and confused nomenclature; yet gratifying success has attended efforts to resolve the amphibian tissue and discover mammalian homologies or their primordia. It is not our purpose in this chapter to give a comprehensive list of the tracts of Amblystoma that have been identified and named. References are given to the literature in which incomplete lists have been published and to pages of this book where particular systems of fibers are described. To these there are added descriptions of a few other systems of fibers of special importance and complexity.

Lists of tracts of Necturus have been published for the medulla oblongata ('14a, '30), midbrain and thalamus ('17), and forebrain ('33b). At the end of the latter paper is an alphabetical list of abbreviations which includes many tracts, with references to previous descriptions and synonyms. For Amblystoma a similar list of abbreviations ('36, pp. 309-12) includes many tracts of the cerebral peduncle, with page references to the text. Other lists of tracts of Amblystoma have been published ('27, '36, '39, '39b, '42), and Bindewald ('14) gave in tabular form a useful summary of previous descriptions of the parts of the amphibian forebrain and related fiber tracts, with homologies. The tracts of several species of urodeles have been described in the literature cited on page 11. The following systems of fibers of Amblystoma have been more or less completely described: peripheral nerves (Coghill, '02), olfactory tracts ('27, p. 282; Bindewald, '14; compare Necturus, '33b, p. 124, '34b), optic tracts and tectal connections ('25, '41, '42; compare Necturus, '41a), postoptic (supra-optic) commissures ('42, p. 219; Necturus, '41a, p. 513), the visceral-gustatory system ('44a, '44b), cerebellum (Larsell, '20, '32), tegmental fascicles ('36).

In this work under appropriate headings there are lists of tracts related with the several parts of the brain, accompanied by diagrams, and some classified lists, including the lemniscus and visceral-gustatory systems (chap. xi); connections of the interpeduncular nucleus (chap. xiv); some tectal connections (chap. xvi); stria medullaris thalami, fasciculus retroflexus, stria terminalis, and fornix (chap. xviii); some olfactory connections (chap. xix); and in the next following sections the basal forebrain bundles, tegmental fascicles, f. tegmentalis profundus, and the commissural systems.

THE BASAL FOREBRAIN BUNDLES

In a survey of the forebrains of fishes, attention was called to the important part played in morphogenesis by these great systems of longitudinal conduction ('22a, p. 175), and here this theme has received further consideration in chapter vii. These bundles of Necturus were described in comparison with those of other amphibians ('33b, p. 166), and their arrangement in Amblystoma has been illustrated ('27, p. 285; '36, p. 335 and figs. 5, 6; '39b, p. 533 and fig. 1). In sections of these brains the basal bundles are the most obvious landmarks, extending from the hemispheres backward through the brain into the medulla oblongata. Some of their fibers, in other species, have been described as extending into the ventral funiculi of the spinal cord, but our material yields no evidence that in Amblystoma any fibers which descend from the hemispheres in these bundles go back without interruption farther than the level of the roots of the VIII nerve. These bundles contain ascending and descending fibers, many of the latter decussating in the anterior commissure. In addition to these main lines of through descending traffic, many other kinds of fibers enter these bundles, and these connections will now be summarized.

The fibers of these bundles are arranged in three groups of fascicles as indicated in figures 19, 20, 21, 101; but these groups are not sharply separated, for there is much interchange of fibers among them. Their descending fibers are roughly comparable with the subcortical components of the mammalian extra-pyramidal systems, and the more dorsal ascending fibers of the thalamo-frontal tract (figs.19, 101, *tr.th.f.*) correspond with the thalamo-striatal system. The dorsal group of fascicles, the lateral forebrain bundle (f. lateralis telencephali, *f.lat.t.*), connects with the lateral wall of the hemisphere and has dorsal and ventral components with different connections. The ventral group, the medial forebrain bundle (f. medialis telencephali, *f.med.t.*), connects with the medial and ventral walls and also has dorsal and ventral components related, respectively, with the dorsal and ventral parts of the hypothalamus. The third group of fascicles comprises the tractus olfacto-peduncularis (*tr.ol.ped.*), which lies between the lateral and the medial bundles, connecting the anterior olfactory nucleus and the head of the caudate with the hypothalamus, peduncle, and interpeduncular nucleus.

Lateral forebrain bundle.—Most of these fibers are connected with the primordial corpus striatum and amygdala, and some of them with the piriform area. The descending fibers correspond with the human ansa lenticularis, and most of them are well myelinated. They appear earlier in ontogeny than do the ascending fibers. The dorsal and ventral fascicles of this bundle interchange fibers and in Necturus are not clearly separable ('33b, p. 170). They are connected mainly, though not exclusively, with the dorsal and ventral nuclei, respectively, of the corpus striatum, as described in chapter vii. When first studied, this relationship could not be demonstrated ('27, p. 286), but subsequent examination of sagittal sections ('36, p. 335) convinces me that this incomplete separation marks the beginning of differentiation of the globus pallidus. In 1927 the ventral fascicles were named tr. strio-peduncularis ('27, p. 287) and the dorsal fascicles tr. strio-tegmentalis, but these terms are inappropriate because both of them terminate in the peduncle and also in the tegmentum, though in different areas and evidently with different physiological import. The dorsal fascicles enter group (9) of the tegmental fascicles and the ventral fascicles enter group (10), as described below.

The descending fibers of the ventral fascicles (*f.lat.t.v.* and (*10*) of the figures) arise from the large cells distributed throughout the striatal gray. Their terminals are widely spread in the ventral parts of the peduncle and isthmic tegmentum. They do not extend so far spinalward as do some fibers of the dorsal fascicles. In the analysis of the tegmental fascicles ('36) these fibers comprise group (10). They make their chief synaptic connections with the large cells of the peduncle and tegmentum, the axons of which enter the ventral tegmental fascicles of groups (4), (5), and (6), descending to the medulla oblongata, and some of which continue in the f. longitudinalis medialis into the spinal cord. This connection is regarded as provision for cerebral control of mass movements of the trunk, limbs, and eyeballs.

The descending fibers of the dorsal fascicles (*f.lat.t.d.* and tegmental fascicles of group (9)) extend farther spinalward with different distribution. They give numberless collaterals to the thalamus, the dorsal part of the peduncle, and the dorsal and isthmic tegmentum; and posteriorly they turn laterally into the trigeminal tegmentum dorsally of its motor zone (figs. 29, 30). The chief synaptic connections of these fibers are not with the primary motor column but with areas of intermediate-zone type. They appear later in embryonic development than do the ventral fascicles, and their functions are believed to be cerebral control and conditioning of bulbar reflexes.

Another important descending component of these dorsal fascicles is the recently described tr. strio-tectalis (figs. 11, 101, *tr.st.tec.*; '42, p. 262) and the associated tr. strio-pretectalis (figs. 14, 101, *tr.st.pt.*). These fibers probably arise from the corpus striatum, though this has not been demonstrated. They end by wide arborizations in the optic tectum, pretectal nucleus, and geniculate neuropil of the thalamus.

Some fibers descend from the dorsal striatal nucleus in company with more from the amygdala in that component of the stria terminalis complex known as the dorsal olfactory projection tract (*ol.p.tr.*), as described on page 242.

The most noteworthy ascending system of fibers of the lateral bundle is tr. thalamo-frontalis (*tr.th.f.*). These slender unmyelinated axons arise from cells of the middle sector of the dorsal thalamus and descend in several small compact strands to the dorsal fascicles of the lateral forebrain bundle, within which they turn forward. As in Necturus ('33b, p. 170), they probably all end in the striatal neuropil, though their terminals in Amblystoma have not been described. These are precursors of the ascending thalamic radiations of mammals. They arise from the undifferentiated nucleus sensitivus of the thalamus, and no evidence has been seen of any separation among them of projection tracts related with different functional systems. The course of these fibers, so far as known, has been fully illustrated (figs. 3, 15, 30–34, 75, 95, 101, 102, 103; '39b, figs. 1, 7, 8, 13–17).

Medial forebrain bundle.—Most of the fibers of this large system are unmyelinated, passing in both directions between the olfactory bulb and the medioventral parts of the olfactory area of the hemisphere and the preoptic nucleus and hypothalamus. Many of these fibers decussate in the ventral part of the anterior commissure ridge. Analysis of this complex is possible only with the aid of elective Golgi impregnations, and there are few of these in our collection of Amblystoma sections. The Necturus material has been more instructive, and the following summary is based on data from both species (Amblystoma, figs. 6, 25, 71, 75; '27, p. 285; '39b, p. 534 and figs. 1, 79; Necturus, '33b, pp. 173, 261 and fig. 14; '34b). The medial bundle is incompletely separable into dorsal and ventral fascicles connected, respectively, with the dorsal and ventral parts of the hypothalamus. Both fascicles have extensive connections with the preoptic nucleus, and both contain descending and ascending fibers. Posteriorly of the anterior commissure ridge, thick collaterals of fibers of both groups of fascicles ascend to the habenula in tr. olfacto-habenularis of the stria medullaris (p. 257; '39b, fig. 79).

The dorsal fascicles contain descending fibers arising in the septum (tr. septo-hypothalamicus) and in the primordium hippocampi (precommissural fornix, p. 254), also some components of the stria terminalis system (p. 256). These are accompanied by some ascending fibers the connections of which are not clear.

The ventral fascicles contain secondary olfactory fibers from the bulb and some of higher order from the ventral and medial sectors of the anterior olfactory nucleus. These fibers are distributed to the preoptic nucleus and ventral part of the hypothalamus and are accompanied by many preoptico-hypothalamic fibers. Slender filaments of the nervus terminalis are spread among these fascicles for their entire length. The large hypophysial tract arises from neurons which are widely scattered throughout the hypothalamic field reached by both the ventral and the dorsal fascicles of the ventral bundle. In the floor of the preoptic recess there is a median fascicle of unmyelinated fibers, among which are a few with myelin sheaths. This is tr. preopticus (p. 244).

Very slender unmyelinated fibers arise from small cells at the extreme posterior end of the ventral hypothalamus and ascend for undetermined distances in the ventral fascicles. There are doubtless other ascending fibers, but their courses have not been recorded.

Olfacto-peduncular tract.—This is a well-defined round bundle, lying between the lateral and the medial bundles and less myelinated than the former (figs. 18, 21, 25–30, 53, 54, 59, 72, 95–99, 101; '27, p. 286; '36, p. 336; '39b, p. 534, figs. 1, 79, 80). Its fibers arise from the head of the caudate nucleus and neighboring parts and distribute to the dorsal part of the hypothalamus, ventral part of the peduncle, and interpeduncular nucleus. The hypothalamic connection allies this tract with the dorsal fascicles of the medial forebrain bundle, the peduncular connection with the ventral fascicles of the lateral forebrain bundle. The connection with the interpeduncular nucleus is extensive and physiologically distinctive. This tract is the most direct connection between the olfactory area and the primary motor field of the peduncle and the interpeduncular nucleus. It is an internuncial of intermediate-zone type, with the olfactory influence predominant.

THE TEGMENTAL FASCICLES

The strio-thalamic and strio-peduncular components of the lateral forebrain bundle activate large neurons of the ventral thalamus and peduncle, and the axons of these cells transmit impulses downward to the tegmentum from the isthmus to the spinal cord. These descending fibers form the most conspicuous components of a series of ventral tegmental fascicles, some of which at lower levels are assembled in the f. longitudinalis medialis (figs. 6, 18, 91). Ventrally and dorsally of these are many other tegmental fascicles, composed chiefly of descending fibers; and still farther dorsally are the ascending lemniscus systems (fig. 14). The analysis of the composition of these longitudinal fascicles has been very difficult, yet this knowledge is essential for an understanding of the brain stem.

The topographical analysis of the brain stem of 1935 was followed in 1936 by a reconnaissance survey of the related fibers in the peduncle and tegmentum. The more obvious and constant bundles of longitudinal fibers were enumerated as ten fascicles or groups of fascicles ('36, p. 303), and the specific tracts represented in these and some other bundles, so far as then known, were listed ('36, pp. 334–46). These fascicles are divided into a ventral and a dorsal series, the former including fascicles (1) to (6) and part of (10); the latter, (7) to (10). Figures 91, 92, 94, and 102–4 are here reproduced from the paper of 1936; compare figures 30–33 and 101.

This classification was arbitrary for descriptive purposes only and included only the most clearly defined and constant fascicles bordering the gray substance. The symbol for each group is an Arabic number inclosed in parenthesis. Externally of these ten groups, in the intermediate alba, there are other less well-defined fascicles with a larger proportion of unmyelinated fibers. These are imbedded in much neuropil and are more variable than the deeper fascicles. Superficially in the subpial neuropil there is another series of fascicles, chiefly of unmyelinated fibers, some of which form recognizable tracts. Dorsally of all these fascicles are the lemniscus systems; these and the ascending secondary visceral-gustatory tract are described elsewhere in this work. Further study has yielded additional details about the composition of the ten groups of tegmental fascicles. These and some other fascicles and tracts of this region are here analyzed as far as their composition is now known. Most of these fibers are descending. There are ascending fibers also, but our material has yielded little information about them.

During the preliminary study it was anticipated that each group of tegmental fascicles would prove to be composed chiefly or wholly of fibers of a single tract or related group of tracts, as are the lemniscus systems. This proves not to be the case, for most of these bundles are mixtures of fibers of diverse sorts from unexpectedly widely separated sources. The reasons for their fasciculation in the pattern observed are not clear. Some specific tracts, like the mamillo-interpeduncular and strio-tegmental, are fairly clearly segregated (in groups (2) and (9) in the cases mentioned), but most of the bundles are heterogeneous mixtures. The pattern of fasciculation seems to be determined more by the ultimate destination of the fibers than by their nuclei of origin. In Weigert and especially in reduced silver preparations these fascicles, particularly those bordering the gray, are clearly defined for long distances; but Golgi sections show that there is much anastomosis among them and that there are numberless unmyelinated fibers which are not fasciculated but spread diffusely in the alba.

As the lateral forebrain bundles recurve dorsally at the antero-dorsal border of the peduncle (figs. 6, 16, 101, 102; '36, figs. 5, 6), sagittal Weigert sections show that the compact bundles of myelinated fibers disintegrate, with diffuse spread of the fibers in the alba of the posterior part of the thalamus, dorsal tegmentum, and peduncle. Some of these fibers are reassembled farther spinalward in fascicles (9) and (10). Golgi sections show a similar dispersal of the thinner unmyelinated fibers. Reduced silver preparations, however, reveal many slender fascicles of unmyelinated fibers which traverse this region without loss of their individuality. It is evident that these bundles, like most of the other tegmental fascicles, are mixtures of fibers of diverse distribution and physiological significance. In transverse Golgi sections of adult brains, in which the myelinated fibers of the deeper fascicles are slightly darkened and the unmyelinated fibers are electively impregnated, it is seen that the proportion of unmyelinated fibers is greater in the dorsal fascicles than in the ventral and that almost all fibers bordering the gray in the ventral fascicles are myelinated.

Ventromedian fascicles (1).—These are limited to the midbrain and isthmus, extending spinalward from the commissure of the tuberculum posterius. Most of their fibers decussate obliquely in the ventral commissure. They comprise 5 to 12 well-defined bundles of rather thick myelinated fibers arranged close to the gray under the ventral angle of the ventricle, with some admixture of unmyelinated fibers. These fibers are derived from various sources, and they are variously distributed. The bundles as definite anatomical entities are assembled only in the space bounded approximately by the levels of the nuclei of the III and IV cranial nerves. This distance is greater in urodeles than in most other vertebrates, and the arrangement of these fascicles found in Amblystoma has not been described in any other species. In Necturus reduction of the optic system involves corresponding shrinkage of these median fascicles ('36, p. 348), though the general plan is similar. Three components of these fascicles have been distinguished.

a) Tractus tecto-bulbaris et spinalis cruciatus, pars anterior (fig. 12, *tr.t.b.c.1.*).—This is by far the largest component of the group. Its fibers, chiefly myelinated, descend from the tectum and turn spinalward in the mid-plane, here decussating in component 4 of the commissure of the tuberculum posterius, and posteriorly of this very obliquely in the ventral commissure ('36, pp. 303, 330, figs. 2, 7). At the level of the nucleus of the IV nerve they turn laterally and descend in tr. tecto-bulbaris et spinalis within the ventromedial alba of the medulla oblongata. The course of this tract, as seen in horizontal sections, is shown in figures 27–36 (for complete description see '36, p. 340, and '42, p. 268). The posterior division of this tract (figs. 27–36, *tr.t.b.c.2.*) does not enter the tegmental fascicles but decussates transversely, at the level of the nucleus of the IV nerve.

b) Tractus pedunculo-bulbaris ventralis cruciatus ('36, figs. 2, 7, *f.m.t.(1)b.*; '39*b*, p. 582 and figs. 23, 24).—These fibers arise from the ventral thalamus and peduncle and enter ventral fascicles (1). Here they decussate obliquely, mingled with those of the crossed tecto-bulbar tract; and after crossing they separate from the latter to descend in the ventrolateral fasciculi of the medulla oblongata, and some of them in the f. longitudinalis medialis, as described in the references cited.

c) Tractus pedunculo-tegmentalis cruciatus ('36, figs. 2, 7, f.m.t.(1)c.).—These fibers enter the ventral fascicles from the peduncle and perhaps also from the ventral thalamus. At the posterior end of these fascicles they do not turn laterally with the others but continue posteriorly and dorsally as one of the components of the f. tegmentalis profundus. They arborize in the deeper layers of neuropil of the isthmic tegmentum.

Ventromedian fascicles (2).—These fascicles of thin unmyelinated fibers comprise tr. mamillo-interpeduncularis, lying laterally and ventrally of those of group (1). These are shown in figure 19 and separating from *tr.mam.teg.* in figure 21. They assemble in the periventricular neuropil of the ventral lobe of the dorsal part of the hypothalamus. Figure 27 (*tr.mam.inp.(2)*) shows them converging into a ventricular protuberance of this lobe, which, immediately dorsally of this level, extends across the mid-plane to join the corresponding structure of the opposite side at the attenuated ventral border of the commissure of the tuberculum posterius (compare the median section, fig. 2C). Here some of these fibers decussate as component 1 of this commissure, as shown in figure 28. The crossed and uncrossed fibers continue dorsalward and then spinalward, recurving around the cerebral flexure at the extreme ventral surface (figs. 29, 30), to end in open arborizations within the interpeduncular neuropil. Their entire course in the larva is shown in a published diagram ('39b, fig. 22; other details are also illustrated in that paper—figs. 6–9, 35, 41, 42, 57–61). These fibers comprise the whole of group (2) of the tegmental fascicles, described in 1936. They do not form a compact bundle but are rather loosely spread, and their courses can be followed only in favorable Golgi sections. Unlike the other connections of the dorsal part of the hypothalamus, these fibers are aggregated in the deep periventricular neuropil, and evidently their physiological properties are radically different from those of the mamillo-peduncular and mamillo-tegmental tracts.

Ventral fascicles (3).—The chief component is tr. mamillo-tegmentalis (fig. 21; '36, figs. 3, 8; '42, fig. 3). These thin myelinated and unmyelinated fibers pass, probably in both directions, between the dorsal hypothalamus and the tegmentum in close association with similar fibers related with the peduncle and thalamus. The mammalian equivalents of the complex are found in the mamillo-thalamic and mamillo-peduncular tracts and the mamillary peduncle.

Amblystoma has no differentiated corpus mamillare; its primordium is in the dorsal part of the hypothalamus, from which efferent fibers go out dorsalward to the peduncle, forward to the ventral thalamus, and backward to the tegmentum. They are dispersed in the alba of the hypothalamus, and as they leave it those for the thalamus and tegmentum accumulate rostrally of those for the peduncle. Some of them decussate in components 1 and 2 of the commissure of the tuberculum posterius ('36, fig. 2). Afferent fibers to the hypothalamus are known to be present in the mamillo-peduncular and mamillo-thalamic tracts, and this may be true also of the mamillo-tegmental.

Mamillo-peduncular fibers have wide distribution in the alba of the nucleus of the tuberculum posterius, including the neuropil of the area ventrolateralis pedunculi (figs. 6, 18, 27-30, tr.mam.ped.; '36, p. 338, figs. 3, 8; '39b, p. 338, figs. 6-12, 22, 35, 42, 89; '42, fig. 39; Necturus, '33b, p. 246; '34b, p. 422). The accompanying pedunculo-mamillary fibers transmit visceral sensory, gustatory, and optic impulses received by the area ventrolateralis pedunculi to the hypothalamus. The mamillo-thalamic tract and the accompanying thalamo-mamillary tract put the mamillary region into reciprocal relations with the anterior part of the thalamus, as has been fully illustrated ('39b, figs. 22, 35; '42, fig. 39; Necturus, '33b, p. 247; '34b, p. 423, figs. 2, 8, 9).

The fibers of tr. mamillo-tegmentalis arise in company with those of the two preceding systems and form a loose fascicle rostrally of those of tr. mamillo-peduncularis (figs. 27-31; '36, p. 338, figs. 3, 8-21; '39b, p. 552, fig. 43; '42, fig. 3). Most of these fibers enter ventral tegmental fascicles of group (3), as shown in figures 92 and 94, some of them first decussating in the two ventral components of the commissure of the tuberculum posterius. These fascicles lie ventrolaterally of those of group (4), which contains thicker and more heavily myelinated fibers from the ventral hypothalamus. They terminate in the alba of the isthmic tegmentum, chiefly through the f. tegmentalis profundus. Mingled with these hypothalamic fibers are some thicker well-myelinated fibers of tr. pedunculo-tegmentalis. Some of the latter, and probably some of the hypothalamic fibers also, descend for long distances in the f. longitudinalis medialis.

Ventral fascicles (4).—These are mixed bundles derived chiefly from the postoptic commissure. The largest component is tr. hypothalamo-peduncularis et tegmentalis (figs. 21, 23), which contains both descending and ascending fibers. In the first description these were designated simply as "fibers from the postoptic commissure" ('36, p. 304, figs. 3, 8, 19, *po.(4)*); but now their connections are better known ('42, p. 226 and fig. 3), thanks to the fact that they mature very early in ontogeny and so can be seen in young stages, despite their dispersed arrangement. A few elective Golgi impregnations confirm these findings in the adult.

Fibers related with the entire ventral part of the hypothalamus converge into the commissura tuberis in the caudal part of the postoptic commissure complex. After crossing here in diffuse arrangement, they recurve around the tuberculum posterius, where most of them spread and end in the peduncle (tr. hypothalamo-peduncularis). Others descend in ventral tegmental fascicles of group (4) as tr. hypothalamo-tegmentalis. Only the longer fibers of the latter tract are entered on the drawings of the horizontal sections (figs. 25–32, marked *tr.hy.teg.(4)*, or simply (4)). Figure 32 cuts these fibers (4) at the most dorsal level of their arched course through the peduncle; compare their projection on the sagittal plane ('36, fig. 3, *po.(4)*). Other similar fibers extend dorsally from both ventral and dorsal parts of the hypothalamus to enter the midbrain without decussation in the chiasma ridge ('42, figs. 22, 23, 39). Some of these decussate in component 2 of the commissure of the tuberculum posterius, but most of them are uncrossed.

This system of fibers is evidently the main descending pathway from the ventral hypothalamus and neuropil of the chiasma ridge to the motor field of the peduncle and tegmentum. The associated fibers connected with the dorsal part of the hypothalamus probably should be classed with the mamillo-peduncular system, though they are not included in the bundles so designated. This more dorsal system, and perhaps the ventral system also, contain some fibers which are afferent to the hypothalamus, though most of them evidently are efferent. Most of the longer fibers of this system which reach the tegmentum descend in bundles of group (4), and here they are joined by thalamo-tegmental and pedunculo-tegmental fibers ('42, p. 225 and fig. 4). The longest fibers enter the f. longitudinalis medialis (figs. 27, 28), and it is uncertain whether these are of hypothalamic or of peduncular origin.

Most of the hypothalamic fibers of groups (3) and (4) pass into f. tegmentalis profundus, and most of these end in the neuropil related with the central nucleus of the isthmic tegmentum. These are doubtless precursors of the hypothalamic component of the mammalian f. longitudinalis dorsalis of Schütz.

Tegmental fascicles of groups (5) to (10) contain fibers of diverse origin, and there is much anastomosis of their finer fibers. The coarser fibers, however, are well fasciculated in an arrangement which seems to be determined primarily by the terminal distribution of the descending systems. Their analysis has been clarified by the previously published embryological studies; see particularly the general survey ('39b) and for the decussating systems the description of the postoptic commissure of the adult ('42). The data upon which this summary of the composition of the several fascicles is based are to be found mainly in the two papers just cited.

Ventral fascicles (5).—This is a group of large fascicles mainly composed of fibers (chiefly myelinated) descending from the ventral thalamus and peduncle. Most of them are uncrossed, but some decussate in company with those of tr. tecto-bulbaris cruciatus in the ventromedial fascicles of group (1) ('39b, p. 546 and fig. 23, $f.v.t.(5)$). They arise from all parts of the ventral thalamus and peduncle, and most of them terminate in the alba of the isthmic and trigeminal tegmentum, where they are in synaptic contact with the large cells of this region. The thicker and more heavily myelinated fibers are closely fasciculated laterally of the ventral angle of the ventricle, and many of these descend for undetermined distances in the f. longitudinalis medialis. The origin of these fibers from the ventral thalamus is shown in figures 31 and 32. At those levels similar thick fibers enter fascicles (5) from large cells of the peduncle, but they are not drawn in these figures. Figures 6 and 18 show fibers entering the f. longitudinalis medialis from large cells of the dorsal and ventral parts of the peduncle. The former (shown but not named in fig. 32) are in synaptic connection with terminals of the posterior commissure and doubtless correspond with the mammalian nucleus of Darkschewitsch (p. 217). The more ventral large cells of the peduncle (fig. 31) correspond with the interstitial nucleus of Cajal. In Amblystoma, unlike Necturus and most other vertebrates, the f. longitudinalis medialis is definitely organized only spinalward from the isthmus ('36, p. 334).

The fascicles of group (5) also receive some fibers from other systems, as will appear below. For their courses as seen in horizontal sections see figures 27-32; in transverse sections, figures 91-94 and '36, figures 9–16; in sagittal sections, '36, figures 3, 18–21. The thalamo- and pedunculo-tegmental fibers of this group are similar in many respects to the crossed and uncrossed fibers of tr. thalamo-tegmentalis ventralis of groups (4), (6), and (8). Those of group (5) arise chiefly from the peduncle, the others from the thalamus. In the aggregate these thick descending paths comprise the chief final common paths from the cerebrum to the peripheral neuromotor apparatus of the primary activities of the skeletal musculature, notably those of locomotion and feeding. In early larval development these long fibers from the peduncle and ventral thalamus are among the first to appear in the cerebrum. Their adult distribution in the several tegmental fascicles seems to be determined primarily not by the arrangement in space of the groups of cells from which they arise but by the lower motor fields into which they discharge their nervous impulses. Those in groups (4), (5), and (6) descend more medially and ventrally, the longest in the f. longitudinalis medialis. These longer fibers evidently activate the trunk and limbs. Collaterals of these fibers and accompanying shorter fibers end throughout the isthmic and bulbar tegmentum ('39b, figs. 84, 93), thus insuring coordination of head movements with those of the trunk and limbs. These more ventromedial fibers, accordingly, comprise the final common paths of fundamental mass movements, total patterns of behavior.

The more dorsal fibers of group (8), accompanied by the striotegmental fibers of group (9), descend along the outer border of the isthmic gray (figs. 29, 30) to end far laterally in the white substance of the isthmic and trigeminal tegmentum. Here the fundamental reflexes of the musculature of the head, concerned primarily with feeding, are organized. These more dorsal fibers, crossed and uncrossed, have innumerable terminals and collaterals in the peduncle, dorsal tegmentum, and lower tegmental fields ('39b, figs. 79–82).

Ventral fascicles (6).—These fascicles, like those of the fourth group, were originally designated as fibers from the postoptic commissure ('36, figs. 4, 8, 18, *po.*(6)). They are now known to be composed chiefly of two systems of thick myelinated and unmyelinated fibers which decussate in the postoptic commissure, derived, respectively, from the tectum and the ventral thalamus—tr. tecto-thalamicus et hypothalamicus cruciatus posterior ('42, p. 221 and fig. 5) and tr. thalamo-tegmentalis ventralis cruciatus ('42, p. 225 and fig. 4). The coarsest fibers of these fascicles, probably including some of both the systems just mentioned, traverse the whole length of the peduncle and isthmic tegmentum, and some of them enter the f. longitudinalis medialis. Figure 32 cuts these fascicles a few sections ventrally of their most dorsal course as they arch across the peduncle; compare the sagittal section (fig. 103).

The thickest fibers of tegmental fascicles (4), (6), and (8) decussate in the dorsal and posterodorsal part of the postoptic commissure, those of group (8) being the most dorsal fibers of this complex at their crossing in the mid-plane. Most of the latter enter group (8), but some enter group (6). These are fibers of tr. tecto-thalamicus et hypothalamicus cruciatus posterior of my former descriptions, but these longer fibers are properly called tr. tecto-peduncularis et tegmentalis cruciatus. Posteriorly and ventrally of these at the decussation are similar coarse fibers of tr. thalamo-tegmentalis ventralis cruciatus, which cross in more dispersed arrangement posteriorly and ventrally of the preceding system. These enter the three groups of fascicles, (4), (6), and (8). In the horizontal sections here illustrated the courses of these fibers are in some places not very clearly seen, for they are not separately fasciculated. Their courses are indicated on the drawings as determined (where doubt arises) by comparison with sections in other planes and with the early larvae, where their courses are clear. It should be noted that this crossed system of ventral thalamo-tegmental fibers is accompanied by similar uncrossed fibers, most of which enter fascicles of group (8). These were first described in the larva ('39b, p. 546, figs. 13–16, 81) and in the adult are seen here in figures 30–33 as fibers joining bundle (8) from the thalamus. There is also a broad uncrossed connection from both ventral and dorsal parts of the thalamus to the tegmentum by tr. thalamo-tegmentalis rectus (figs. 30–34, *tr.th.teg.r.*). These fibers pass from the thalamic gray to the pial surface of the tegmentum and end here in the superficial neuropil and obviously have physiological properties different from the deeper fibers of groups (4), (6), and (8), which make synaptic contacts with the large cells of the tegmentum in the intermediate and deep neuropil.

Dorsal fascicles (7).—Two systems of fibers have been identified in these fascicles: (*a*) one from the dorsal thalamus and dorsal tegmentum and (*b*) one from the tectum.

a) Tractus tegmento-isthmialis (fig. 21).—This tract was first described as the chief component of fascicles of group (7) under the name tr. tegmento-bulbaris ('36, p. 334, figs. 4, 6, *f.d.t.(7)*; '39b, p.

584). That name is inappropriate for two reasons, first, because few of its fibers reach the bulb and, second, because there is a large tegmento-bulbar tract more ventrally (figs. 27, 28, 29, *tr.teg.b.*). The tract here under consideration passes from the posterior part of the dorsal thalamus, eminence of the posterior commissure, and dorsal tegmentum into fascicles of group (7). Its fibers enter f. tegmentalis profundus and end in relation with the small cells of the central nucleus of the isthmic tegmentum. Its entire course is shown in figures 29–33, here marked (7). The isthmic tegmentum is regarded as the pool within which the bulbar reflexes concerned with feeding are organized, and this tract probably plays a critical role in bringing to this center the appropriate afferents from the intermediate zone. This tract carries only part of the efferent fibers from the dorsal tegmentum. Many of these fibers pass directly ventrally to enter the peduncle and lower tegmental levels uncrossed or with decussation in the ventral commissure. Others ascend or descend in various other tegmental fascicles.

b) Tractus tecto-bulbaris rectus.—It has recently been found that these fibers leave the tectum by various courses and that many of those from the anterior part of the tectum enter fascicles of group (7), from which they separate in the isthmus to enter the medulla oblongata in company with other tectal fibers (p. 225). These are the fibers which in 1936 were followed from fascicles (7) into the bulb.

Dorsal fascicles (8).—This conspicuous group contains thick and thin fibers, many of which are well myelinated. Those which come from the postoptic commissure ('36, pp. 304, 338, figs. 4, 8, 17, 18, *po.(8)*) comprise two quite separate systems, (*a*) from the tectum and (*b*) from the ventral thalamus. The latter are joined by many uncrossed fibers of the same system.

a) Tractus tecto-thalamicus et hypothalamicus cruciatus posterior.—The coarsest and most heavily myelinated fibers of the complex so named in earlier papers, after decussation in the postoptic commissure, enter fascicles of group (8) and distribute to the dorsal, isthmic, and trigeminal tegmentum ('42, p. 221). Other shorter fibers enter fascicles (6) to reach the isthmus, and others end in the peduncle and hypothalamus. These longer and coarser fibers to the tegmentum were variously interpreted and named in my earlier papers, but good elective impregnations have now clarified their connections as tr. tecto-tegmentalis cruciatus.

b) Tractus thalamo-tegmentalis ventralis.—As previously mentioned, this large and complicated system of crossed and uncrossed fibers was overlooked until early larval stages revealed its essential features ('39b, p. 546). These thick fibers arise in all parts of the ventral thalamus. Some of them decussate in the postoptic commissure and enter fascicles (4), (6), and (8). Others descend uncrossed in all these groups and also in group (5). Golgi sections show that these tracts have numberless terminals and collaterals throughout their lengths. The thicker fibers of group (8) course more dorsally and terminate more laterally in the isthmic and trigeminal tegmentum.

The composition of group (8) as it leaves the postoptic commissure is shown in figures 25 and 26 and its further course in figures 27–33. Figures 30–33 show uncrossed fibers from the ventral thalamus joining the bundles of crossed fibers (compare '39b, fig. 2, and see the concluding comment about fascicles of group (9)).

Dorsal fascicles (9).—These are composed chiefly of tr. strio-tegmentalis (fig. 101; '27, p. 287; '36, pp. 304, 335; '39b, fig. 79), which is the dorsal component of the lateral forebrain bundle, passing from the dorsal nucleus of the primordial corpus striatum to the trigeminal tegmentum. Some of these fibers decussate in the anterior commissure, and a small number descend as far as the level of the VII nerve roots, with terminals and collaterals along the way.

These fascicles are easily followed in both horizontal and transverse sections (figs. 28–33, *f.lat.t.d.* and (9)), and especially clearly in sagittal sections (figs. 16, 19, 21, 72, 102; '36, fig. 5; '39, figs. 3, 7–12; '39b, figs. 1, 79). There is interchange of fibers between these fascicles and those of groups (7a) and (8), and the thicker fibers of both groups turn laterally in the trigeminal tegmentum close to the gray and spray out dorsally of the V and VII nuclei, where they apparently activate both large and small tegmental cells. As the fascicles of group (9) reach the posterior end of the ventral thalamus, they turn sharply dorsad, parallel with the limiting sulcus, *s*, around the anterodorsal border of the nucleus of the tuberculum posterius. Here those of (9) lie ventrolaterally of (8) (figs. 16, 91–94; '36, fig. 5), with much anastomosis between them. The finer fibers of both groups spread widely in the neuropil of the peduncle and dorsal and isthmic tegmentum, and some descend in fascicles of group (7). The primary function of groups (7), (8), and (9) appears to be the control of the feeding musculature of the jaws and hyoid from the corpus striatum, thalamus, and tectum.

Fascicles (10).—These comprise the greater part of the ventral fascicles of the lateral forebrain bundles (figs. 26–32, 101, 102, *f.lat.t.v.* and *(10)*). In 1927 (p. 287) this was termed tr. strio-peduncularis, but it now appears that these fibers arborize not only in the peduncle but also throughout the length of the isthmic tegmentum, though they do not extend as far spinalward as those of group (9). This tr. strio-peduncularis et tegmentalis arises from the ventral nucleus of the corpus striatum, decussates partially in the anterior commissure, traverses the ventral thalamus, and in the peduncle breaks up into a number of small fascicles which are widely distributed among the ventral and dorsal tegmental fascicles. Its fibers are less myelinated than those of group (9), and, in general, they distribute in the alba more ventrally and laterally (figs. 92, 102, 103; '36, pp. 304, 336, fig. 6; '39b, fig. 79; '42, figs. 2, 17); compare the preceding description of the lateral forebrain bundles.

Most of the numbered groups of fascicles which have just been listed lie in the deeper layers of the alba close to the gray. Superficially of these and in part mingled with them are many dispersed fibers, loosely fasciculated in variable arrangements. These include many of the shorter tracts mentioned in the descriptions of the several regions, such as thalamo-tegmental, pedunculo-tegmental, tecto-tegmental, and tegmento-bulbar tracts. The well-defined tr. olfacto-peduncularis also belongs in this series. The thalamo-tegmental fibers are very numerous, and some of them are well fasciculated. From both dorsal and ventral thalamus uncrossed fibers stream backward to the isthmus, some deeply and some superficially—tr. thalamo-tegmentalis rectus (figs. 15, 31–34, *tr.th.teg.r.*). These are in addition to the thick fibers from the ventral thalamus, which enter tegmental fascicles (6) and (8), some uncrossed and some decussating in the postoptic commissure (fig. 17, *tr.th.teg.v.r.* and *c.* '39b, p. 546). Analysis of the fibers from the dorsal thalamus which decussate in this commissure is very difficult. The earlier accounts are confused and inaccurate, but the essential features are now clear, as described in the next chapter.

FASCICULUS TEGMENTALIS PROFUNDUS

This name has been given to a loosely arranged sheet of fibers which borders the gray in the tegmental region of the midbrain and the isthmus. Most of these fibers are unmyelinated and are part of an

THE SYSTEMS OF FIBERS

extensive tegmento-peduncular and tegmento-bulbar system of internuncials, passing obliquely from the tegmental to the motor field; but fibers of various other systems are mingled with these. They comprise important components of the periventricular, deep and intermediate tegmental neuropil, with a dorsoventral trend. At the outer border of the gray they are especially numerous, and here some of the components are fasciculated as anatomically separate tracts for part of their courses. Other more dispersed components also are named as tracts in cases in which their terminal connections are revealed by elective impregnations. Many of these fibers decussate in the ventral commissure in the tuberculum posterius and spinalward of it through the medulla oblongata. In the midbrain and upper rhombencephalon there are numberless short crossed and uncrossed fibers, some strictly commissural between the tegmental fields and larger numbers passing from the tegmentum obliquely rostrad or caudad, with or without decussation, to other parts of the motor field.

The list below includes all the tracts which have been identified in this complex, and in addition to these there are diffuse connections through the neuropil with all neighboring regions. This summary is based upon what has been seen in both Amblystoma and Necturus.

Brachium conjunctivum (p. 176 and fig. 10).—This is one of the largest and most compact components of the complex, though rarely impregnated in our material. From the cerebellum to its decussation it accompanies the isthmic sulcus, as is well shown in figure 71.

Tertiary visceral-gustatory tract (p. 169 and figs. 8, 23).—These fibers accompany those of the brachium conjunctivum and are indistinguishable from them except in elective impregnations.

Tegmento-interpeduncular and interpedunculo-tegmental fibers (pp. 199, 201 and figs. 19, 60–68, 79, 80, 83, 84).—These very numerous fibers, passing in both directions between the interpeduncular nucleus and the isthmic and trigeminal tegmentum, are diffusely spread in the neuropil of the gray and the deeper layers of alba.

Tractus tegmento-peduncularis (p. 215 and fig. 18).—These fibers pass from the dorsal and isthmic tegmentum to the peduncle, dispersed at all depths of gray and white substance. Some of them decussate in the ventral commissure in the vicinity of the fovea isthmi. They accompany similar fibers of tr. tecto-peduncularis (fig. 24).

Tractus tegmento-isthmialis (p. 283 and fig. 21).—These fibers from the dorsal to the isthmic tegmentum form a large part of the dorsal tegmental fascicles of group (7), but many of them are mingled with more dispersed fibers of tr. tecto-tegmentalis.

Tractus tecto-tegmentalis (fig. 21).—Fibers from all parts of the superior and inferior colliculus stream backward into the isthmic tegmentum. Most of the former enter dorsal tegmental fascicles of group (7) in company with longer fibers of tr. tecto-bulbaris rectus (p. 225), but many of them descend dispersed in the neuropil (fig. 79).

Tractus pedunculo-tegmentalis cruciatus.—These, as described above (p. 278), enter the ventral median tegmental fascicles of group (1) from the peduncle and, after decussation here, pass to the isthmic tegmentum by way of f. tegmentalis profundus.

Tractus mamillo-tegmentalis.—Some fibers from the dorsal hypothalamus reach the isthmic tegmentum accompanying the superficial tr. mamillo-cerebellaris (p. 170). Others take a deeper course in ventral tegmental fascicles of group (3) and f. tegmentalis profundus, as described on page 279.

CHAPTER XXI

THE COMMISSURES

GENERAL CONSIDERATIONS

THE commissures are in two series, dorsally and ventrally of the ventricles. The fibers which cross the mid-plane are of two sorts: (1) some are strictly commissural, connecting corresponding regions of the two sides; (2) most of them are decussations, like the optic chiasma, connecting dissimilar regions. Those of the dorsal series include fibers of correlation which arise and terminate within the sensory zone and also connections between the sensory zone and other zones. Those of the ventral series are concerned in the main with motor co-ordination on the two sides of the body, some passing from sensory and intermediate zones to the motor zone, others lying wholly within the motor zone, and both sorts being accompanied by uncrossed fibers.

In addition to the crossed systems of correlation and co-ordination to which reference has just been made, it is a noteworthy fact that the main lines of fore-and-aft ascending and descending conduction in the brains of all vertebrates decussate, so that the adjusting centers for organs on the right side of the body are in the left side of the brain and vice versa. This applies especially to the apparatus of somatic adjustment, but not so generally to the visceral systems. Most of the fibers of the ascending secondary visceral-gustatory tract ($tr.v.a.$) and of the olfacto-hypothalamic tracts are uncrossed. The proprioceptive systems are both crossed and uncrossed.

The reason for the decussation of the major conduction pathways of the somatic sensori-motor systems has puzzled neurologists for many years, and fantastic theories have been expressed, some bearing the names of great masters—Wundt, Flechsig, Cajal, and others. This extensive literature has been reviewed by Jacobsohn-Lask ('24), who, instead of elaborating a theory in terms of the highly specialized human brain, like his predecessors, reviews the entire history of the evolution of the nervous system, from its first appearance in coelenterates, in relation to the bilateral symmetry of the body. None of these speculations have yielded satisfying conclusions.

A survey of the commissures and decussations of all vertebrate brains shows that some of them, like the posterior commissure and optic chiasma, hold constant positions throughout the series, while others vary widely in position and composition. It is evident that in the latter cases the site of crossing of a particular system of fibers is determined by the topographic arrangement of the parts to be connected. These arrangements are widely diversified in the lower groups of vertebrates and the crossing fibers tend to take the shortest available pathways. But when the course of a particular decussating tract has been established in an ancestral species, the original site of the decussation may be retained, despite great changes in the relative sizes of parts of the brain in phylogenetic descendants of the ancestral form, so that the tract may take a circuitous course to reach its decussation, as illustrated by some components of the postoptic complex in fishes. This conservatism may be accounted for by the fact that the decussating fibers may have collateral connections at any part of the course both before and after crossing. These considerations suggest that great caution should be observed in using the actual sites of crossing of the various systems of fibers as criteria of homology. Some of them are very stable; in other instances fibers with similar origins and terminations may cross in surprisingly different places, as illustrated by the various courses taken by fibers of the hippocampal commissure.

Primitively the roof plate of the brain, unlike the floor plate, was membranous, and the locations of commissures which invade it are determined by the functional requirements of the organs differentiated below it in the various species of animals. In the telencephalon the topographic arrangements of parts are extremely variable. In all cases there is a wide interruption of the commissures at the site of the stem-hemisphere fissure, paraphysis, dorsal sac, and their derivatives. In the diencephalon the habenular commissure is separated by the pineal recess from the commissura tecti, and the latter is continuous spinalward with the posterior commissure and the com. tecti of the mesencephalon. In cyclostomes the middle part of the mesencephalic tectum is a membranous chorioid plexus, and the com. tecti is here interrupted.

At the posterior end of the tectum its commissure is continuous with a thin sheet of crossed and uncrossed fibers in the anterior medullary velum, and this, in turn, with the massive cerebellar commissures. Between the cerebellum and the calamus scriptorius the roof is a membranous chorioid plexus except in some fishes, in which the large acousticolateral areas fuse over the ventricle, with some decussating fibers. In the region of the calamus there is a dorsal crossing of visceral sensory fibers (com. infima of Haller) and of somatic sensory fibers between the funicular nuclei. This dorsal commissure, reduced in size, is present through the entire length of the spinal cord.

The ventral series of commissures is concentrated in the anterior commissure of the telencephalon, the chiasma ridge of the diencephalon, the so-called ansulate commissure of the midbrain; and posteriorly of this it extends continuously as the ventral commissure of the rhombencephalon and spinal cord. The ventral decussations of Necturus from the tuberculum posterius to the spinal cord were described in 1930 ('30, p. 89) and the other commissures in papers before and after that date.

The arrangement of the telencephalic commissures of the Amphibia differs radically from that of all animals higher in the scale. As seen in median section (fig. 2), the anterior and hippocampal commissures do not cross in or above the lamina terminalis but in a high anterior commissure ridge which projects upward from the floor posteriorly of the interventricular foramina. This ridge is separated from the lamina terminalis by the wide precommissural recess, or aula, which is the vestibule of the interventricular foramina (fig. 1B). Both pallial and subpallial parts of the hemispheres contribute fibers to the commissures in this ridge, and many fibers from the olfactory and hippocampal areas also cross in the habenular commissure.

The peculiar arrangement of the anterior and hippocampal commissures, the lamina terminalis, and the related chorioid plexuses of Amblystoma has been described and illustrated ('35). It is explained by the topography at the di-telencephalic junction and the presence of unusually wide interventricular foramina. The explanation of this topography, in turn, is to be sought in the phylogenetic ancestors of existing Amphibia. In Protopterus as described by Rudebeck ('45), pallial formation is confined to the lateral wall of the hemisphere, and the hippocampal commissure passes down behind the foramen to cross in close association with the anterior commissure in essentially the same way as in amphibians. In view of the close similarity of development of the brains of amphibians and lungfishes, it is probable that the ancestral amphibian resembled Protopterus in this region.

The arrangement of myelinated fibers in all the commissures of Amblystoma as seen in the median plane is shown in figure 2C. The components of these commissures will now be summarized, with references to more detailed descriptions.

THE DORSAL COMMISSURES

Commissura hippocampi.—These fibers converge from all parts of the hippocampal area to its ventral border, from which they descend behind the foramen to cross in the dorsal part of the anterior commissure ridge (figs. 96–99). Many of these fibers join a mixed longitudinal fascicle, termed "fimbria," which borders the primordium hippocampi for its entire length at the ventromedial margin. Another fascicle, composed exclusively of commissural fibers, assembles at the ventrolateral border of the primordium. Most of these fibers are unmyelinated. Weigert sections show a few brilliantly stained myelinated fibers scattered among them.

Accompanying this commissural bundle as it leaves the hippocampus are other similar fibers of tractus cortico-thalamicus medialis and tr. cortico-habenularis medialis. Some of the latter cross in the habenular commissure and return to the hippocampal area of the opposite side, this being the com. pallii posterior (figs. 32, 34, 71, 72, 76).

Commissura habenularum (com. superior).—The analysis of the stria medullaris as detailed in chapter xviii reveals the following components of the habenular commissure: (1) The most anterior member is the com. pallii posterior (fig. 20, no. 8). (2) Posteriorly of this the com. superior telencephali includes components *3, 4, 5,* and *6* of the diagram, viz., tr. olfacto-habenularis anterior, tr. cortico-habenularis lateralis, and tr. amygdalo-habenularis. (3) Still more posteriorly are crossing fibers of uncertain connections, some of which may come from other components of the stria medullaris. There are doubtless strictly commissural fibers connecting the habenulae of the two sides. The commissural fibers from the hemisphere are accompanied by uncrossed fibers. It is possible that these tracts are accompanied by crossed and uncrossed fibers, which pass from the habenular nuclei forward into the hemispheres, but this has not been demonstrated.

Commissura tecti diencephali.—Most of these fibers are commissural, connecting the two pretectal nuclei. Some tecto-habenular and habenulo-tectal fibers decussate here.

Commissura posterior.—This is the primary pathway from the anterior part of the optic tectum to the motor zone of the midbrain. Its decussating fibers appear very early in embryogenesis, accompanied by uncrossed fibers from the tectum and eminence of the posterior commissure. In the adult many of the commissural fibers end in this eminence. Crossed and uncrossed fibers spread widely in the posterior part of the thalamus, the nucleus of the tuberculum posterius, and the dorsal tegmentum. The largest fascicles connect with the big cells of the nucleus of Darkschewitsch (p. 217).

Commissura tecti mesencephali.—This thin sheet of crossing fibers is continuous between the posterior commissure and the anterior medullary velum. It contains thin and thick fibers (many of the latter myelinated) which spread widely through all layers of the tectum. Most of these seem to be commissural between the tecti of the two sides, but no satisfactory analysis has been recorded. Some fibers of the mesencephalic root of the V nerve apparently decussate here, but, if so, the number is small.

Commissures of the anterior medullary velum.—Most of the fibers in the velum are longitudinal—tr. tecto-cerebellaris—and some of these may decussate here. The most constant and noteworthy component is the decussation of the IV nerve roots ('36, p. 342; '42, p. 255). The velum contains cells and fibers of the mesencephalic V root, and some of these may decussate here. Our preparations give no clear evidence of crossed fibers of this root; if present, the number is certainly not large.

Cerebellar commissures.—Larsell's analysis of these commissures is confirmed. The two systems are quite distinct. (1) The com. cerebelli is related with the median body of the cerebellum, including decussating fibers of tr. spino-cerebellaris, sensory root fibers of the V nerve, secondary trigeminal fibers from the superior sensory V nucleus in the auricle, and commissural fibers between these nuclei and between the two sides of the corpus cerebelli. (2) The com. vestibulo-lateralis cerebelli is a more dispersed system of fibers related with the vestibular and lateral-line centers of adjustment in the auricles. It is composed of root fibers of the VIII nerve and secondary fibers of both vestibular and lateral-line systems. There are doubtless also commissural fibers between the two auricles. None of these fibers make significant connections with the median body of the cerebellum through which they pass. Their terminal relations are with auricular tissue which is the primordium of the floccular part of the mammalian flocculonodular lobe.

Commissura infima Halleri.—This is a decussation of the fasciculi solitarii at the calamus scriptorius, containing both root fibers and secondary fibers of the visceral-gustatory system and doubtless also commissural fibers between the two commissural nuclei of this system.

Commissure of the funicular nuclei.—Intimately associated with the preceding are commissural fibers between the nuclei of the dorsal funiculi in the calamus region, with which decussating fibers are mingled. This dorsal commissure of somatic sensory fibers is extended, reduced in size, downward through the entire length of the spinal cord. There is some evidence that the visceral sensory com. infima is also represented in the cord (p. 125).

This completes the summary of the dorsal commissures. We now turn to the ventral series, beginning, as before, at its anterior end.

THE VENTRAL COMMISSURES

COMMISSURA ANTERIOR

The complex com. anterior occupies the entire anterior commissure ridge except its dorsal border. Its largest components are the partial decussations of the medial forebrain bundles below and the lateral forebrain bundles above (figs. 25–28). Associated with these fibers are others, including the com. amygdalarum (which is part of the stria terminalis system, p. 256), some fascicles of the nervus terminalis, and a dispersed decussation between the olfactory fields of the anterior parts of the hemispheres ('39b, fig. 21).

The crossing fibers of the anterior commissure ridge are enveloped by a thin layer of gray which expands laterally as the large bed-nuclei of the anterior commissure. The thin floor of the long preoptic recess between the anterior commissure ridge and the chiasma ridge contains the longitudinal fibers of tr. preopticus, some of which decussate here.

CHIASMA OPTICUM

All fibers of the optic nerves decussate in the chiasma opticum. The crossing occupies the anterior border of the chiasma ridge (figs. 2B and 2C). At the posterior border of the chiasma there is some mingling of optic fibers with those of the postoptic commissure, but in some of our Golgi preparations the optic fibers are electively impregnated and can be separated from the others ('41, '42).

COMMISSURA POSTOPTICA

The complex com. postoptica is represented in mammals by the supra-optic commissures, but its composition is so different in urodeles and mammals that exact homologies cannot be established. Further analysis of intervening species is requisite before these relationships can be clarified.

In Amblystoma this complex includes decussating fibers derived from the superior and inferior colliculi, the entire diencephalon, the amygdala, and the subpallial olfactory field of the cerebral hemisphere. Many of these tracts have collateral connections along their courses, both before and after crossing, and are accompanied by uncrossed fibers. There are also strictly commissural fibers connecting some of these regions (for evidence of such fibers in the frog see '25, p. 480). These decussating systems connect, after crossing, with extensive fields of the intermediate and motor zones, including the strio-amygdaloid area, preoptic nucleus, hypothalamus, ventral thalamus, geniculate neuropil, dorsal tegmentum, peduncle, isthmic tegmentum, and bulbar tegmentum. The posterior part of the postoptic commissure is comparable with the com. tuberis of some other vertebrates and consists mainly of decussating fibers from the ventral part of the hypothalamus to the peduncle and interpeduncular nucleus. The direction of conduction of most of these fibers has not been clearly determined, though most of the larger systems evidently converge from the sensory zone into the motor field.

The fibers of few of these components are assembled in well-organized tracts; most of them are so dispersed and commingled that analysis is very difficult. Some of them are myelinated, and these tend to be assembled in recognizable tracts. These myelinated tracts, as seen in Weigert sections, were the first to be described, but their distribution after crossing baffled analysis. These tracts are accompanied by much larger numbers of unmyelinated fibers in more dispersed arrangement. The courses of some of these have been revealed by elective Golgi impregnations, and other systems have been clarified by study of the sequence of their development as published in a series of papers from 1937 to 1941. In view of the complexity of these connections and the technical difficulties encountered in their study, it is not surprising that the earlier descriptions were incomplete and not free from error. It is believed that now it is possible to present an analysis of this complex which, though still incomplete, reveals its major features.

This generalized arrangement as seen in urodeles is probably primitive and may be taken as the point of departure in the study of the postoptic systems of more specialized brains of both fishes and higher vertebrates. These commissures have been analyzed in Necturus ('41a, p. 513), where they are still more generalized. Our present knowledge of these systems of Amblystoma was summarized on pages 219–28 of the paper of 1942, with diagrams illustrating the connections of the principal tracts. All known components are assembled in the following list. Here some of the earlier names of tracts have been replaced by more accurate terms; some others are retained, though now known to be inappropriate or inadequate. In this list there are included, first, the groups of fibers which descend to the chiasma ridge from the tectum and pretectal nucleus, followed by those descending from the thalamus, next, the systems arising in the hypothalamus, and, finally, a heterogenous group with hypothalamic connections. For additional details about some of these tracts in preceding chapters consult the Index.

1. *Tractus tecto-thalamicus et hypothalamicus cruciatus anterior* (fig. 12, *tr.t.th.h.c.a.*).—This *anterior tectal fasciculus* is a mixture of myelinated and unmyelinated fibers from the dorsomedial part of the tectum opticum and the pretectal nucleus which descend across the thalamus in company with the more anterior fascicles of the optic tract. After partial crossing in the anteroventral part of the chiasma ridge, its fibers spread in the neuropil of the chiasma ridge and hypothalamus. Some of them may reach beyond this region. This cumbersome name was applied in my earlier papers to a mixed fascicle which had not been analyzed. It can now be replaced by the names of the several tracts of which it is composed. Some of these are uncrossed (notably tr. pretecto-thalamicus), and some fibers of two of them decussate in the chiasma ridge, nos. 2 and 3 below. The arrangement of these components as seen in horizontal sections is shown in figures 25–36.

2. *Tractus tecto-hypothalamicus anterior* (*tr.t.hy.a.*).—This is the tectal component of the preceding fasciculus; it is evidently an optic pathway to the hypothalamus (p. 224). It passes through the pretectal nucleus and is accompanied by tecto-pretectal fibers. It probably is physiologically related with no. 3.

3. *Tractus pretecto-hypothalamicus* (fig. 15, *tr.pt.hy.*).—This tract as

it leaves the pretectal nucleus is accompanied by a large tr. pretecto-thalamicus (p. 234 and figs. 35, 36).

4. *Tractus tecto-thalamicus et hypothalamicus cruciatus posterior* (fig. 12, *tr.t.th.h.c.p.*).—This, like no. 1, is a mixed fascicle which has been analyzed. Both these names may now be discarded in favor of shorter terms—*anterior and posterior tectal fascicles*. The posterior fascicle arises chiefly from the nonoptic nucleus posterior tecti and the adjoining ventrolateral margin of the optic tectum, the latter region in Necturus receiving few terminals of the optic tract ('41a, p. 516). This fascicle is probably activated primarily by lemniscus, rather than optic, fibers or by a combination of the two. The tracts of which it is composed have collateral connections with the geniculate neuropil and ventral thalamus both before and after crossing. These fibers descend from the tectum parallel with those of the lateral optic tract and internally of them. So far as known they have a common origin in the tectum, but, after crossing, they take widely divergent courses. The tracts, consequently, are named according to their terminal distribution. The most important components are the following, nos. 5 and 6.

5. *Tractus tecto-hypothalamicus posterior.*—These are finer fibers which terminate in the postoptic neuropil and neighboring regions of the hypothalamus. They are more clearly seen in Necturus ('41a, p. 516) than in Amblystoma.

6. *Tractus tecto-tegmentalis cruciatus* (fig. 12, *tr.t.teg.c.*).—The course and distribution of the thicker fibers of the posterior tectal fascicle were not clarified until they were identified in larval stages, in which they were electively impregnated because they mature precociously. These were first recognized in early feeding larvae ('39, p. 106) and later in adult Necturus ('41a, p. 516) and Amblystoma ('42, p. 222). Some erroneous descriptions of these fibers in my earlier papers have been corrected ('39, p. 110).

The more heavily myelinated fibers of this tract decussate in the dorsal part of the chiasma ridge (figs. 2C, 12, 16, 25, *tr.t.th.h.c.p.*), and after crossing they enter tegmental fascicles of groups (8) and (6), and in smaller numbers they are dispersed in other fascicles. The dispersed fibers spread in the peduncle. Those which enter fascicles numbered (8) take a longer and more dorsal course, distributing to the dorsal, isthmic, and trigeminal tegmentum, some of them extending as far as the level of the V nerve roots. The entire course of these fibers can be followed in the horizontal sections, figures 25–35, where they are marked *tr.t.th.h.c.p.* before their decussation and (*8*) beyond the crossing.

There are two strong systems of crossed tecto-peduncular and tecto-tegmental fibers, both of which are drawn in figure 12. The system just described descends chiefly from the nonoptic part of the tectum (*tr.t.th.h.c.p.*) and, after crossing, spreads in the peduncle by way of tegmental fascicles (6) and throughout the tegmentum by way of fascicles (8). The second system is tr. tecto-peduncularis cruciatus (*tr.t.p.c.*), which arises in the optic part of the tectum, crosses in the commissure of the tuberculum posterius, and then spreads out in the peduncle. This well-myelinated tract is accompanied by similar fibers from the pretectal nucleus and dorsal thalamus. These two tecto-peduncular systems evidently have quite different physiological significance.

Attempts to analyze the postoptic components arising in the thalamus were unsuccessful until the sequence of development of these fibers was revealed by embryological studies. These findings were then confirmed by elective Golgi impregnations of older larvae and adults. Some errors in the earlier descriptions have been corrected, and now it is possible to give a fairly complete account of both the crossed and the uncrossed fibers which diverge from the thalamus. Since the direct and crossed fibers are evidently intimately related physiologically, both series are included in the following description. Efferent fibers from the thalamus are arranged in two sharply contrasted series, which arise, respectively, from the dorsal thalamus and the ventral thalamus.

The efferent series from the dorsal thalamus includes uncrossed fibers to the tectum, habenula, cerebral hemisphere, ventral thalamus, hypothalamus, and peduncle which need not be further considered here; but some of the other uncrossed tracts, which evidently are in reciprocal physiological relation with the crossed tracts, should be specifically mentioned. Decussating fibers from the dorsal thalamus are in two groups. The first includes thick myelinated fibers of tr. thalamo-peduncularis cruciatus (*tr.th.p.c.*), which, as already mentioned, joins tr. tecto-peduncularis cruciatus (*tr.t.p.c.*) to decussate in the commissure of the tuberculum posterius as described under that caption below. The second group is a much larger number of thin unmyelinated or lightly myelinated fibers which decussate in the postoptic commissure and will next be described.

7. *Tractus thalamo-hypothalamicus et peduncularis cruciatus* (*tr.th.h.p.c.*).—In the earlier descriptions of both Amblystoma and Necturus this name was given to a large collection of unmyelinated and lightly myelinated fibers which descends in dispersed arrangement from the dorsal thalamus to the postoptic commissure. Their distribution beyond the decussation could not be clearly followed, and some of those descriptions now require correction. As elsewhere pointed out ('42, p. 223), this name should now be discarded because at least three quite distinct tracts are represented here and relatively few of these fibers have any connection with the peduncle. The three tracts represented in this complex (nos. 8, 9, and 10) have a common origin in the dorsal thalamus, chiefly its middle sector, and, after crossing, take widely different courses. Their decussation is posterior to that of the tectal components of the commissure in a band which is narrow dorsally and spreads ventrally through a wide area of the neuropil of the chiasma ridge (fig. 2C, *tr.th.h.d.c.*).

8. *Tractus thalamo-hypothalamicus dorsalis cruciatus* (*tr.th.h.d.c.*).— These fibers, most of which are unmyelinated, cross in the middle region of the chiasma ridge (figs. 2C, 15, 25) and then spread in the hypothalamus. In figures 27–33, 95, 102, and 103 the symbol *tr.th.h.d.c.* refers to the mixture of fibers of tracts 8, 9, and 10 in their descending course from the thalamus to the commissure. The fibers of nos. 9 and 10 cross dorsally of those of no. 8, though there is mingling of the fibers of the three tracts with one another and with those of surrounding decussations.

The remaining fibers of this complex, after crossing, are distributed to the tegmentum in two tracts which take parallel courses, one superficially, the other at the border of the gray. These are designated components A and B, respectively. In figures 15 and 21 the components A and B are not separately designated.

9. *Tractus thalamo-tegmentalis dorsalis cruciatus A* (*tr.th.teg.d.c.A.*). —This tract was first described in the early feeding larva ('39, p. 116) and later in the adult ('42, p. 224). After decussation, these unmyelinated fibers ascend from the chiasma ridge parallel with the course of the descending uncrossed limb of this commissure and more superficially. Their further course is shown in figures 26–34, here marked A. In the dorsal tegmentum they turn spinalward along the ventrolateral border of the tectum. Here they lie close to the pial surface and immediately ventrally of the lateral optic tract (fig. 94, A). In this part of their course they join an uncrossed tract with similar

origin from the dorsal thalamus and similar distribution in the tegmentum—tr. thalamo-tegmentalis rectus (figs. 15, 31–34, 94, *tr.th.teg.r.*). This latter tract arises from both dorsal and ventral thalamus, but only the dorsal component of it is under consideration here (fig. 21, *tr.th.teg.d.r.*). In some Golgi preparations there is evidence that axons from the dorsal thalamus may divide, with branches entering both the uncrossed and the crossed thalamo-tegmental tracts ('42, p. 224). Evidently, the crossed and uncrossed tracts are reciprocally related physiologically.

10. *Tractus thalamo-tegmentalis dorsalis cruciatus B.*—This is a deep component of the same system as the preceding, receiving nearly all the myelinated fibers of no. 7 of this list. In the chiasma ridge these fibers separate from the others of the group and cross at the dorsal border of the postoptic commissure (marked B in figs. 26 and 95). Beyond the decussation they scatter widely, and most of them enter the well-myelinated tegmental fascicles of group (8), within which they may descend as far as the V nerve roots.

The two crossed tracts to the tegmentum, nos. 9 and 10 of this list, seem to have the same origin and about the same field of distribution, except that one of them (A) terminates in the superficial tegmental neuropil and the other (B) arborizes in the deep neuropil. The physiological properties of these zones of neuropil evidently are different.

11. *Tractus thalamo-tegmentalis ventralis cruciatus* (figs. 2C, 17, *tr.th.teg.v.c.*).—These thick fibers (many of them well myelinated) converge into the postoptic commissure from all parts of the ventral thalamus. In this part of their course they are not fasciculated but are scattered among other similar fibers so that their courses could not be followed until they were studied embryologically. They mature early and may be impregnated with reduced silver at stages when few other fibers respond to this treatment. We also have good elective Golgi impregnations of them in later larval stages ('39, pp. 98, 120; '39*b*, p. 546; '42, p. 225).

These fibers cross in the posterodorsal part of the chiasma ridge, mingled with those of other systems. The thickest of the myelinated fibers are crowded together at the dorsal margin of the postoptic complex. After crossing, they spread widely in the peduncle and tegmentum. Most of those from the posterior part of the ventral thalamus enter ventral tegmental fascicles of group (4), and some of these may descend in the f. longitudinalis medialis. Many fibers from the middle and anterior parts of the ventral thalamus reach the tegmentum through fascicles of groups (6) and (8). These crossed fibers are accompanied by many others that take similar courses without decussation.

The hypothalamic components of the postoptic commissure include, in addition to the terminals of extrinsic tracts already described, a few well-defined tracts and several other less-well-known connections.

12. *Tractus hypothalamo-peduncularis et tegmentalis* (figs. 18, 23, *tr.hy.ped.*; fig. 21, *tr.hy.teg.*).—This is the most noteworthy component originating within the hypothalamus. Its fibers assemble from the whole of the ventral part of the hypothalamus and comprise the main pathway from this region to the motor field of the peduncle and tegmentum. Some of them decussate in the posterior part of the chiasma ridge; others are uncrossed; still others decussate in the commissure of the tuberculum posterius. They connect by terminals or collaterals with the dorsal part of the hypothalamus, ventral part of the peduncle (including the neuropil of the area ventrolateralis pedunculi), and isthmic tegmentum. The last-mentioned group includes thick fibers, some of which are well myelinated, which enter ventral tegmental fascicles (4); and some of these may take long courses in the f. longitudinalis medialis (for further description see p. 280, and '42, p. 226).

13. *Olfactory projection tract* (fig. 19, *ol.p.tr.*).—This name has been given to a thin strand of unmyelinated fibers which pass in both directions between the strio-amygdaloid area and a specific nucleus at the posterior border of the chiasma ridge ('21, p. 247; '27, p. 304; '36, fig. 5). Some of these fibers decussate here.

14. *Tractus pedunculo-hypothalamicus.*—A large component in the posterodorsal part of the postoptic commissure was provisionally given this name, though the exact connections of its fibers could not be determined ('42, p. 227).

15. *Medial forebrain bundle.*—The fibers of the medial forebrain bundle are interlaced with all components of the postoptic commissure laterally of the chiasma ridge. Many of these fibers of both descending and ascending systems enter the postoptic neuropil and participate in its formation. This implies a transfer of more or less of this activity to the opposite side of the brain, but no details of specific decussational or commissural pathways have been revealed.

Postoptic neuropil.—In the mid-plane, the postoptic decussations occupy most of the chiasma ridge, and these fibers are enveloped on all sides except ventrally by a gray layer, the bed-nucleus of the postoptic commissure. This nucleus is expanded laterally. From this gray layer, richly arborized ependymal elements and dendrites of neurons are spread among the decussating fibers. Similar long dendrites enter it from all surrounding parts, including the nucleus magnocellularis, from which tr. hypophysius arises ('42, fig. 51). The entire chiasma ridge is also permeated with dense axonic neuropil which is continuous with that of surrounding parts. This neuropil receives terminals and collaterals of axons of hypothalamic neurons, medial forebrain bundles, tractus preopticus, and most of the decussating systems. The chief outflow from it seems to be by axons of the nucleus of the postoptic commissure directed into tr. hypothalamo-peduncularis. Other fibers enter the medial forebrain bundle (for further details see '42, p. 219; Necturus, '33b, p. 251; '34b, p. 383).

This neuropil is clearly one of the major adjusting centers of the urodele brain. Situated in the center of the great olfacto-visceral field, its connections indicate that it may be activated from every correlation center of the cerebrum. Undoubtedly it plays an important part in all general visceral activities. It is equally evident that there is no provision here for localization of specific functions. This is probably the undifferentiated primordium from which some of the specialized hypothalamic nuclei of mammals have been elaborated.

COMMISSURA TUBERCULI POSTERIORIS

The commissure of the tuberculum posterius was defined in 1917 (p. 224) as the ventral mesencephalic decussations between the infundibulum and the fovea isthmi. Posteriorly of the latter the ventral tegmental commissure extends backward without interruption through the rhombencephalon and spinal cord. In Necturus there is a very short interruption of the ventral commissural system at the fovea isthmi ('30, p. 89), but in most urodeles this gap does not appear.

The composition and arrangement of these ventral commissures are very diversified in different vertebrates. The hypothalamic connections at the anteroventral end of the commissure of the tuberculum posterius of Amblystoma are in some other vertebrates widely separated as the retro-infundibular decussations, or decussatio hypothalamicus posterior. The remainder of the commissure of the tuberculum and the anterior part of the ventral tegmental commissure comprise the ansulate commissure of the literature.

In Amblystoma I have recognized four chief components of the commissure of the tuberculum posterius ('36, p. 305 and fig. 2), arranged as shown in sagittal section in figures 81, 104; in horizontal sections in figures 28, 29, 30, 31 (component 4 being here marked, *tr.t.b.c.1.*); and in transverse section in figure 94 (components 2 and 3 shown here, components 1 and 4 being shown in a neighboring section, '36, fig. 11). The connections of these components are as follows:

1. This is the partial decussation of unmyelinated fibers from the dorsal (mamillary) part of the hypothalamus to the interpeduncular nucleus ('36, p. 338 and figs. 3, 8, 11–13). Most of them enter ventral tegmental fascicles of group (2) as marked *tr.mam.inp.*, on figures 19, 27–30, 71, 81, 92, 103.

2. This component contains partly myelinated fibers from the dorsal hypothalamus to the peduncle and tegmentum, accompanied by many uncrossed fibers and by fibers conducting in the reverse direction from the peduncle to the hypothalamus (p. 278; figs. 8, 21, *tr.mam.teg.*, 23, *tr.mam.ped.*). Most of them spread in the alba of the peduncle or enter ventral tegmental fascicles of group (3). Another system is tr. hypothalamo-peduncularis et tegmentalis from the ventral part of the hypothalamus to the peduncle and tegmentum (fig. 21, *tr.hy.teg.* and fig. 23, *tr.hy.ped.*). There is interchange of all these fibers with those of component 1 and with other tegmental fascicles. They are marked *tr.mam.ped.*, *tr.mam.teg.*, and *(3)* on figures 18, 21, 23, 27–31, 79, 82, 92, 94, 103.

3. The third component contains heavily myelinated fibers from the tectum, pretectal nucleus, and dorsal thalamus to the peduncle, accompanied by many uncrossed fibers—tr. tecto-peduncularis cruciatus (*tr.t.p.c.*) and tr. thalamo-peduncularis cruciatus (*tr.th.p.c.*), as elsewhere described (p. 223; '42, p. 267). These are shown on figures 12, 15, 18, 22, 29–36, 94, 103.

4. The dorsal component of this commissure is larger than the others and is composed chiefly of tecto-bulbar and tecto-spinal fibers from the anterior part of the tectum opticum, marked *tr.t.b.c.1.* on the figures. Its fibers after oblique decussation enter ventral tegmental fascicles of group (1) (see p. 277 and figs. 12, 27–36, 93, 94).

In addition to these four well-defined components, there are unmyelinated fibers which descend from the tectum and dorsal tegmentum through the gray and deeper layers of the alba to the region of the nucleus of the III nerve (fig. 22, *tr.t.ped.*). Some of these tectopeduncular fibers decussate in the ventral commissure both before and behind the fovea isthmi. In Necturus some fibers of the nervus terminalis probably decussate in the commissure of the tuberculum posterius (McKibben, '11), and this may be true in Amblystoma.

COMMISSURA VENTRALIS

The ventral decussations of Necturus were analyzed in the paper of 1930 (p. 89), and those of Amblystoma are similar. For the details the reader is referred to that description and to a later contribution on larval Amblystoma ('39b).

BIBLIOGRAPHY

BIBLIOGRAPHY

This list includes only the works cited in the text. It is not a systematic bibliography of the amphibian nervous system. Additional references are cited in the author's papers, in the comprehensive work by Ariëns Kappers, Huber, and Crosby ('36), and in numerous other publications.

ADDENS, J. L. 1946. The nucleus of Bellonci and adjacent cell groups in selachians. II, Proc. kon. Akad. Wetensch., Amsterdam, **49**:94–100. Also earlier papers cited.

ADELMANN, H. B. 1929. Experimental studies on the development of the eye. I, J. Exper. Zoöl., **54**:249–90.

———. 1936. The problem of cyclopia, Quart. Rev. Biol., **11**:161–82; 284–304.

AGAR, W. E. 1943. A contribution to the theory of the living organism. Melbourne and London: Melbourne University Press and Oxford University Press.

ARIËNS KAPPERS, C. U. 1929. The evolution of the nervous system in invertebrates, vertebrates and man. Haarlem: Bohn.

ARIËNS KAPPERS, C. U.; HUBER, G. CARL; and CROSBY, E. C. 1936. The comparative anatomy of the nervous system of vertebrates including man. New York: Macmillan Co.

ARONSON, LESTER R., and NOBLE, G. K. 1945. The sexual behavior of Anura. II. Neural mechanisms controlling mating in the male leopard frog, Rana pipiens, Bull. Am. Mus. Nat. Hist., **86**:83–140, article 3.

BAGLEY, CHARLES, JR., and LANGWORTHY, O. R. 1926. The forebrain and midbrain of the alligator, etc., Arch. Neurol. & Psychiat., **16**:154–66.

BAILEY, P., and BONIN, G. VON. 1946. Concerning cytoarchitectonics. Trans. Am. Neurol. Assoc., 71st Meeting, Pp. 89–93.

BAILEY, P., and DAVIS, E. W. 1942. Effects of lesions of the periaqueductal gray matter in the cat, Proc. Soc. Exper. Biol. & Med., **51**:305–6.

———. 1942a. The syndrome of obstinate progression in the cat, ibid., p. 307.

BAKER, R. C. 1927. The early development of the ventral part of the neural plate of Amblystoma, J. Comp. Neurol., **44**:1–27.

BAKER, R. C., and GRAVES, G. O. 1932. The development of the brain of Amblystoma (3 to 17 mm. body length), J. Comp. Neurol., **54**:501–59.

BARNARD, J. W. 1936. A phylogenetic study of the visceral afferent areas, etc., J. Comp. Neurol., **65**:503–602.

BELLONCI, J. 1888. Ueber die centrale Endigung des Nervus opticus bei den Vertebraten, Ztschr. f. wissensch. Zool., **47**:1–46.

BENEDETTI, E. 1933. Il cervello e i nervi cranici del Proteus anguineus Laur., Mem. Ist. ital. di speleologia, ser. biol., Mem. III, pp. 1–80.

BENZON, A. 1926. Die markhaltigen Faserzüge im Vorderhirn von Cryptobranchus japonicus, Ztschr. f. mikr.-anat. Forsch., **5**:285–314.

BERITOFF, J. S. (ed.). 1943. Trans. of the J. Beritashvili Physiologica Institute, No. 5. Pp. xiv+532. Tbilisi (Tiflis), Georgia, U.S.S.R. (with abstracts in English).

BINDEWALD, C. A. E. 1914. Das Vorderhirn von Amblystoma mexicanum, Arch. f. mikr. Anat., Abt. 1, **84**:1–74.

BISHOP, S. C. 1943. Handbook of salamanders: the salamanders of the United States, of Canada and of Lower California. Ithaca, N.Y.: Comstock Pub. Co.

BODIAN, D. 1937. The structure of the vertebrate synapse, J. Comp. Neurol., **68**:117–59.

BODIAN, D. 1942. Cytological aspects of synaptic function, Physiol. Rev., **22**:146–69.
BONIN, G. VON. 1945. The cortex of Galago; its relation to the pattern of the primitive cortex. Illinois Monog. M. Sc. Vol. **5,** No. 3. Urbana: University of Illinois Press.
BRICKNER, RICHARD M. 1930. A new tract in Herrick's gustatory system in certain teleosts, J. Comp. Neurol. **50**:153–57.
BRODAL, A. 1940. Experimentelle Untersuchungen über die olivo-cerebellare Lokalisation, Ztschr. f. d. ges. Neurol. u. Psychiat., **169**:1–153.
BRODAL, A., and JANSEN, J. 1946. The ponto-cerebellar projection in the rabbit and cat, J. Comp. Neurol., **84**:31–118.
BUCY, PAUL C. (ed.). 1944. The precentral motor cortex. Illinois Monog. M. Sc., Vol. **4,** Urbana: University of Illinois Press.
BURR, H. S. 1922. The early development of the cerebral hemispheres in Amblystoma, J. Comp. Neurol., **34**:277–301.
CALDERON, LUIS. 1928. Sur la structure du ganglion interpédonculaire, Trav. du lab. de recherches biol. de l'Univ. de Madrid, **25**:297–306.
CAMPION, GEORGE G., and SMITH, G. ELLIOT. 1934. The neural basis of thought. New York: Harcourt, Brace & Co.
CHEZAR, H. H. 1930. Studies on the lateral-line system of Amphibia. II, J. Comp. Neurol., **50**:159–75.
CHILD, C. M. 1941. Patterns and problems of development. Chicago: University of Chicago Press.
CLARK, W. E. LE GROS. 1933. The medial geniculate body and the nucleus isthmi, J. Anat., **67**:536–48.
———. 1943. The anatomy of cortical vision, Tr. Ophth. Soc., **62**:229–45 (for 1942).
COGHILL, G. E. 1902. The cranial nerves of Amblystoma tigrinum, J. Comp. Neurol., **12**:205–89.
———. 1913. The primary ventral roots and somatic motor column of Amblystoma, ibid., **23**:121–43.
———. 1914–36. Correlated anatomical and physiological studies of the growth of the nervous system of Amphibia, Papers I–XII, ibid., Vols. **24–64.**
———. 1929. Anatomy and the problem of behaviour. Cambridge: Cambridge University Press.
———. 1930. The structural basis of the integration of behavior, Proc. Nat. Acad. Sc., **16**:637–43.
———. 1933. The neuro-embryological study of behavior: principles, perspective and aim, Science, **78**:131–38.
———. 1936. Integration and motivation of behavior as problems of growth, J. Genet. Psychol., **48**:3–19.
———. 1940. Early embryonic somatic movements in birds and in mammals other than man. Monog. Soc. Research in Child Development, Vol. **5,** No. 2. Washington, D.C.: National Research Council.
———. 1943. Flexion spasms and mass reflexes in relation to the ontogenetic development of behavior, J. Comp. Neurol., **79**:463–86.
CONEL, J. LEROY. 1929. The development of the brain of Bdellostoma stouti. I. External growth changes, J. Comp. Neurol., **47**:343–403.
———. 1931. The development of the brain of Bdellostoma stouti. II. Internal growth changes, ibid., **52**:365–499.
CORBIN, K. B. 1940. Observations on the peripheral distribution of fibers arising in the mesencephalic nucleus of the fifth cranial nerve, J. Comp. Neurol., **73**:153–77.

CRAIGIE, E. HORNE. 1938. The vascularization of the hypophysis in tailed amphibians, Tr. Roy. Soc. Canada, ser. 3, **32**:43–50.

———. 1938a. The blood vessels of the brain substance in some amphibians, Proc. Am. Phil. Soc., **78**:615–49.

———. 1939. Vascularity in the brains of tailed amphibians. I. Amblystoma tigrinum (Green), ibid., **81**:21–27.

———. 1940. The cerebral cortex in palaeognathine and neognathine birds, J. Comp. Neurol., **73**:179–234.

———. 1945. The architecture of the cerebral capillary bed, Biol. Rev., **20**:133–46.

CROSBY, ELIZABETH C. 1917. The forebrain of Alligator mississippiensis, J. Comp. Neurol., **27**:325–402.

CROSBY, ELIZABETH C., and WOODBURNE, R. T. 1938. Certain major trends in the development of the efferent systems of the brain and the spinal cord, Univ. Hosp. Bull., Ann Arbor, Mich., **4**:125–28.

———. 1940. The comparative anatomy of the preoptic area and the hypothalamus, A. Research Nerv. & Ment. Dis., Proc., **20**:52–169.

CUSHING, H. 1903. The taste fibers and their independence of the n. trigeminus, Bull. Johns Hopkins Hosp., **14**:71–78.

DEMPSTER, W. T. 1930. The morphology of the amphibian endolymphatic organ, J. Morphol., **50**:71–126.

DETWILER, S. R. 1945. The results of unilateral and bilateral extirpation of the forebrain of Amblystoma, J. Exper. Zoöl., **100**:103–17.

———. 1946. Experiments upon the midbrain of Amblystoma embryos, Am. J. Anat., **78**:115–38.

DEWEY, JOHN. 1896. The reflex arc concept in psychology, Psychol. Rev., **3**:357–70. See also Dewey's later statement in J. Phil., **9**:664–68, 1912, especially the footnote on p. 667.

Dow, R. S. 1942. The evolution and anatomy of the cerebellum, Biol. Rev., **17**:179–220.

ECONOMO, C. 1926. Ein Koeffizient für die Organisationshöhe der Grosshirnrinde, Klin. Wchnschr., 5 Jhrg., I Halbjahres, pp. 593–95.

———. 1929. The cytoarchitectonics of the cerebral cortex. Translated by S. PARKER. Oxford University Press.

EDINGER, L. 1911. Vorlesungen über den Bau der nervösen Zentralorgane, Vol **1**. 8th ed. Leipzig: F. C. W. Vogel.

EMERSON, ALFRED E. 1942. The modern naturalist, Bull. Transylvania Coll., **15**:71–77.

———. 1943. Ecology, evolution and society, Am. Nat., **77**:97–118.

ESTABLE, C. 1924. Terminaisons nerveuses branchiales de larve du Pleurodeles waltlii et certaines données sur l'innervation gustative, Trav. du lab. de recherches biol. de l'Univ. de Madrid, **22**:369–84.

EVANS, F. GAYNOR. 1944. The morphological status of the modern Amphibia among the Tetropoda, J. Morphol., **74**:43–100.

FERRARO, A., and BARRERA, S. E. 1935. Posterior column fibers and their termination in Macacus rhesus, J. Comp. Neurol., **62**:507–30.

FOREL, A. 1877. Untersuchungen über die Haubenregion u.s.w., Arch. f. Psychiat., **7**:393–495.

Fox, C. A. 1941. The mammillary peduncle and ventral tegmental nucleus in the cat. J. Comp. Neurol. **75**:411–25.

FRANCIS, E. T. B. 1934. The anatomy of the salamander. Oxford University Press.

FREY, EUGEN. 1938. Studien über die hypothalamische Opticuswurzel der Amphibien. II. Proteus anguineus, Proc. kon. Akad. Wetensch., Amsterdam, **41**:1015–21.

FULTON, J. F. 1943. Physiology of the nervous system. 2d ed. Oxford Univ. Press.

GANSER, S. 1882. Vergleichend-anatomische Studien über das Gehirn des Maulwurfs, Morphol. Jahrb., **7**:591–725.

GAUPP, E. 1899. Anatomie des Frosches, Abt. 2, Auf. 2. Braunschweig.

GEHUCHTEN, A. VAN. 1894. Contribution à l'étude du système nerveux des téleostéens, Cellule, **10**:255–95.

———. 1895. Le faisceau longitudinal postérieur, Bull. Acad. roy. de méd. de Belgique, **9**:1–40.

———. 1897. La moelle épinière des larves des batraciens (Salamandra maculosa), Arch. de biol., **15**:251–71.

———. 1897a. Le ganglion basal et la commissure habénulaire dans l'encéphale de la salamandre, Bull. Acad. roy. de Belgique, ser. 3, **34**:38–67.

———. 1900. Anatomie du système nerveux de l'homme. 2 vols. Louvain.

GEIRINGER, MARTHA. 1938. Die Beziehungen der basalen Optikuswurzel zur Hypophyse und ihre Bedeutung für den Farbwechsel der Amphibien, Anat. Anz., **86**: 202–7.

GESELL, ROBERT, and HANSEN, E. T. 1945. Anticholinesterase activity of acid as a biological instrument of nervous integration, Am. J. Physiol., **144**:126–63.

GILLILAN, LOIS A. 1941. The connections of the basal optic root (posterior accessory optic tract) and its nucleus in various mammals, J. Comp. Neurol., **74**:367–408.

GLASSER, OTTO (ed.). 1944. Medical physics. Chicago: Year Book Publishers.

GREGORY, WILLIAM K. 1943. Environment and locomotion in mammals. Nat. Hist., **51**:222–27.

GRIGGS, L. 1910. Early stages in the development of the central nervous system of Amblystoma punctatum, J. Morphol., **21**:425–83.

GUDDEN, B. VON. 1881. Mitteilung über das Ganglion interpedunculare, Arch. f. Psychiat., **11**:424–27.

HAMBURGER, V., and KEEFE, E. L. 1944. The effects of peripheral factors on the proliferation and differentiation in the spinal cord of chick embryos, J. Exper. Zoöl., **96**:223–42.

HERRICK, C. JUDSON. 1899. The cranial and first spinal nerves of Menidia: a contribution upon the nerve components of the bony fishes, J. Comp. Neurol., **9**: 153–455.

———. 1900. A contribution upon the cranial nerves of the cod fish, ibid., **10**: 265–316.

———. 1903. The organ and sense of taste in fishes, Bull. U.S. Fish Com., **22**:237–72.

———. 1903a. On the morphological and physiological classification of the cutaneous sense organs of fishes, Am. Nat., **37**:313–18.

———. 1903b. On the phylogeny and morphological position of the terminal buds of fishes, J. Comp. Neurol., **13**:121–38.

———. 1905. The central gustatory paths in the brains of bony fishes, ibid., **15**: 375–456.

———. 1908. The morphological subdivision of the brain, ibid., **18**:393–408.

———. 1909. The nervus terminalis (nerve of Pinkus) in the frog, ibid., **19**:175–90.

———. 1910. The morphology of the forebrain in Amphibia and Reptilia, ibid., **20**:413–547.

———. 1913. Brain, anatomy of the, in Reference handbook of the medical sciences, **2**:274–342. 3d ed. New York: William Wood & Co.

HERRICK, C. JUDSON. 1913a. Some reflections on the origin and significance of the cerebral cortex, J. Anim. Behavior, 3:222–36.

———. 1914. The cerebellum of Necturus and other urodele Amphibia, J. Comp. Neurol., 24:1–29.

———. 1914a. The medulla oblongata of larval Amblystoma, ibid., pp. 343–427.

———. 1917. The internal structure of the midbrain and thalamus of Necturus, ibid., 28:215–348.

———. 1920. Irreversible differentiation and orthogenesis, Science, 51:621–25.

———. 1921. A sketch of the origin of the cerebral hemispheres, J. Comp. Neurol., 32:429–54.

———. 1921a. The connections of the vomeronasal nerve, accessory olfactory bulb and amygdala in Amphibia, ibid., 33:213–80.

———. 1922. What are viscera? J. Anat., 56:167–76.

———. 1922a. Functional factors in the morphology of the forebrain of fishes, Libro en honor de D. Santiago Ramón y Cajal, 1:143–204. Madrid.

———. 1922b. Some factors in the development of the amphibian nervous system, Anat. Rec., 23:291–305.

———. 1924. Origin and evolution of the cerebellum, Arch. Neurol. & Psychiat., 11:621–52.

———. 1924a. The amphibian forebrain. I. Amblystoma, external form, J. Comp. Neurol., 37:361–71.

———. 1924b. The amphibian forebrain. II. The olfactory bulb of Amblystoma, ibid., pp. 273–396.

———. 1924c. Neurological foundations of animal behavior. New York: Henry Holt & Co.

———. 1924d. The nucleus olfactorius anterior of the opossum, J. Comp. Neurol., 37:317–59.

———. 1925. The amphibian forebrain. III. The optic tracts and centers of Amblystoma and the frog, ibid., 39:433–89.

———. 1925a. Morphogenic factors in the differentiation of the nervous system, Physiol. Rev., 5:112–30.

———. 1925b. The innervation of palatal taste buds and teeth of Amblystoma, J. Comp. Neurol., 38:389–97.

———. 1926. Brains of rats and men. Chicago: University of Chicago Press.

———. 1927. The amphibian forebrain. IV. The cerebral hemispheres of Amblystoma, J. Comp. Neurol., 43:231–325.

———. 1929. Anatomical patterns and behavior patterns, Physiol. Zoöl., 2:439–48.

———. 1930. The medulla oblongata of Necturus, J. Comp. Neurol., 50:1–96.

———. 1930a. Localization of function in the nervous system, Proc. Nat. Acad. Sc., 16:643–50.

———. 1931. The amphibian forebrain. V. The olfactory bulb of Necturus, J. Comp. Neurol., 53:55–69.

———. 1931a. An introduction to neurology. 5th ed. Philadelphia: W. B. Saunders Co.

———. 1933. The functions of the olfactory parts of the cerebral cortex, Proc. Nat. Acad. Sc., 19:7–14.

———. 1933a. Morphogenesis of the brain, J. Morphol., 54:233–58.

———. 1933b. The amphibian forebrain. VI. Necturus, J. Comp. Neurol., 58: 1-288.

———. 1933c. The amphibian forebrain. VII. The architectural plan of the brain, ibid., pp. 481–505.

———. 1933d. The evolution of cerebral localization patterns, Science, 78:439–44.

HERRICK, C. JUDSON. 1933e. The amphibian forebrain. VIII. Cerebral hemispheres and pallial primordia, J. Comp. Neurol., **58**:737–59.

———. 1934. The amphibian forebrain. IX. Neuropil and other interstitial nervous tissue, ibid., **59**:93–116.

———. 1934a. The amphibian forebrain. X. Localized functions and integrating functions, ibid., pp. 239–66.

———. 1934b. The hypothalamus of Necturus, ibid., pp. 375–429.

———. 1934c. The interpeduncular nucleus of the brain of Necturus, ibid., **60**:111–35.

———. 1934d. The endocranial blood vascular system of Amblystoma, Ztschr. f. mikr.-anat. Forsch., **36**:540–44.

———. 1935. The membranous parts of the brain, meninges and their blood vessels in Amblystoma, J. Comp. Neurol., **61**:297–346.

———. 1935a. A topographic analysis of the thalamus and midbrain of Amblystoma, ibid., **62**:239–61.

———. 1936. Conduction pathways in the cerebral peduncle of Amblystoma, ibid., **63**:293–352.

———. 1937. Development of the brain of Amblystoma in early functional stages, ibid., **67**:381–422.

———. 1938. Development of the cerebrum of Amblystoma during early swimming stages, ibid., **68**:203–41.

———. 1938a. Development of the brain of Amblystoma punctatum from early swimming to feeding stages, ibid., **69**:13–30.

———. 1938b. The brains of Amblystoma punctatum and A. tigrinum in early feeding stages, ibid., pp. 391–426.

———. 1939. Internal structure of the thalamus and midbrain of early feeding larvae of Amblystoma, ibid., **70**:89–135.

———. 1939a. The cerebrum of Amblystoma tigrinum in midlarval stages, ibid., pp. 249–66.

———. 1939b. Cerebral fiber tracts of Amblystoma tigrinum in midlarval stages, ibid., **71**:511–612.

———. 1941. Development of the optic nerves of Amblystoma, ibid., **74**:473–534.

———. 1941a. Optic and postoptic systems of fibers in the brain of Necturus, ibid., **75**:487–544.

———. 1941b. The eyes and optic paths of the catfish, Ameiurus, ibid., **75**:255–86.

———. 1942. Optic and postoptic systems in the brain of Amblystoma tigrinum, ibid., **77**:191–353.

———. 1943. The cranial nerves: a review of fifty years, Denison Univ. Bull., J. Sc. Lab., **38**:41–51.

———. 1944. The incentives of science, Scient. Monthly, **58**, 462–66.

———. 1944a. Apparatus of optic and visceral correlation in the brain of Amblystoma, J. Comp. Psychol., **37**:97–105.

———. 1944b. The fasciculus solitarius and its connections in amphibians and fishes, J. Comp. Neurol., **81**:307–31.

———. 1947. The proprioceptive nervous system, J. Nerv. & Ment. Dis., **106**:355–58.

———. 1948. George Ellett Coghill, naturalist and philosopher. To be published by University of Chicago Press.

HOAGLAND, H. 1933. Electrical responses from the lateral-line nerves of catfish. I, J. Gen. Physiol., **16**:695–714.

———. 1933b. Quantitative analysis of responses from lateral-line nerves of fishes. II, ibid., pp. 715–31.

HOLMGREN, NILS. 1922. Points of view concerning forebrain morphology in lower vertebrates, J. Comp. Neurol., **34**:391–459.

———. 1946. On two embryos of Myxine glutinosa, Acta Zool., **27**:1–90.

HOLMGREN, NILS, and VAN DER HORST, C. J. 1925. Contribution to the morphology of the brain of Ceratodus, Acta Zool., **6**:59–165.

HOLTFRETER, J. 1945. Differential inhibition of growth and differentiation by mechanical and chemical means, Anat. Rec., **93**:59–74.

HOOKER, DAVENPORT. 1944. The origin of overt behavior. Ann Arbor: University of Michigan Press.

HOWELL, A. BRAZIER. 1944. Speed in animals, their specialization for running and leaping. Chicago: University of Chicago Press.

HUBER, G. C., and CROSBY, ELIZABETH C. 1933. A phylogenetic consideration of the optic tectum. Proc. Nat. Acad. Sc., **19**:15–22.

———. 1934. The influences of afferent paths on the cytoarchitectonic structure of the submammalian optic tectum, Psychiat. en Neurol. Bl, pp. 459–74.

———. 1943. A comparison of the mammalian and reptilian tecta, J. Comp. Neurol., **78**:133–68.

HUMPHREY, TRYPHENA. 1944. Primitive neurons in the embryonic human central nervous system, J. Comp. Neurol., **81**:1–45.

JACOBSOHN-LASK. 1924. Die Kreuzung der Nervenbahnen und die bilaterale Symmetrie des tierschen Körpers. Berlin: S. Karger.

JANSEN, JAN. 1930. The brain of Myxine glutinosa, J. Comp. Neurol., **49**:359–507.

JESERICH, MARGUERITE W. 1945. The nuclear pattern and the fiber connections of certain non-cortical areas of the telencephalon of the mink (Mustela vison), J. Comp. Neurol., **83**:173–211.

JOHNSTON, J. B. 1901. The brain of Acipenser, Zool. Jahrb., **15**:59–260.

———. 1902. The brain of Petromyzon, J. Comp. Neurol., **12**:1–86.

———. 1916. Evidence of a motor pallium in the forebrain of reptiles, ibid., **26**: 475–79

JONES, D. S. 1945. The origin of the ciliary ganglia in the chick embryo, Anat. Rec., **92**:441–47.

KABAT, H. 1936. Electrical stimulation of points in the forebrain and midbrain: the resultant alterations in respiration, J. Comp. Neurol., **64**:187–208.

KATO, G. 1934. The microphysiology of nerve. Tokyo: Maruzen Co., Ltd.

KINGSBURY, B. F., 1895. On the brain of Necturus maculatus, J. Comp. Neurol., **5**:139–205.

———. 1930. The developmental significance of the floor-plate of the brain and spinal cord, ibid., **50**:177–207.

KODAMA, S. 1929. Ueber die sogenannten Basalganglien. B. Ueber die Faserverbindungen den Basalganglien u.s.w., Schweiz. Arch. f. Neurol. u. Psychiat., **23**:179–265.

KOSTIR, W. J. 1924. An analysis of the cranial ganglia of an embryo salamander, Amblystoma jeffersonianum (Green), Ohio J. Sc., **24**:230–63.

KREHT, HANS. 1930. Ueber die Faserzüge im Zentralnervensystem von Salamandra maculosa L., Ztschr. f. mikr.-anat. Forsch., **23**:239–320.

———. 1931. Ueber die Faserzüge im Zentralnervensystem von Proteus anguineus Laur., ibid., **25**:376–427.

KUHLENBECK, H. 1921. Zur Morphologie des Urodelenvorderhirns, Jenaische Ztschr. f. Naturwissensch., **57**, N.F., **50**:463–90.

———. 1922. Zur Morphologie des Gymnophionengehirns, ibid., **58**:453–84.

KUPFFER, C. VON. 1906. Die Morphogenie des Zentralnervensystems, Hertwig's Handb. f. Entw. d. Wirbeltiere, **2**, Teil 3, 1–272. Jena.

LANDACRE, F. L. 1921. The fate of the neural crest in the head of the urodeles, J. Comp. Neurol., 33:1–43.

———. 1926. The primitive lines of Amblystoma jeffersonianum, ibid., 40:471–95.

LARSELL, O. 1920. The cerebellum of Amblystoma, J. Comp. Neurol., 31:259–82.

———. 1923. The cerebellum of the frog, ibid., 36:89–112.

———. 1924. The nucleus isthmi of the frog, ibid., 36:309–22.

———. 1925. The development of the cerebellum in the frog (Hyla regilla) in relation to the vestibular and lateral-line systems, ibid., 39:249–89.

———. 1929. The nerve terminations in the lateral-line organs of Amblystoma, ibid., 48:465–70.

———. 1931. The cerebellum of Triturus torosus, ibid., 53:1–54.

———. 1932. The development of the cerebellum in Amblystoma, ibid., 54:357–435.

———. 1934. The differentiation of the peripheral and central acoustic apparatus in the frog, ibid., 60:473–527.

———. 1934a. Morphogenesis and evolution of the cerebellum, Arch. Neurol. & Psychiat., 31:373–95.

———. 1937. The cerebellum: a review and interpretation, ibid., 38:580–607.

———. 1945. Comparative neurology and present knowledge of the cerebellum, Bull. Minnesota M. Foundation, 5:73–85.

———. 1947. The cerebellum of myxinoids and petromyzonts, including developmental stages in the lampreys, J. Comp. Neurol., 86:395–445.

———. 1947a. The nucleus of the IV nerve in petromyzonts, ibid., pp. 447–66.

LASHLEY, K. S. 1934. The mechanism of vision. VII, J. Comp. Neurol., 59:341–73.

———. 1934a. The mechanism of vision. VIII, ibid., 60:57–79.

———. 1941. Thalamo-cortical connections of the rat's brain, ibid., 75:67–121.

LASHLEY, K. S., and CLARK, GEORGE. 1946. The cytoarchitecture of the cerebral cortex of Ateles: a critical examination of architectonic studies, J. Comp. Neurol., 85: 223–305.

LIGGETT, J. R. 1928. An experimental study of the olfactory sensitivity of the white rat, Genet. Psychol. Monog., 3:1–64.

LILLIE, RALPH S. 1945. General biology and philosophy of organism. Chicago: University of Chicago Press.

LOO, Y. T. 1931. The forebrain of the opossum, Didelphis virginiana. II, J. Comp. Neurol., 52:1–148.

MCKIBBEN, PAUL S. 1911. The nervus terminalis in urodele Amphibia, J. Comp. Neurol., 21:261–309.

———. 1913. The eye-muscle nerves in Necturus, ibid., 23:153–72.

MAGOUN, H. W. 1944. Bulbar inhibition and facilitation of motor activity, Science, 100:549–50.

MARBURG, O. 1944. The structure and fiber connections of the human habenula, J. Comp. Neurol., 80:211–33.

METTLER, FRED A. 1945. Fiber connections of the corpus striatum of the monkey and baboon, J. Comp. Neurol., 82:169–204.

METTLER, FRED A., and METTLER, CECILIA C. 1941. Role of the neostriatum, Am. J. Physiol., 133:594–601.

MONCRIEFF, R. W. 1944. The chemical senses. London: Leonard Hill, Ltd.

MURPHY, J. P., and GELLHORN, E. 1945. Multiplicity of representation versus punctate localization in the motor cortex, Arch. Neurol. & Psychiat., 54:256–273.

NEIMANIS, EMMA. 1931. Individual variation of form of the brain of Triton cristatus Laur. and its relation to the specific variation of the brain of Urodela, Bull. Soc. biol. de Lettonie, 2:67–92.

NOBLE, G. KINGSLEY. 1931. The biology of the Amphibia. New York: McGraw-Hill Book Co.

O'NEILL, H. M. 1898. Hirn- und Rückenmarkshüllen bei Amphibien, Morphol. Arb., Schwalbe, **8**:47-64.

OSBORN, H. F. 1888. A contribution to the internal structure of the amphibian brain, J. Morphol., **2**:51-96.

PALAY, S. L. 1944. The histology of the meninges of the toad (Bufo), Anat. Rec., **88**:257-70.

———. 1945. Neurosecretion. VII. The preoptico-hypophysial pathway in fishes. J. Comp. Neurol., **82**:129-43.

PAPEZ, JAMES W. 1936. Evolution of the medial geniculate body. J. Comp. Neurol., **64**:41-61.

———. 1944. Structures and mechanisms underlying the cerebral functions, Am. J. Psychol., **57**:291-316.

PARKER, G. H. 1912. The relation of smell, taste, and the common chemical sense in vertebrates, J. Philadelphia Acad. Nat. Sc., ser. 2, **15**:219-34.

———. 1918. A critical survey of the sense of hearing in fishes, Proc. Am. Phil. Soc., **57**:1-30.

———. 1922. Smell, taste and allied senses in the vertebrates, Philadelphia: J. B. Lippincott Co.

PARKER, G. H., and VAN HEUSEN, A. P. 1917. The reception of mechanical stimuli by the skin, lateral-line organs and ears in fishes, especially in Amiurus, Am. J. Physiol., **44**:463-89.

PEARSE, A. S. (ed.). 1936. Zoological names: a list of phyla, classes and orders. Durham, N.C.: Duke University Press.

PEARSON, A. A. 1945. Further observations on the intramedullary sensory type neurons along the hypoglossal nerve, J. Comp. Neurol., **82**:93-100.

PIATT, JEAN. 1945. Origin of the mesencephalic V root cells in Amblystoma, J. Comp. Neurol., **82**:35-53.

———. 1946. The influence of the peripheral field on the development of the mesencephalic V nucleus in Amblystoma, J. Exper. Zoöl., **102**:109-41.

POLYAK, S. L. 1941. The retina. Chicago: University of Chicago Press.

RAMÓN Y CAJAL, P. 1922. El cerebro de los batracios. Libro en Honor de D. S. Ramón y Cajal, **1**:13-150. Madrid.

RAMÓN Y CAJAL, S. 1911. Histologie du système nerveux de l'homme et des vertébrés, Vol. **2**. Paris.

RASMUSSEN, G. L. 1946. The olivary peduncle and other fiber projections of the superior olivary complex, J. Comp. Neurol., **84**:141-219.

REISER, OLIVER L. 1946. The world sensorium, New York: Avalon Press.

RÖTHIG, PAUL. 1911. Zellanordnungen und Faserzüge im Vorderhirn von Siren lacertina, Abh. kgl. Preuss. Akad. Wissensch., Anhang, pp. 1-23.

———. 1911a. Beiträge zum Studium des Zentralnervensystems der Wirbeltiere. IV. Die markhaltigen Faserzüge im Vorderhirn von Necturus maculatus, Arch. f. Anat. (u. Physiol.), pp. 48-56.

———. 1912. Beiträge..... V. Die Zellanordnungen im Vorderhirn der Amphibien, Verhandl. Kon. Akad. Wetensch., Amsterdam, sec. 2, Deel 17, pp. 1-23.

——— 1923. Beiträge..... VIII. Ueber das Zwischenhirn der Amphibien, Arch. f. mikr. anat., **98**:616-45.

———. 1924. Beiträge..... IX. Ueber die Faserzüge im Zwischenhirn der Urodelen, Ztzchr. f. mikr.-anat. Forsch., **1**:5-40.

———. 1927. Beiträge..... XI. Ueber die Faserzüge im Mittelhirn, Kleinhirn und der Medulla oblongata der Urodelen und Anuren, ibid., **10**:381-472.

Romer, A. S. 1946. The early evolution of fishes, Quart. Rev. Biol., **21**:33–69.
Roofe, P. G. 1935. The endocranial blood vessels of Amblystoma tigrinum, J. Comp. Neurol., **61**:257–93.
———. 1937. The morphology of the hypophysis of Amblystoma. J. Morphol., **61**:485–94.
———. 1938. The blood vascular system of the hypophysis of Amblystoma tigrinum, J. Comp. Neurol., **69**:249–54.
Rudebeck, Birger. 1945. Contributions to forebrain morphology in Dipnoi, Acta Zool., **26**:9–156.
Scharrer, E. 1932. Experiments on the function of the lateral-line organs in the larvae of Amblystoma punctatum, J. Exper. Zoöl., **61**:109–14.
Scharrer, E., and Scharrer, B. 1940. Secretory cells within the hypothalamus, A. Research Nerv. & Ment. Dis. Proc., **20**:170–94.
Scharrer, E.; Smith, S. W.; and Palay, S. L. 1947. Chemical sense and taste in the fishes, Prionotus and Trichogaster, J. Comp. Neurol., **86**:183–98.
Schriever, H. 1935. Aktionspotentiale des N. lateralis bei Reizung der Seitenorganie von Fischen, Arch. f. d. ges. Physiol., **235**:771–84.
Shanklin, W. M. 1933. The comparative neurology of the nucleus opticus tegmenti with special reference to Chameleon vulgaris, Acta Zool., **14**:163–84.
Sheldon, R. E. 1909. The reactions of the dogfish to chemical stimuli, J. Comp. Neurol., **19**:273–311.
Sherrington, C. S. 1906. The integrative action of the nervous system. New York: Charles Scribner's Sons.
Söderberg, Gertie. 1922. Contributions to the forebrain morphology in amphibians, Acta Zool., **3**:65–121.
Sosa, Julio Maria. 1945. Collateral nerve fibers within septum dorsale of the spinal cord and medulla oblongata and their connections, J. Comp. Neurol., **83**:157–71.
Speidel, C. C. 1946. Prolonged histories of vagus nerve regeneration patterns, sterile distal stumps, and sheath cell outgrowths (abstr.), Anat. Rec., **94**:499.
Sperry, R. W. 1943. Effect of 180 degree rotation of the retinal field on visuomotor coordination, J. Exper. Zoöl., **92**:263–79.
———. 1944. Optic nerve regeneration with return of vision in anurans, J. Neurophysiol., **7**:57–70.
———. 1945. Restoration of vision after crossing of optic nerves and after contralateral transplantation of eye, ibid., **8**:15–28.
———. 1945a. Centripetal regeneration of the 8th cranial nerve root with systematic restoration of vestibular reflexes, Am. J. Physiol., **144**:735–41.
———. 1945b. The problem of central nervous reorganization after nerve regeneration and muscle transposition, Quart. Rev. Biol., **20**:311–69.
Stensiö, E. A. 1927. The Downtonian and Devonian vertebrates of Spitsbergen. I. Family Cephalaspidae. Det Norske Vidensk.-Akad. i Oslo.
Stone, L. S. 1922. Experiments on the development of the cranial ganglia and the lateral-line sense organs in Amblystoma punctatum, J. Exper. Zoöl., **35**:421–96.
———. 1926. Further experiments on the extirpation and transplantation of mesectoderm in Amblystoma punctatum, ibid., **44**:95–131.
———. 1944. Functional polarization in retinal development and its reestablishment in regenerating retinae of rotated grafted eyes, Proc. Soc. Exper. Biol. & Med., **57**:13–14.
Stone, L. S., and Ellison, F. S. 1945. Return of vision in eyes exchanged between adult salamanders of different species, J. Exper. Zoöl., **100**:217–27.
Stone, L. S., and Zaur, I. S. 1940. Reimplantation and transplantation of adult eyes in the salamander (Triturus viridescens) with return of vision, J. Exper. Zoöl., **85**:243–69.

STRÖER, W. F. H. 1939. Ueber den Faserverlauf in den optischen Bahnen bei Amphibien, Proc. kon. Akad. Wetensch., Amsterdam, **42**:649–56.

———. 1939a. Zur vergleichenden Anatomie des primären optischen Systems bei Wirbeltieren, Ztschr. f. Anat. u. Entwcklngsgesch., **110**:301–21.

———. 1940. Das optische System beim Wassermolch (Triturus taeniatus), Acta néerl. morphol., **3**:178–95.

STRONG, O. S. 1895. The cranial nerves of the Amphibia, J. Morphol., **10**:101–230.

SUMI, R. 1926. Ueber die Morphogenese des Gehirns von Hynobius nebulosus, Folia Anat. Japon., **4**:171–270.

———. 1926a. Ueber die Sulci und Eminentiae des Hirnventrikels von Diemictylus pyrrhogaster, ibid., pp. 375–88.

TAYLOR, A. C. 1944. Development of the innervation pattern in the limb bud of the frog, Anat. Rec., **87**:379–413.

THOMPSON, D'ARCY WENTWORTH. 1944. On growth and form. New York: Macmillan Co.

THOMPSON, ELIZABETH L. 1942. The dorsal longitudinal fasciculus in Didelphis virginiana, J. Comp. Neurol., **76**:239–81.

TUGE, H. 1932. Somatic motor mechanisms in the midbrain and medulla oblongata of Chrysemys elegans (Wied), J. Comp. Neurol., **55**:185–271.

WALLENBERG, A. 1931. Beiträge zur vergleichenden Anatomie des Hirnstammes, Deutsche Ztschr. f. Nervenh., Vols. **117, 118, 119,** pp. 677–98.

WARREN, JOHN. 1905. The development of the paraphysis and the pineal region in Necturus maculatus, Am. J. Anat., **5**:1–27.

WEISS, PAUL. 1936. Selectivity controlling the central-peripheral relations in the nervous system, Biol. Rev., **11**:494–531.

———. 1939. Principles of development. New York: Henry Holt & Co.

———. 1941. Self-differentiation of the basic patterns of coordination, Comp. Psychol. Monog., **17**, No. 4, 1–96.

WHITMAN, C. O. 1892. The metamerism of Clepsine, Festschr. f. Rudolf Leuckharts, pp. 384–95. Leipzig: Wilhelm Engelmann.

———. 1899. Animal behavior. Biological lectures from the Marine Biological Laboratory, Woods Hole, for 1898, pp. 285–338. Boston.

WOODBURNE, R. T. 1936. A phylogenetic consideration of the primary and secondary centers and connections of the trigeminal complex in a series of vertebrates, J. Comp. Neurol., **65**:403–501.

———. 1939. Certain phylogenetic anatomical relations of localizing significance for the mammalian central nervous system, ibid., **71**:215–57.

YNTEMA, C. L. 1937. An experimental study of the origin of the cells which constitute the VIIth and VIIIth cranial ganglia and nerves in the embryo of Amblystoma punctatum, J. Exper. Zoöl., **75**:75–101.

———. 1943. An experimental study on the origin of the sensory neurones and sheath cells of the IXth and Xth cranial nerves in Amblystoma punctatum, ibid., **92**:93–119.

———. 1943a. Deficient efferent innervation of the extremities following removal of neural crest in Amblystoma, ibid., **94**:319–49.

YOUNGSTROM, K. A. 1938. Studies on the developing behavior of Anura, J. Comp. Neurol., **68**:351–79.

———. 1940. A primary and a secondary somatic motor innervation in Amblystoma, ibid., **73**:139–51.

———. 1944. Intramedullary sensory type ganglion cells in the spinal cord of human embryos, ibid., **81**:47–53.

ILLUSTRATIONS

ILLUSTRATIONS

GENERAL STATEMENT

All of the 113 figures are of adult or late larval Amblystoma tigrinum, except figures 86B, 86C, 111, 112, and 113 of Necturus and figure 107 of Rana pipiens. The indicated magnifications of the drawings show wide diversity in actual sizes of the specimens due to two things: in the first place, the fresh specimens vary greatly in size, and, in the second place, shrinkage during preparation may amount to as much as one-fourth of the original linear dimensions. Figures 2–85 are original drawings, and figures 1 and 86–113 are selected from previous publications, with minor alterations in some of them.

The internal structure of the brain of Necturus has been illustrated by sections drawn at close intervals in the three conventional planes ('30, '33b) and by many drawings of detail in several other publications. The general plan of the brain of Amblystoma is so similar to that of Necturus that comparisons are readily made. The sections of Amblystoma have been under investigation for nearly forty years, and the exigencies of the study and of publication have required concentration upon particular topics rather than description of the brain as a whole, and no comprehensive atlas of sections has been prepared. It is the aim of this book to supply a general view, but, unfortunately, the figures necessary to illustrate it are scattered in many publications. Some figures of especial value for general orientation are included here, and references are given to others in the literature.

Three series of Weigert sections of adult A. tigrinum, prepared by the late Dr. P. S. McKibben in 1910, were selected as standards of reference. These are his numbers IIC, transverse, IC, horizontal, and C, sagittal. All were fixed in a formalin-bichromate mixture, imbedded in paraffin, cut serially at 12 μ, and stained on the slides. These show minimum distortion of form, excellent histological preservation, and brilliant stain of myelinated fibers, with decolorization arrested at a stage which leaves all cell bodies clearly stained. The transverse series was chosen as the type specimen, and this has been quite fully illustrated, though these drawings were published at different times. To assist the reader who may wish to assemble an atlas of this specimen, a list of all available figures of it was published ('35a, p. 241), with serial numbers of the sections pictured, and to this list the following numbers may now be added:

Section
98.... Olfactory bulb. '24b, fig. 5
570.... Nucleus posterior tecti. '36, fig. 15
595.... Cerebellum. Figure 91
635.... Posterior border of V nerve roots. Figure 90
730.... Level of IX nerve roots. Figure 89
929.... Anterior root of first spinal nerve. Figure 88
956.... Calamus scriptorius. '44b, fig. 4
975.... Commissura infima. Figure 87
990.... Nuclei of dorsal funiculi. '44b, fig. 6

Figure 2C of this work is the median section of the middle part of the brain stem, reconstructed from this series of sections; and figures 1–4 of the paper of 1935 are similar reconstructions, showing chorioid plexuses and blood vessels.

A paper model of this transverse series was prepared, × 75, each section being drawn on cardboard seventy-five times the thickness of the section and then cut out along external and ventricular surfaces. These sheets, when properly stacked, show the external and ventricular configuration, and upon them details of internal structure were drawn in colored inks. The relevant data about this specimen were published ('35a) in connection with the preparation of the diagram of the median section, shown here as figure 2C (cf. fig. 1B). Scales accompanying the diagram give the section numbers, so that any section figured can be accurately located on the projection. This outline has been used as the basis for many diagrams of internal structure in this work and previous papers. Figures 2A and 2B are similar diagrams, made from the series of horizontal sections illustrated in figures 25–36. A wax model was made from the sagittal Weigert series C (figs. 1A and 85).

Two series of transverse sections of the adult forebrain based mainly on Golgi sections were published in 1927; six of these are here reproduced as figures 95–100. Six sagittal sections of the middle part of the adult brain stem by Rogers' reduced silver method were published in 1936, figures 17–22; and three of these are shown here as figures 102, 103, and 104. Seven pictures of the chiasma region from the sagittal Weigert series C are in figure 11 of 1941. A series of twelve horizontal Cajal sections of the adult follows in this book (figs. 25–36). Serial sections of larvae of 38 mm., prepared by Cajal's reduced silver method, have been illustrated (horizontal, '39b, figs. 2–20; transverse, '14a, figs. 4–14, and '39a, figs. 2–14). Some details of the development of the external form of the brain and the internal structures are reported in the papers of 1937–41. For meninges and blood vessels see pages 24–27, my papers ('34d and '35), Roofe ('35), and Dempster ('30).

FIGURES AND DESCRIPTIONS

Figure 1A, B, and C.—Lateral and median views of the brain of adult A. tigrinum, reproduced from '24a, figures 1, 2, and 3.

A.—Drawn from a wax model made from sagittal Weigert sections, no. C, collected at Chicago, Ill. × 15. The lateral wall of the cerebral hemisphere has been cut away to open the lateral ventricle; compare figure 85 drawn from the same model.

B.—Median section of a specimen from Colorado. × 10. The dissection was prepared by Dr. P. S. McKibben and drawn by Katharine Hill. Brains of several specimens were exposed, fixed *in situ* in formalin-Zenker, then removed from the head, washed in water, and hardened in graded alcohols before being cut in the mid-sagittal plane. The shrinkage in alcohol accentuates the ventricular sculpturing.

C.—Key drawing to accompany figure 1B.

Figure 2A, B, and C.—Three drawings of the median section of the brain stem of the adult prepared by graphic reconstruction from sections.

A.—This median section is reconstructed from the series of horizontal Cajal sections illustrated in figures 25–36. × 23. Compare figure 2C made from a specimen differently prepared.

The shape of the two specimens is somewhat different, owing chiefly to slight dorsoventral compression of the midbrain and other distortions of the Cajal specimen. Despite these defects, the correspondence of the two median sections is fairly close. The figures of the two specimens are drawn on the same plan, with the important exception that in the drawing of 2C the ventricular sulci are projected upon the median plane, while in 2A and B the outlines of the chief cellular areas are thus projected. In general, the sulci mark the boundaries of these areas, but this correspondence is not exact, and there is great individual variation in both these features. In the posterior part of the tegmentum isthmi a vertical dotted line marks, somewhat arbitrarily, the boundary between its central nucleus and a posterior sector composed of larger cells and transitional to the large-celled component of the trigeminal tegmentum.

Two drawings based on this reconstruction are shown, with emphasis on different features. In figure 2A the larger subdivisions of this part of the brain stem are demarcated so as to assist

ILLUSTRATIONS

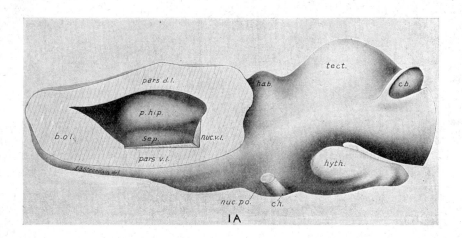

1 A

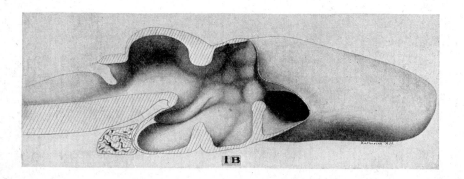

1 B

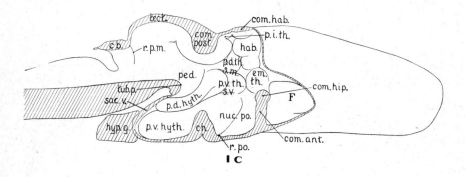

1 C

324 THE BRAIN OF THE TIGER SALAMANDER

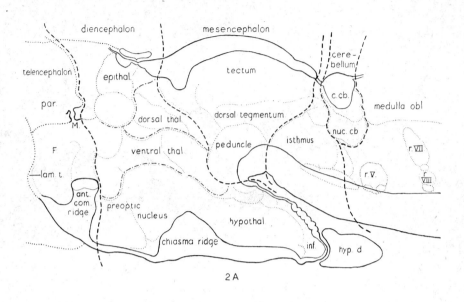

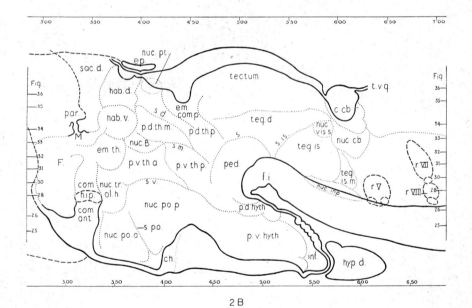

ILLUSTRATIONS 325

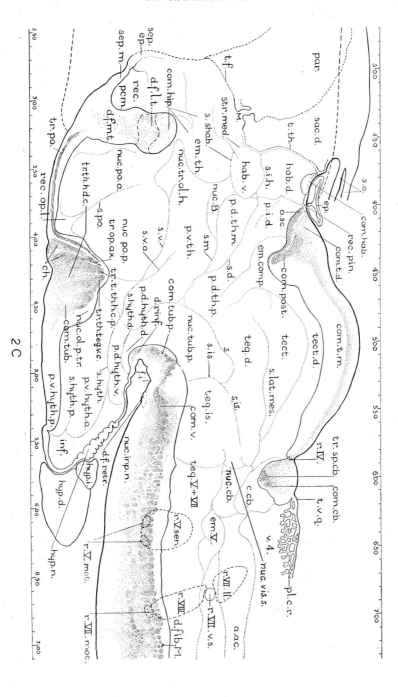

the reader in comparing the amphibian topography with the conventional mammalian analysis. Median structures, that is, the cut surfaces of the section, are outlined with full lines, paramedian structures in dotted lines (compare the gross section, fig. 1B; Necturus, '17, fig. 63; and McKibben, '11, fig. 5). Heavy broken lines mark the boundaries of the larger divisions as here defined, and dot-and-dash lines mark the subdivisions of the diencephalon.

B.—The same outline, bearing names of some of the smaller subdivisions. The scales at right and left indicate the approximate planes of the sections shown in figures 25–36. These levels are not exact, for unequal shrinkage during preparation produced some irregularities which have been smoothed in the diagram. The scales at top and bottom show section numbers of the type specimen, no. IIC, from which figure 2C was drawn, as indicated on that figure. Here again the correspondence is not exact, though sufficiently close to facilitate orientation of the published transverse sections with reference to the two median sections.

C.—This is the median section of the type specimen, IIC (p. 321), copied from 1935a, figure 1, with some changes in the lettering. × 30.

Figure 3.—Outline drawing of the dorsal surface of the adult brain. × 10. The paraphysis and chorioid plexus of the fourth ventricle have been removed. The courses of the sensory fibers of the trigeminal and dorsal spinal nerve roots and of the spino-bulbar, spino-cerebellar, and spinal lemniscus secondary tracts are diagrammatically indicated. The general bulbar lemniscus (*lm.* of the other figures) arises from the entire sensory zone of the medulla oblongata and ascends approximately parallel with the spinal lemniscus.

Figures 4 and 5.—Two diagrams illustrating the extent of the sensory and motor zones as seen in horizontal sections of the adult brain (chap. v). × 15. The outlines are from a series of Cajal sections cut in a slightly different plane from those illustrated in figures 25–36, with the anterior end more dorsal.

Fig. 4.—This passes through the hippocampal commissure and shows on the left side the extent of the motor zone, as here defined, by oblique hatching and the sensory zone in the olfactory area by hatching in the reverse direction. The plane of section is about the same as that of figures 28 anteriorly and 27 posteriorly. It passes below the tuberculum posterius at the cerebral flexure, above which the motor zone of the peduncle and isthmus is continuous between the medulla oblongata and the ventral thalamus (fig. 30).

Fig. 5.—This shows the extent of the sensory zone at a more dorsal level, the plane being about that of figures 35 anteriorly and 34 posteriorly. The undifferentiated anterior olfactory nucleus encircles the base of the olfactory bulb, and through it passes the very large fasciculus postolfactorius (cf. fig. 105), from which the various olfactory tracts of the hemisphere are distributed (p. 55; '27, p. 283).

Figure 6.—Selected examples of long pathways of conduction leading toward the skeletal musculature of the trunk and limbs, seen as projected upon the lateral aspect of the brain. × 10. The only afferent systems drawn are the olfactory and optic, and from these receptive fields only a few lines of descending conduction are indicated as typical representatives of through fore-and-aft transmission.

Figure 7.—Diagram of the central courses of the sensory components of cranial nerves V to X of larval Amblystoma seen as projected upon the lateral surface of the medulla oblongata. × 36. The drawing is based on figure 3 of 1914. The general cutaneous component is drawn in dashed lines, the vestibular in thick unbroken lines, the visceral-gustatory in dotted lines, the three lateral-line VII roots in thick dash lines, the two lateral-line X roots in dot-and-dash lines, and correlation tracts *a* and *b* in thin continuous lines. Some fibers of the ascending roots of the general cutaneous and vestibular systems decussate in the cerebellum, and some visceral root fibers of the f. solitarius decussate in the commissura infima at the commissural nucleus of Cajal.

Figure 8.—Diagram of the central connections of the visceral-gustatory system seen as projected upon the lateral surface of the adult brain. × 10. A probable direct connection from the superior visceral-gustatory nucleus (*nuc.vis.s.*) to the hypothalamus is indicated by the dotted line.

Figure 9.—Diagrammatic transverse section of the larval medulla oblongata near the level of the IX nerve roots, showing four types of neurons of the sensory zone. × 100. Drawn from preparations illustrated in 1914. Figure 89 shows a section of the adult brain in about

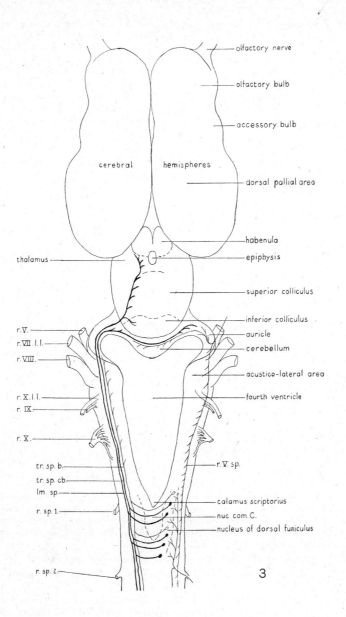

the same plane. The arrangement of the fascicles of nerve root fibers is indicated. Neuron 1 is in synaptic relation with all components of these cranial nerves. Neuron 2 makes its chief connection with visceral-gustatory fibers of the f. solitarius and less intimate connection with vestibular and trigeminal fibers. Neuron 3 connects only with root fibers of the trigeminus. Neuron 4 connects with the trigeminus and also with the reticular formation and motor zone. Similar elements have been seen to connect also with fascicles of the VIII and other nerve roots. Axons of all these types decussate in the ventral commissure and may ascend in the general bulbar lemniscus. The axon of neuron 2 divides, one branch ascending in the secondary visceral tract (*tr.v.a.*) of the same side and the other crossing to the opposite side. Some similar neurons connect only with the f. solitarius and have unbranched axons entering *tr.v.a.* only. Numberless permutations of the various types of connection here shown have been observed.

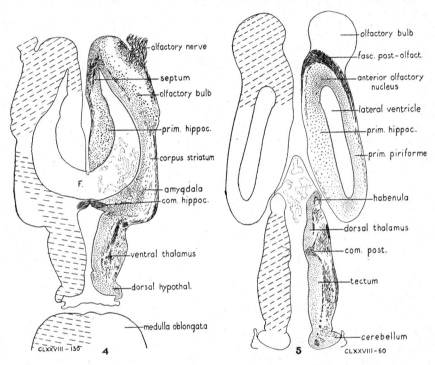

Figure 10.—Diagram of the chief afferent connections of the body of the cerebellum and of the brachium conjunctivum seen as projected on the median section of the brain (chaps. iv and xii). × 18. The more lateral vestibular connections are not drawn. The outline is that of figure 2C, and dotted lines mark ventricular sulci and the boundaries of the chief subdivisions of the brain wall.

Figure 11.—Diagram of the chief afferent tracts to the tectum (pp. 48, 220).

Figure 12.—Diagram of the chief efferent tracts from the tectum. Many shorter connections are omitted (p. 223).

Figure 13.—Diagram of the connections of the mesencephalic nucleus of the V nerve and of some tecto-bulbar and tegmento-bulbar tracts probably concerned with feeding reflexes (p. 140).

Figure 14.—Diagram of the chief afferent tracts to the dorsal thalamus and of some other connections of the optic tracts (pp. 49, 236).

Figure 15.—Diagram of the chief efferent tracts from the dorsal thalamus (pp. 49, 237) and pretectal nucleus (p. 39).

ILLUSTRATIONS

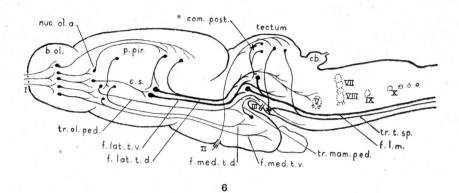

6

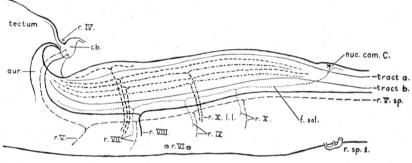

7

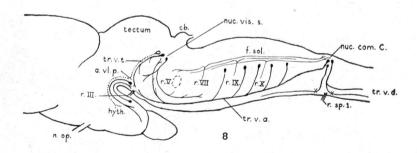

8

330   THE BRAIN OF THE TIGER SALAMANDER

Figure 16.—Diagram of the chief afferent tracts to the ventral thalamus (p. 239).

Figure 17.—Diagram of the chief efferent tracts from the ventral thalamus (p. 240).

Figure 18.—Diagram of the chief connections of the "peduncle" (nucleus of the tuberculum posterius, pp. 50, 217). Many shorter connections are omitted. For the connections of the ventrolateral neuropil see chapter iii and figure 23.

Figure 19.—Diagram of the diencephalic connections of the amygdala (pp. 52, 248) and of some of the connections of the interpeduncular nucleus (chap. xiv).

Figure 20.—Diagram of the connections of the habenula. In the stria medullaris thalami the components are arranged in the fore-and-aft order in which they ascend, as also are the

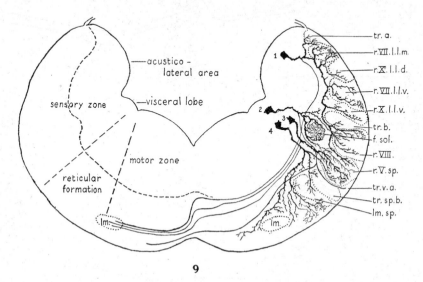

decussations in the habenular commissure (compare the horizontal sections, figs. 25–36). Efferent fibers from the habenula enter three tracts (*tr.hab.t.*, *f.retr.*, and *tr.hab.th.*). The afferent fibers are numbered as in the analysis of the stria medullaris in chapter xviii:

1. Tr. olfacto-habenularis medialis
2. Tr. olfacto-habenularis lateralis
3. Tr. olfacto-habenularis anterior, ventral division
4. Tr. olfacto-habenularis anterior, dorsal division
5. Tr. cortico-habenularis lateralis
6. Tr. amygdalo-habenularis
7. Tr. septo-habenularis
8. Tr. cortico-habenularis medialis
9. Tr. cortico-thalamicus medialis
10. Tr. olfacto-thalamicus
11. Tr. thalamo-habenularis
12. Tr. tecto-habenularis

Not shown in this figure is the tr. pretecto-habenularis and a probable tr. strio-habenularis.

Figure 21.—Diagram of the chief afferent connections of the tegmentum isthmi (p. 179). Most of the tracts here indicated end in both the isthmic and the trigeminal tegmentum. For the connections of the interpeduncular nucleus and its neuropil see chapter xiv.

Figure 22.—Diagram of the most direct connections between the retina and the peduncle. Optic tracts and efferent paths from the peduncle are drawn in full lines, internuncial connections in broken lines, and thick myelinated fibers in heavier lines (chap. xvi).

ILLUSTRATIONS

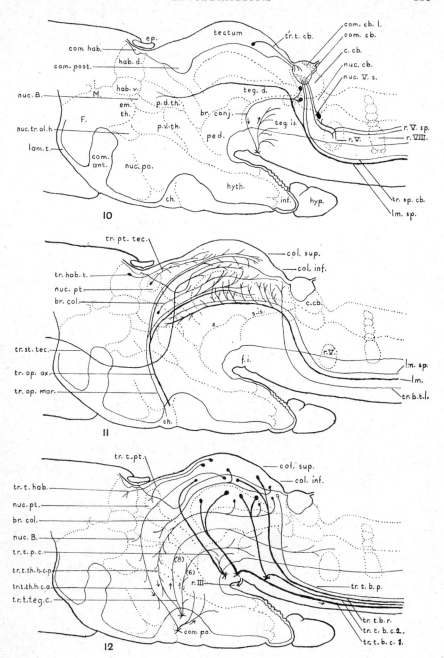

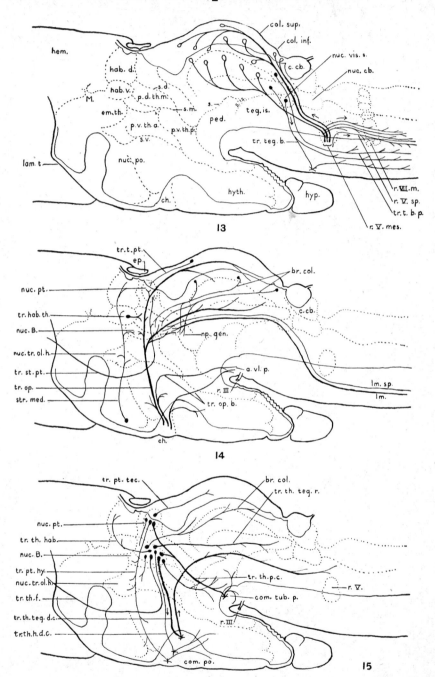

ILLUSTRATIONS

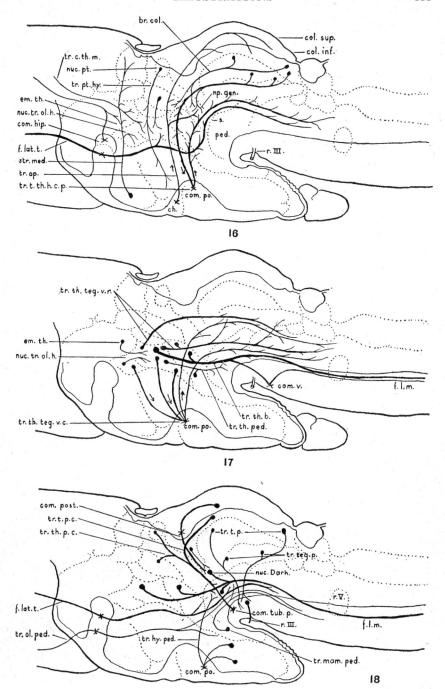

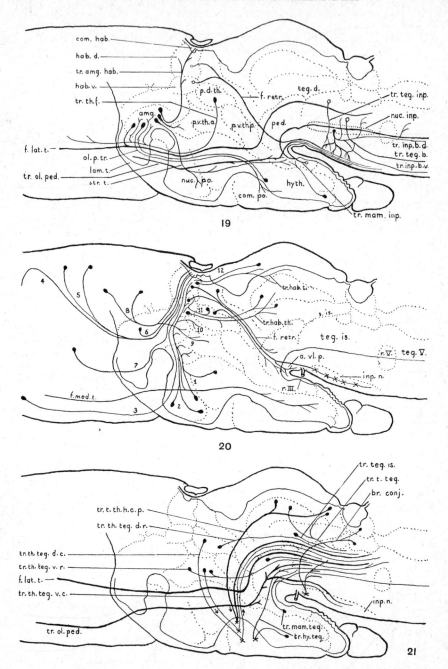

ILLUSTRATIONS 335

Figure 23.—The chief nonvisual afferent connections of the ventrolateral peduncular neuropil (chap. iii) seen as projected upon the median section. Most of these fibers are unmyelinated, of thin or medium size. The visceral-gustatory system has more fibers than any of the others.

Figure 24.—Diagrammatic thick transverse section through the middle of the optic tectum and the oculomotor nucleus, illustrating typical connections between the tectum and the

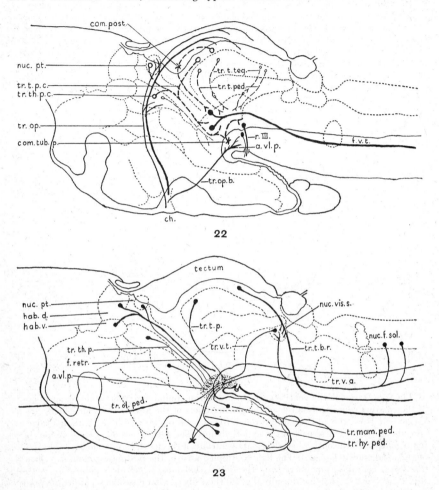

cerebral peduncle. The gray is outlined by a broken line, and the ventral border of the tectum by dotted lines. On the right side a typical neuron of the peduncle is drawn, with axon entering the ventral tegmental fascicles (*f.v.t.*). Four outlying cell bodies are seen at the border of the ventrolateral peduncular neuropil (*a.vl.p.*). On the left side is a neuron of the oculomotor nucleus, the dendrites of which connect with terminals of the basal optic tract in the ventrolateral neuropil. Compare figure 93.

Figures 25 to 36.—These semidiagrammatic drawings are made from sections selected from a horizontal series of the adult brain prepared by P. S. McKibben, Cajal's reduced silver method after fixation in alcohol. × 35. For approximate planes of section see figure 2B. The sections are exactly horizontal right and left.

336   THE BRAIN OF THE TIGER SALAMANDER

In this specimen the cell bodies are not blackened, and no dendrites are visible. All nuclei of cells are stained, and the gray pattern is clearly shown. There is scanty impregnation of the neuropil. In the cerebral hemispheres only the thickest axons are impregnated; elsewhere the thick and medium fibers are brilliantly differentiated. In some places where the boundaries of fascicles and tracts are obscure, interpretation has been aided by comparison with other specimens by methods of Cajal, Rogers, Golgi, and Weigert.

Horizontal sections are not so favorable for analysis of the tegmental fascicles (chap. xx) as are those cut in transverse and sagittal planes, and it is difficult to follow individual bundles as they recurve around the tuberculum posterius in the peduncle and tegmentum; but, by comparison with sections prepared by other methods and cut in various planes, the courses of most of the tracts and of the tegmental fascicles of groups (1) to (10) can be followed. The limits of the numbered groups of fascicles are not always clear, but their identification by number on the drawings is believed to be substantially correct.

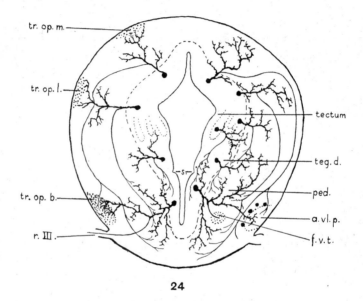

24

Fig. 25.—Through the ventral part of the anterior commissure ridge and the dorsal border of the chiasma ridge. Only a few of the thicker fibers of the medial forebrain bundle are impregnated. These thread their way through the decussating fascicles of the chiasma ridge and spread out in the ventral part of the hypothalamus. The dense neuropil of these regions is not impregnated. All the optic fibers decussate ventrally of this level, except the most dorsal fibers of the axial tract (*tr.op.ax.*). The lightly stippled area in the postoptic commissure, marked *tr.th.h.d.c.*, contains thin fibers from the dorsal thalamus, which decussate more ventrally (fig. 2C, *tr.th.h.d.c.*). Farther dorsally (fig. 26) these fibers, after crossing, separate into fascicles A and B (p. 299).

Fig. 26.—Ten sections more dorsally, the section passes through the dorsal fascicles of the medial forebrain bundle, tr. olfacto-peduncularis, and above the chiasma ridge. The mixed system of thalamo-hypothalamic and tegmental fibers (fig. 25, *tr.th.h.d.c.*) has separated into superficial (A) and deep (B) tracts for the tegmentum, as described on page 299. The course of the superficial tract (A) can be followed in figures 27–34; most of the deeper fibers (B) join the dorsal tegmental fascicles of group (8).

Fig. 27.—This cuts the decussation of the dorsal fascicles of the lateral forebrain bundle in the anterior commissure and the motor roots and nucleus of the V cranial nerve. The fibers descending from the dorsal thalamus to the postoptic commissure (*tr.th.h.d.c.*) lie deep in the alba. The same fibers after decussation appear in the superficial tract, A, and in the deeper fascicles of group (8). Rostrally of the V roots, most of the thick fibers of the ventral commissure belong to tr. tegmento-bulbaris cruciatus. Among these the f. longitudinalis medialis is assembling from fascicles of groups (4), (5), and (6) (p. 281). Medially and dorsally of these are finer fibers of the dorsal division of tr. interpedunculo-bulbaris (*tr.inp.b.d.*).

Fig. 28.—Through the hippocampal commissure and the middle of the sensory root of the V nerve. The section cuts the ramus communicans posterior of the arterial system (*r.c.p.*) and the decussation of the posterior division of tr. tecto-bulbaris cruciatus (*tr.t.b.c.2.*), also the most ventral fibers of the commissure of the tuberculum posterius.

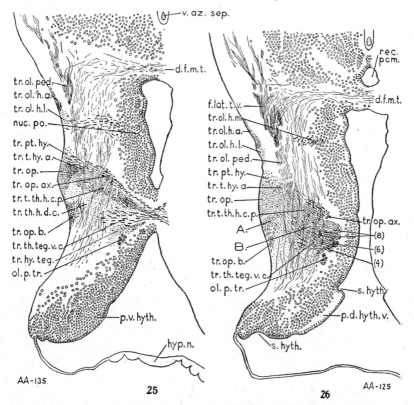

Fig. 29.—This passes immediately below the cerebral flexure at the tuberculum posterius and the ventral border of the peduncle. Mingled with the crossing fibers of *tr.t.b.c.2.* are the most ventral crossing fibers (unmarked) of the anterior division of tr. tecto-bulbaris cruciatus. The central nucleus of the isthmus (*nuc.is.c.*) is here well defined, and laterally of it are terminals of the dorsal tegmental fascicles of groups (7) to (10).

Fig. 30.—This section cuts the ventral border of the junction of the peduncle with the isthmus, a locus marked by the superficial origin of the III nerve. The anterior division of tr. tecto-bulbaris cruciatus (*tr.t.b.c.1.*) is decussating within the ventral tegmental fascicles of group (1) (p. 277). More posteriorly, the section cuts the dorsal border of the sensory V root and the visceral sensory roots of the VII and IX nerves, showing the entire prevagal part of the f. solitarius (pp. 148, 166 and figs. 37, 38).

Fig. 31.—Through the nucleus of the III nerve and the most dorsal component of the commissure of the tuberculum posterius containing decussating fibers of tr. tecto-bulbaris cruciatus 1. Farther forward the section cuts the ventral thalamus at its widest **part** and

338 THE BRAIN OF THE TIGER SALAMANDER

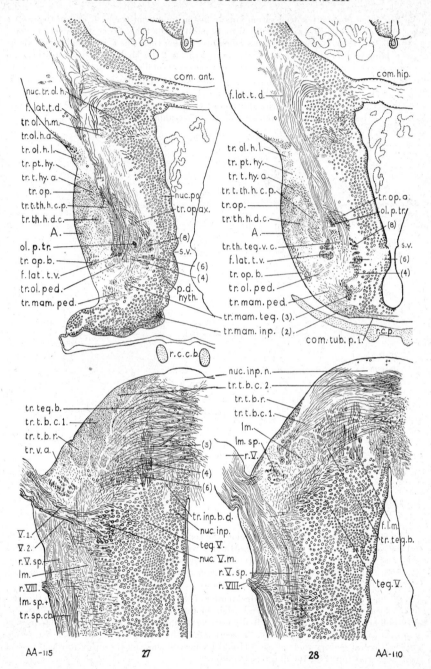

ILLUSTRATIONS

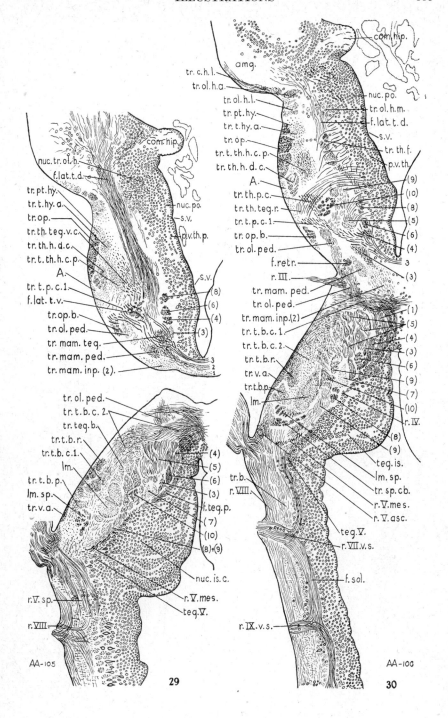

340   THE BRAIN OF THE TIGER SALAMANDER

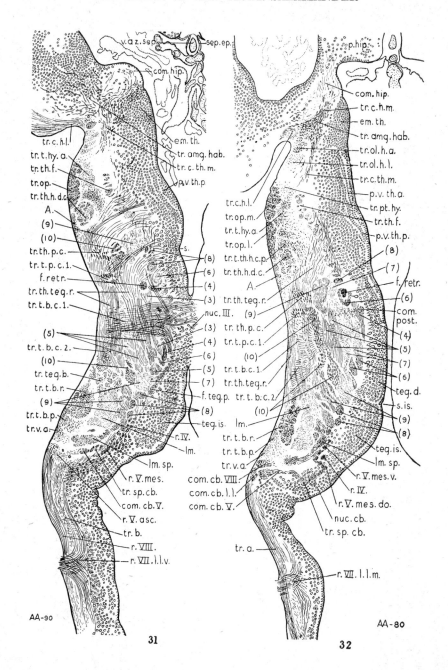

ILLUSTRATIONS

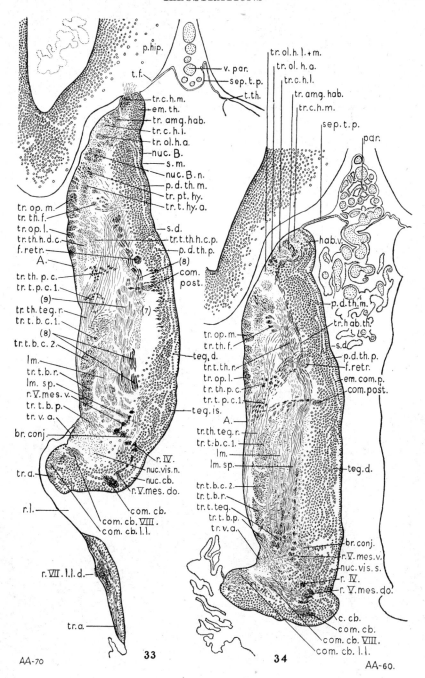

342   THE BRAIN OF THE TIGER SALAMANDER

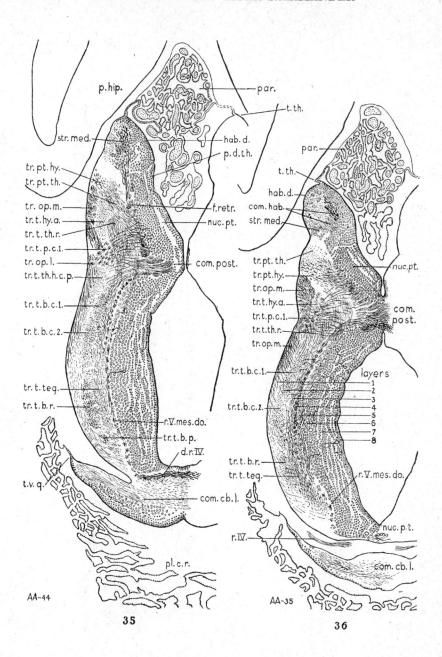

passes through the ventral border of the eminentia thalami. Large numbers of thick fibers pass from the ventral thalamus to the peduncle and isthmic tegmentum (*tr.th.teg.r.*), both superficially and deeper in the alba. Similar fibers (not drawn) arise from the large cells of the peduncle, which comprise a primordium of the mammalian interstitial nucleus of the f. longitudinalis medialis. Many of the thalamic and most of the peduncular fibers enter ventral tegmental fascicles of group (5) (see p. 281). This section cuts tegmental fascicles of group (3) at the dorsal convexity of their course as they recurve around the tuberculum posterius.

Fig. 32.—Section passing just above the interventricular foramen and through the middle of the eminentia thalami. Accompanying fibers of the hippocampal commissure are those of tr. cortico-habenularis medialis for the stria medullaris and tr. cortico-thalamicus medialis (primordium of the column of the fornix). Thick fibers stream backward from the ventral thalamus in tr. thalamo-tegmentalis rectus, some superficially, some at middle depth, and some at the inner border of the alba, the latter entering dorsal tegmental fascicles of group (8). Similar fibers arise from the larger cells of the peduncle, but to simplify the pictures these have not been drawn in figures 31 and 32. Most of them enter tegmental fascicles of group (5), and some of these continue spinalward in f. longitudinalis medialis. The more ventral large cells of the peduncle (fig. 31) correspond with the interstitial nucleus (Cajal) of f. longitudinalis medialis, the more dorsal of these cells (fig. 32) to the nucleus of Darkschewitsch (p. 217 and figs. 6, 18). For further details see the analysis of the tegmental fascicles in chapter xx and references there given.

At the level of figure 32, tegmental fascicles of groups (4), (5), and (10) are cut at the dorsal convexity of their courses through the peduncle, and here they are insinuated among the fibers of the anterior and posterior divisions of tr. tecto-bulbaris cruciatus, descending from the tectum toward their decussations. Farther back in the posterior isthmic neuropil the bulbar and spinal lemniscus systems (*lm.* and *lm.sp.*) are turning upward and forward to ascend in the dorsal tegmentum (figs. 33, 34). The wide auricle contains terminals of vestibular and lateral-line root fibers, other bulbar connections, and nucleus cerebelli.

Fig. 33.—This passes through the floor of the stem-hemisphere fissure, cutting the most dorsal fibers of tr. cortico-habenularis medialis as they enter the stria medullaris and the dorsal border of the eminentia thalami. Fibers of tr. thalamo-frontalis are seen emerging from the ventral border of the dorsal thalamus. Large cells of the nucleus of Darkschewitsch occupy the dorsal border of the peduncle. Laterally of these, tegmental fascicles (7) and (9) are recurving over the peduncle, and two sections farther dorsally, fascicles of group (8) are cut at the top of their convexity. More posteriorly the section passes through the dorsal border of the central nucleus of the isthmic tegmentum, laterally of which a few fibers of the brachium conjunctivum are impregnated. Some of these are seen to arise from the nucleus cerebelli. Externally of this nucleus is the posterior neuropil of the isthmus containing dendrites from the superior visceral-gustatory nucleus, terminals of the ascending secondary visceral tract (*tr.v.a.*), and many other components. Behind the auricle the section passes through the ventral part of the lateral recess of the ventricle and spinalward of this through the "dorsal island" of Kingsbury.

Fig. 34.—Ten sections farther dorsally, all components of the stria medullaris can be identified in the ventral habenular nucleus. The tr. habenulo-thalamicus (*tr.hab.th.*) contains fibers passing in both directions between this nucleus and the thalamus and regions posteriorly of it. The bulbar lemniscus (*lm.*) and the spinal lemniscus (*lm.sp.*) traverse the dorsal tegmentum to reach the dorsal thalamus. More posteriorly the section passes through the gray of the superior visceral-gustatory nucleus (*nuc.vis.s.*) and the junction of the body of the cerebellum (*c.cb.*) with the auricle.

Fig. 35.—Through the ventral borders of the dorsal habenular nucleus, nucleus pretectalis, and tectum. Posteriorly it passes through the superior medullary velum, showing the decussation of the IV nerve roots, and through the body of the cerebellum dorsally of the cerebellar commissure. Fibers of com. vestibulo-lateralis cerebelli are crossing at this level. Compare 1942, figures 78 and 79, cut in about the same plane, Weigert and Golgi.

Fig. 36.—Through the com. posterior and the tectum at its widest part. In figures 35 and 36 the brachia of the superior and inferior colliculi are designated tr. tecto-thalamicus rectus (*tr.t.th.r.*).

Figures 37, 38, and 39.—These semidiagrammatic sketches show some details of the medulla oblongata from an advanced larva. The sections are obliquely horizontal, with the right side and the anterior end more dorsal. × 30. The series contains 27 thick Golgi sections, numbered from dorsal to ventral surfaces, and the data here assembled are selected from sections 11–24. The outlines of sections 13, 17, and 20 are drawn, and upon these there are sketched some details from these and neighboring sections. The impregnation is scanty on a

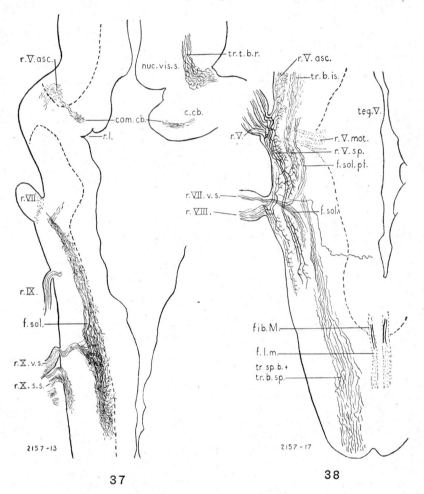

clear ground, with long courses of individual fibers clearly shown. There is no impregnation of the lemniscus systems and but little of fibers descending from above the isthmus, so that the intrinsic bulbar structures impregnated are well defined. In figures 37 and 38 the boundary between gray and white layers is marked by a broken line; in figure 39 the gray of the interpeduncular nucleus is stippled.

Fig. 37.—On the left side this section cuts through the base of the auricle and the floor of the lateral recess. The right side is more dorsal and includes the superior visceral-gustatory nucleus and its neuropil, within which are collaterals from tr. tecto-bulbaris rectus (p. 225). On the left side, V root fibers arborize in the ventral part of the isthmic neuropil, and some of them continue into the cerebellar commissure. Almost the entire course of the postfacial f. solitarius is projected from several sections.

Fig. 38.—This section on the left cuts the superficial origins of the V, VII, and VIII nerve roots and, farther forward, of the III root. The sensory V fibers are seen to bifurcate with more slender ascending branches, most of which arborize in a neuropil, which is the primordium of the superior sensory V nucleus of mammals. The prefacial f. solitarius fibers are smooth and unbranched to their terminals rostrally and internally of the entering V root. Some impregnated fibers of the vestibular root (r. VIII.) extend forward to end within and rostrally of the superior V neuropil, but these are not drawn. Internally of the trigeminal neuropil and confluent with it are terminals of ascending crossed and uncrossed fibers termed tr. bulbo-isthmialis (tr.b.is.), which is continuous with tr. spino-bulbaris (tr.sp.b.). These fibers ascend and descend

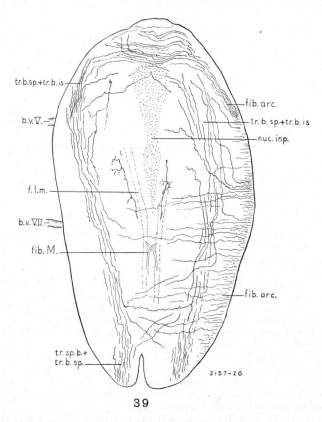

from all levels of the medulla oblongata, many of them first decussating as internal or external arcuate fibers. Many of them bifurcate into ascending and descending branches with or without decussation. They accompany the spinal lemniscus and comprise a mixed bulbo-spinal and spino-bulbar system.

Fig. 39.—This section, parallel with the ventral surface, illustrates the courses of some of the external arcuate fibers, many of which are impregnated on the right side and none on the left. Useful landmarks are provided by the decussation of Mauthner's fibers (fib.M.) and two blood vessels, which, farther dorsally, are related with the roots of the V (b.v.V.) and VII (b.v.VII.) nerves. No fibers of lemniscus and secondary visceral systems are impregnated. The visible fibers evidently are concerned chiefly with bulbar and spinal adjustments.

Figure 40.—Detail from a horizontal Golgi section of a late larva, showing the entrance of the sensory V root and three neurons of the motor V nucleus. × 50. The dendrites extend laterally to engage collaterals of the sensory root fibers, and in the adjoining section ventrally other dendrites of these cells ramify downward into the ventral alba, engaging collaterals and terminals of the descending and ascending fibers of the ventral funiculus, here unstained. Nothing except the elements drawn is impregnated in this region. Presumably, the axons of these cells enter the motor V root, the unimpregnated myelinated fibers of which are here darkened by the Golgi fluid.

Figures 41 to 44.—These semidiagrammatic sketches are made from horizontal Golgi sections of an advanced larva, with elective impregnation of the spinal lemniscus and a few other details. $\times$ 30. The series contains 26 thick sections, numbered from ventral to dorsal. Sections 2, 3, and 4, illustrating innervation of the hypophysis, have been published ('42, figs. 56, 57, 58). Here sections 7, 10, 12, and 14 are drawn, with some additions in each case from intervening sections.

Fig. 41.—The section passes through the V and VIII roots, with impregnation of superficial fascicles of descending fibers of both roots and ascending VIII fibers, some of which extend as far as the cerebellum. No ascending sensory V fibers are impregnated. The loci of the III, IX, and X roots farther dorsally are indicated by blood vessels which accompany these roots. The thick arcuate fibers from the region of the calamus scriptorius enter a mixed tract marked *lm.sp.*, which here contains bulbo-spinal, spino-bulbar, spino-cerebellar, and spinal lemniscus fibers. Most of the thin unmyelinated fibers here impregnated come from lower levels of the spinal cord, and their further ascending course is shown in figures 42, 43, and 44. A few thick fibers of the ventral funiculi are impregnated, some of which decussate in the ventral commissure below the auricle.

Fig. 42.—At this level, thick arcuate fibers in the vicinity of the first spinal roots descend in the calamus region and after crossing join the spinal lemniscus. Some of these bifurcate with a thick ascending branch and a slender descending. Farther forward under the auricle the lemniscus fibers turn dorsalward in the ventral part of the isthmic neuropil, and here many of them end. This is tr. bulbo-isthmialis of figures 38 and 39. The lemniscus fibers which continue rostrad have collateral endings here. Other elective Golgi preparations from this lot show the origin of spinal lemniscus fibers from the nuclei of the dorsal funiculus ('44b).

Fig. 43.—This section cuts the III nerve root and its nucleus and, in the isthmic neuropil, dispersed fibers of the spinal lemniscus and ascending VIII root. Medially of these are axons emerging from the gray of the trigeminal tegmentum, which turn forward and arborize within the gray of the isthmic tegmentum (p. 185).

Fig. 44.—At this level the few remaining impregnated fibers of the spinal lemniscus are turning forward across the dorsal tegmentum to reach the dorsal thalamus. The crude drawing gives a very inadequate picture of the delicacy of these widely branched terminal arborizations. The more numerous fibers of this tract, which arborize in the tectum, are not impregnated in this preparation (compare fig. 101 and '39b, fig. 26). For the patterns of vascular supply sketched in the left thalamus see page 27.

Figure 45.—Terminals of the dorsal lateral-line root of the facial nerve, from the right side of a horizontal Golgi section of a late larva. $\times$ 50. Only three fibers of this root are impregnated, and each of these ramifies through almost the entire extent of the "dorsal island" of the area acusticolateralis immediately spinalward of the auricle, which is the exclusive central field reached by fibers of this root (compare fig. 33).

Figures 46 and 47.—Details from horizontal Golgi sections of an advanced larva. $\times$ 75.

Fig. 46.—A neuron of the anterodorsal part of the trigeminal tegmentum in the ventral part of the auricle (p. 185). The plane of section is approximately that of figure 31.

Fig. 47.—A small neuron at the anteroventral border of the body of the cerebellum, where it merges with the nucleus cerebelli. The slender dendrites extend outward into the dorsal part of the isthmic neuropil, and the axon is directed forward and downward into the brachium conjunctivum. Fibers of the trigeminal component of the cerebellar commissure (*com.cb.*) are approaching their decussation immediately dorsally of this level, as also are more scattered fibers of the vestibulo-lateral component (*com.cb.l.*).

Figure 48.—Detail from a transverse Golgi section previously pictured ('42, fig. 43). $\times$ 50. The small neuron of the dorsal tegmentum has a short axon, which ramifies in the deep neuropil of the gray near the cell body and also dorsalward in the alba of the anterior end of the nucleus posterior tecti. On the vascular loops (*b.v.*) seen here see page 27.

ILLUSTRATIONS

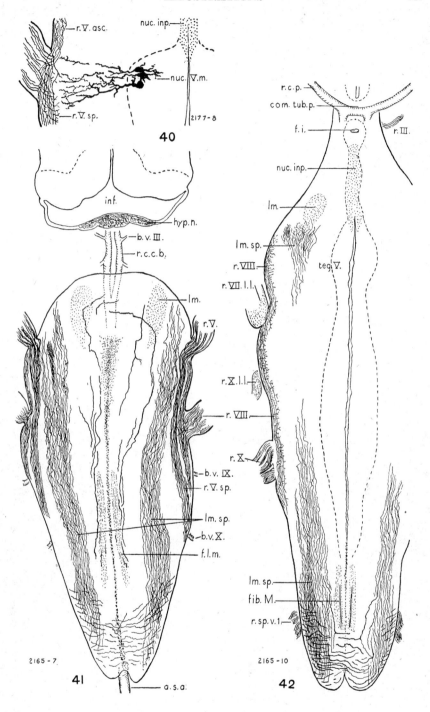

348   THE BRAIN OF THE TIGER SALAMANDER

Figure 49.—A typical neuron of the dorsal tegmentum from a horizontal Golgi section of an advanced larva. × 75. The ependymal surface is marked by a thin line at the left, the pial surface by a thicker line at the right, and the outer border of the gray by a broken line. The axon arises from the dendrite and is directed spinalward (downward in the figure) and probably enters tegmental fascicle (7). A similar neuron *in situ* is shown in figure 44 of the paper of 1942.

Figure 50.—Horizontal section through the decussation and spiral endings of the f. retroflexus (p. 197). Golgi method. × 50. The semidiagrammatic drawing is based on several adult specimens, in each of which the fasciculus of both sides is massively impregnated from the habenula to the interpeduncular neuropil. All details drawn are visible in two sections of the specimen outlined (no. 2229), which is, however, poorly preserved. Several other series of sec-

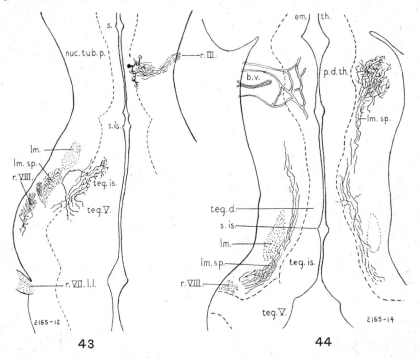

tions show the same relations more clearly, though the oblique planes of some of the sections are less convenient. The considerable lateral obliquity of one of these gives an interesting view of the interior of the spiral. In these specimens the spiral endings are impregnated for about two-thirds of the distance between the decussation and the level of the V roots.

Figures 51 and 52.—Two adjoining horizontal Golgi sections of a late larva, illustrating the decussation and spiral terminals of the f. retroflexus, which is impregnated on only one side. The right side is more ventral. × 50.

Figures 53 and 54.—Two adjacent thick horizontal Golgi sections from an advanced larva. × 50.

Fig. 53.—At the level of the fovea isthmi (*f.i.*) and the decussation of the f. retroflexus. Laterally of the interpeduncular neuropil are impregnated fibers of tr. olfacto-peduncularis, with endings by open arborizations in this neuropil; compare figure 59, cut in a similar plane a little more dorsally.

Fig. 54.—Ventrally and posteriorly of the decussation only a few isolated fibers of f. retroflexus are impregnated, and tr. olfacto-peduncularis is terminating in the ventral interpeduncular neuropil in the same area as the spiral. The tufted endings seen here are not derived from either of these tracts.

ILLUSTRATIONS

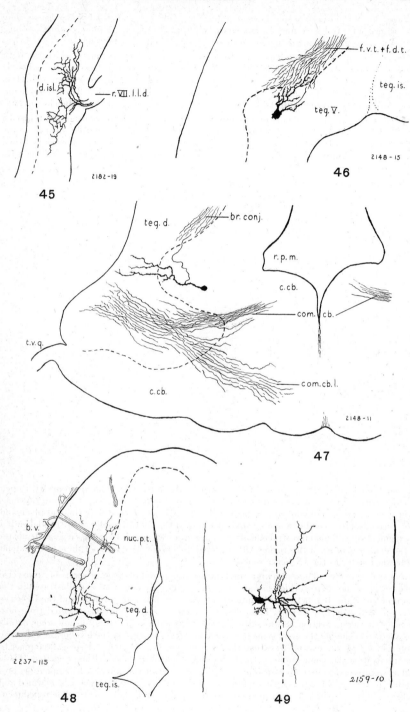

Figure 55.—Detail of the spiral course of fibers of the f. retroflexus below the decussation, from a horizontal Golgi section of a late larva. × 50. In the interpeduncular neuropil nothing but these three fibers is impregnated, so there is no possibility of confusion.

Figures 56, 57, 58.—Three horizontal Golgi sections of an adult brain, in which the right f. retroflexus is unstained and the left fasciculus is abundantly impregnated from the habenula to its decussation. × 50. Some details of the decussation are added to figure 58 from the section adjoining it ventrally. The impregnation fails spinalward of the decussation. The left fasciculus shows an atypical division into two bundles as it enters the alba of the peduncle (p. 262).

Figure 59.—An obliquely horizontal section through the superficial origins of the III roots, advanced larva. × 50. Here it is in about the plane of figure 30, but sharply inclined to the horizontal plane, with the posterior end more ventral and the anterior end more dorsal than that level. Impregnated fibers of the olfacto-peduncular tract pass the region of the fovea isth-

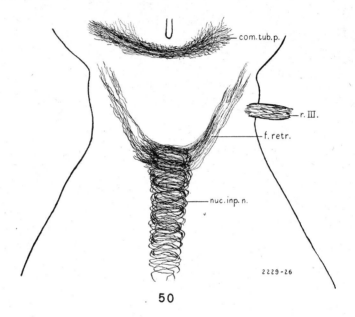

mi and then turn medially. Most of them end in the interpeduncular neuropil with open arborizations, though some extend farther spinalward (compare figs. 53, 54). From this neuropil slender axons of the dorsal interpedunculo-bulbar tract descend near the mid-plane and soon turn laterally to end in a dense axonic neuropil at the ventral border of the caudal end of the tegmentum isthmi and rostral end of the tegmentum trigemini. Here they engage dendrites of the smaller neurons of this region. Other preparations show that some of these fibers extend spinalward as far as the IX nerve roots.

Figures 60 to 64.—These drawings present additional details from a series of transverse Golgi sections (no. 2246), which has already been quite fully described and illustrated in the sector of the brain stem between the interventricular foramen and the nucleus of the IV nerve ('27, pp. 271, 278, figs. 22–40). Those figures were drawn from sections selected from nos. 55–97 of the series. Section 88 was subsequently drawn on a larger scale to show some details of structure, including the arrangement of the tegmental fascicles, at the level of the III nerve roots ('42, fig. 44). Figures 60–64 extend the series spinalward (× 50), with special reference to the interpeduncular connections of the tegmentum, which are here well impregnated. These tegmento-interpeduncular fibers comprise one component of the complex f. tegmentalis profundus (p. 286), and the incomplete references to them made in 1927 can now be clarified and rectified. In this specimen there is no impregnation of the f. retroflexus or tr. olfacto-peduncu-

ILLUSTRATIONS

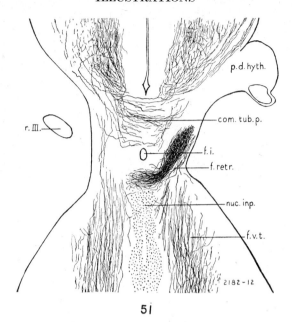

51

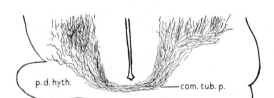

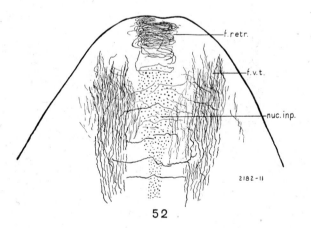

52

352   THE BRAIN OF THE TIGER SALAMANDER

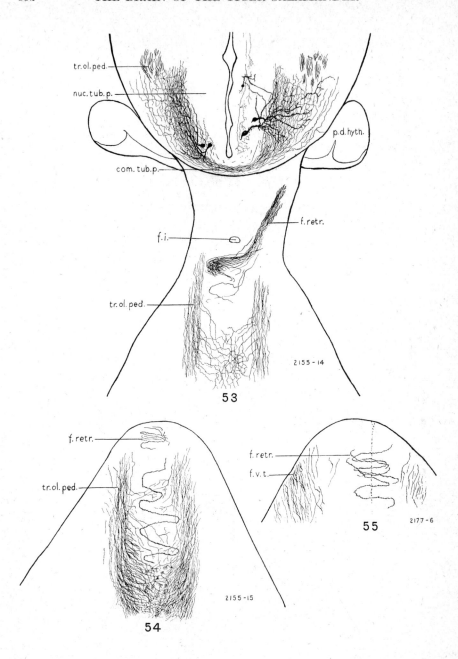

ILLUSTRATIONS

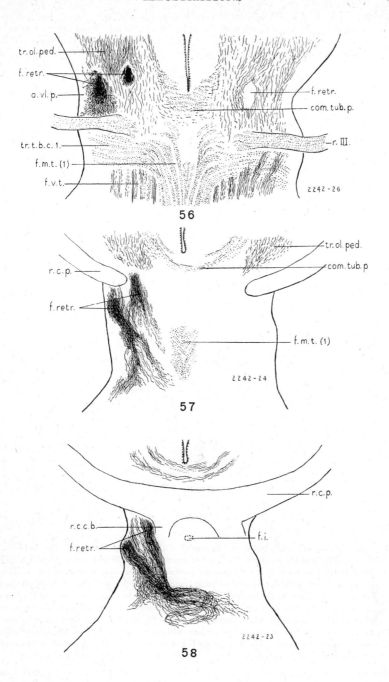

56

57

58

laris. A few neurons of the interpeduncular nucleus are incompletely stained, but apparently no axons from these cells are impregnated. The dense interpeduncular neuropil here seen is composed almost exclusively of axons from the tegmentum and tr. mamillo-interpeduncularis. Terminals of the latter, when separately impregnated, are seen to be of more open texture than the dense vertically arranged terminal tufts of the tegmento-interpeduncular fibers (fig. 60). The two types of terminals are closely interwoven with each other and with tufted terminals derived from axons of neurons of the interpeduncular nucleus (not here impregnated).

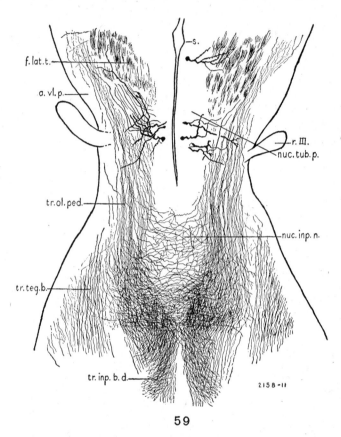

Fig. 60.—This section lies between the nuclei of the III and IV nerves, i.e., between the levels of figures 92 and 93, at about the level of figure 13 of 1936 and two sections rostrally of figure 39 of 1927. The drawing is a composite, containing some details from the two adjoining sections. Axons from both dorsal and isthmic tegmentum descend in the f. tegmentalis profundus, some terminating in the underlying interpeduncular neuropil and some of thicker caliber decussating in the ventral commissure (*tr.teg.b.*), where collaterals separate from them to end in tufts of the interpeduncular neuropil. In figure 39 of 1927 these decussating fibers are called tr. tegmento-peduncularis, and the same fibers are here called tr. tegmento-bulbaris. Both designations are correct. The larger number of these fibers after decussation turn spinalward, some sending collaterals forward also. Others turn rostrad into the peduncle.

Fig. 61.—This section passes through the nucleus of the IV nerve, and the three neurons impregnated probably belong to this nucleus. Approximately the same plane is illustrated in several published figures ('25, figs. 9, 19; '36, fig. 14; '42, fig. 43). At this level the myelinated fibers of the ventral median fascicles (1) are turning laterally to enter tr. tecto-bulbaris cruciatus in the position of the broken lines under the gray of the interpeduncular nucleus. The thick fibers of tr. tegmento-bulbaris shown in figure 60 are here reduced in number, and the thin fibers of tr. tegmento-interpeduncularis are more numerous. These terminals and those of tr. mamillo-interpeduncularis enter into the dense axonic interpeduncular neuropil.

Fig. 62.—This detail is from the lower part of the interpeduncular nucleus rostrally of the V nerve roots, not far from the plane of figure 91. There is impregnation of a typical large neuron of the trigeminal tegmentum and a smaller element at the border of the interpeduncular nucleus. In the interpeduncular neuropil the visible fibers are nearly all derived from the overlying tegmentum, with perhaps some terminals of tr. mamillo-interpeduncularis. At this level the ventral fascicles (4), (5), and (6) have united to form a compact f. longitudinalis medialis, and the dorsal fascicles spread out laterally into the isthmic neuropil and bulbar tegmentum (compare the horizontal sections, figs. 29–32).

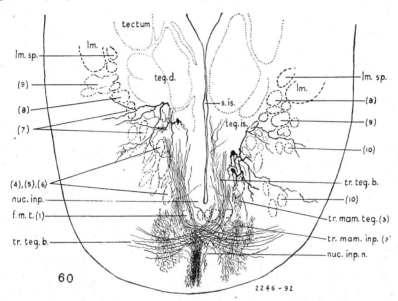

Fig. 63.—Detail of floor-plate structure immediately caudad of the V nerve roots. Though the interpeduncular nucleus is not considered to extend as far spinalward as this level, the lateral and dorsal interpeduncular axonic neuropil is still rather dense (compare figs. 79, 81). The fibers impregnated here are derived chiefly from the overlying tegmentum, and some of them descend as far as the VII roots accompanying the dorsal and ventral tr. interpedunculo-bulbaris, which is not impregnated in this specimen. Two typical ependymal elements of the bulbar tegmentum are drawn.

Fig. 64.—This is a similar sketch of the ventral median raphe at the level of the VII nerve roots. A remnant of the tegmento-interpeduncular neuropil persists, and this extends no farther spinalward, though other preparations show that axons from the interpeduncular nucleus reach at least as far as the IX nerve roots. The ependyma here is more compact and mossy than at the level of the trigeminus.

Figures 65 and 66.—Two transverse sections through the interpeduncular region at the level of transition between the isthmic and the trigeminal tegmentum of the adult. × 50. Each figure is a composite of two adjacent sections, so that four consecutive sections are represented in the two drawings.

Sections taken farther forward from this well-impregnated series have been shown ('42, figs. 45–47). Between the levels of the III and IV nuclei, tr. tegmento-interpeduncularis is

356 THE BRAIN OF THE TIGER SALAMANDER

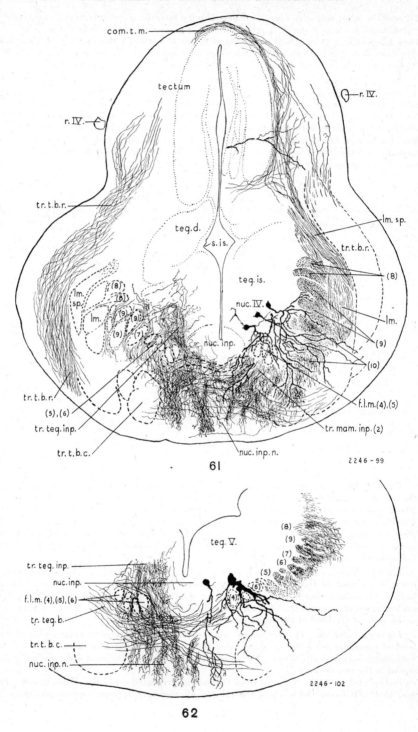

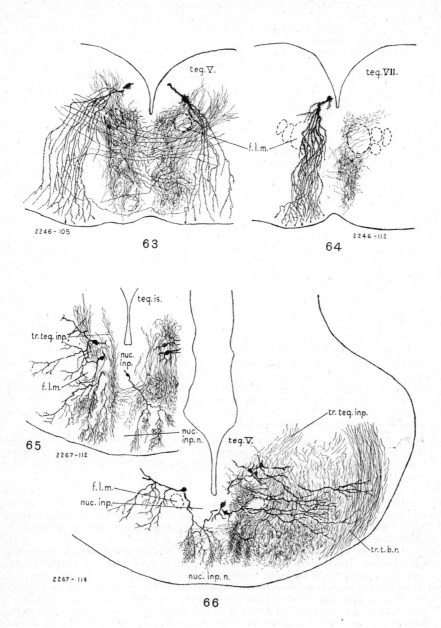

electively stained, with tufted endings in the interpeduncular neuropil similar to those shown in figures 61 and 62. At the level of figure 65 only a few of these fibers are impregnated and also a few neurons of the interpeduncular nucleus with dendrites extending downward into the interpeduncular neuropil.

In figure 66 the neurons of the interpeduncular nucleus are seen to have thick dendrites extending laterally to ramify widely in the alba of the isthmic and trigeminal tegmentum and thinner dendrites directed ventrally into the interpeduncular neuropil. Both dendritic and axonic arborizations enter the glomerulus-like tufts; but, in order to clarify their relations, only the dendritic component is drawn on the left side and the axonic component on the right (compare figs. 83, 84).

Figure 67.—Transverse section through the middle of the interpeduncular nucleus of a half-grown larva. × 75. The broken line marks the outline of the gray substance. The Golgi impregnation is scanty, showing in this region only a few decussating fibers of tr. tegmento-bulbaris on a clear ground. The more dorsal of the two fibers shown is sketched from the adjoining section spinalward. These thick axons probably arise from unimpregnated cells of the isthmic tegmentum, as shown in figure 68. In the ventral commissure, where these fibers decussate, slender collaterals leave them to ramify in the interpeduncular nucleus and its neuropil.

Figure 68.—A diagram based on Golgi sections of a larval Amblystoma from the same lot as figure 67. × 75. The sections are obliquely transverse. The section outlined is in about the same plane as figure 67 on the right and passes through the auricle on the left. The details are from this and the two adjoining sections. Everything drawn was observed, but the assembly is schematic. Thick axons of neurons of the isthmic tegmentum (*tr.teg.b.*) converge into the ventral commissure, where slender collaterals separate from them to arborize in tufted form in the interpeduncular neuropil.

Figures 69 and 70.—These two drawings illustrate details of the interpeduncular region at and immediately rostrally of the nucleus of the IV nerve (compare the diagram, fig. 68). × 50. The impregnation of the transverse Golgi sections of this adult brain is exceptionally good, and it has been quite fully illustrated ('27, figs. 24–30; '42, figs. 24–31).

Fig. 69.—This section is adjacent posteriorly to the one shown in figure 30 of 1942, to which reference may be made for the topographic relations. In both figures the bundles of myelinated fibers of the ventral and ventromedian tegmental fascicles are outlined with broken lines. In the interpeduncular region of the section here shown, there is no impregnation of any nervous elements except a few fibers of tr. tegmento-bulbaris at their decussation in the ventral commissure. Within the commissure slender collaterals descend into the interpeduncular neuropil, where some of them have tufted endings.

Fig. 70.—This is drawn from the second section spinalward of figure 69, through the nucleus of the IV nerve, one neuron of which is impregnated. Thick myelinated axons of the IV nerve root (darkened by the Golgi fluid) ascend from this nucleus in the position indicated by broken lines. Typical ependymal elements of the ventral raphe are drawn.

Figure 71.—Obliquely longitudinal section of the adult brain taken not far from the mid-sagittal plane, with the dorsal and anterior sides inclined somewhat laterally, so that almost the entire length of the f. retroflexus appears in a single thick Golgi section. Golgi method. × 30. The drawing is semidiagrammatic, with some details added to the section outlined from neighboring sections and from the opposite side.

At the point where tr. cortico-habenularis medialis joins the stria medullaris, numberless fine collaterals separate from it to enter the eminentia thalami. The section cuts the optic nerve at its junction with the brain. Peripherally, each optic nerve between the foramen and the brain shows several hundred impregnated unmyelinated fibers, most of which lose the stain before entering the brain. The surviving impregnated fibers decussate in chiasmatic bundles 1 (*ch.1.*) at the extreme anteroventral angle of the chiasma and ascend toward the tectum as the most rostral fibers of the marginal optic tract, losing the stain before reaching the tectum ('42, p. 232).

The abundant impregnation of the f. retroflexus extends only as far as the decussation, below which its fibers are unstained. As these fibers turn medially toward the decussation, many of them separate and descend uncrossed along the lateral margin of the interpeduncular neuropil. The locus of the sulcus isthmi on the ventricular surface is projected upon the drawing as

ILLUSTRATIONS

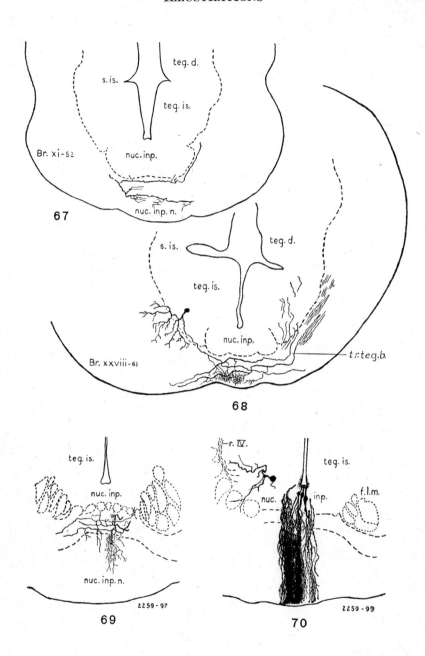

a thick broken line. In the posterior lip of this fissure there is massive elective impregnation of the brachium conjunctivum and the locus of the decussation of these fibers more medially is marked by three crosses. Before and after crossing, many of these fibers turn spinalward in the superficial alba of the isthmic tegmentum.

Figures 72 and 73.—Two semidiagrammatic drawings from an obliquely longitudinal series of Golgi sections of an adult brain. The sections are cut at an angle of about 30° from the sagittal plane, with dorsal and posterior sides more lateral.

Fig. 72.—These details are assembled from several consecutive sections to illustrate some components of the complex of fibers at the di-telencephalic boundary. × 25. Only some of the thicker unmyelinated axons are impregnated in this preparation. The ventral border at the

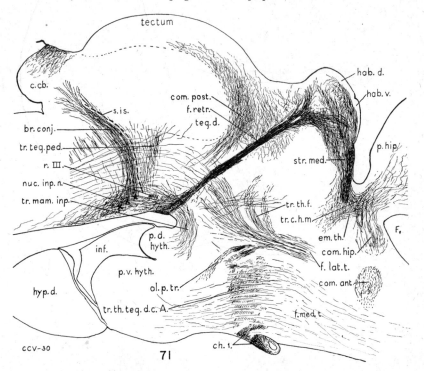

left is near the median plane, and posteriorly of this place the section is inclined laterally. At the posteroventral border of the primordium hippocampi some thick fibers are seen to converge at the thalamic junction, and Weigert sections show that a few of them are myelinated. The larger number of these fibers descend to the hippocampal commissure, and a smaller number ascend in the stria medullaris thalami as tr. cortico-habenularis medialis. Others descend obliquely posteroventrad into the ventral thalamus, passing through and posteriorly of the gray of the eminentia thalami—tr. cortico-thalamicus medialis. This is the precursor of the mammalian columna fornicis, but here apparently none of these fibers reach the mamillary region of the hypothalamus. These thick fibers are accompanied by far more numerous very thin fibers, some of which arise as collaterals from the decussating fibers of the hippocampal commissure (fig. 71). An impregnated neuron of the bed-nucleus of the anterior commissure ridge is sketched from the most lateral of the sections here represented. Its dendrites spread dorsally and ventrally among the hippocampal fibers; the course of the axon is not revealed.

The well-impregnated fibers of tr. thalamo-frontalis arise from the cells of the dorsal thalamus shown in figure 73. In the chiasma ridge, chiasmatic bundles 2 and 3 of optic fibers are impregnated, and their terminals in the lateral part of the tectum are seen at the top of

ILLUSTRATIONS

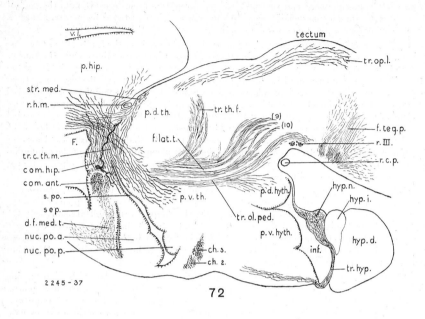

72

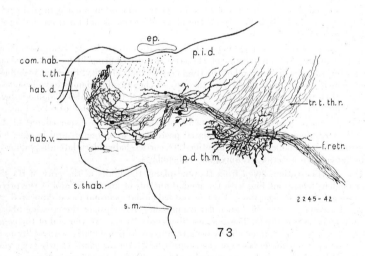

73

the figure. The details of this optic connection have been published ('42, p. 235). The fibers of f. tegmentalis profundus seen at the right are ascending from their decussation in the ventral commissure to spread superficially over the lateral aspect of the isthmic tegmentum. Most of these fibers belong to the brachium conjunctivum.

Fig. 73.—Section more medially, showing details of the origin of the f. retroflexus from the habenula. × 50. The drawing is a combination of three adjoining sections. Only the thicker axons of the fasciculus are here impregnated. These surround a dense fascicle of thinner unstained fibers. The fibers here impregnated arise from neurons of the habenula, from others more posteriorly in the pars intercalaris, and probably also from some of the impregnated neurons of the pars dorsalis thalami. None of these stained fibers reach the interpeduncular nucleus. They spread out in the alba of the posteroventral part of the peduncle, some turning forward into area ventrolateralis pedunculi. The dendrites of the impregnated neurons of the dorsal thalamus extend dorsally among terminals of tr. tecto-thalamicus rectus and also for long distances posteroventrally, accompanying the f. retroflexus into the area of the primordial geniculate bodies (pp. 221, 238). Sections taken more laterally show abundant impregnation of axons from this group of cells which enter tr. thalamo-frontalis (fig. 72).

Figures 74 to 78.—Five sagittal sections of the adult brain, illustrating the composition of the stria medullaris thalami (p. 256 and fig. 20). × 37. The sections were prepared by the reduced silver method of Cajal and cut at 10 μ slightly oblique to the sagittal plane, with the dorsal and rostral sides more lateral (compare '36, figs. 17–79, where the dorsoventral obliquity is in the reverse direction). In these sections the thickest axons, both myelinated and unmyelinated, are black, and the thinner fibers range through gray to brown or yellow, so that components of the stria are well differentiated.

Fig. 74.—This cuts through the di-telencephalic junction near its lateral border, passing through the olfacto-peduncular tract below, and above this through the dorsal and ventral fascicles of the lateral forebrain bundle. The artery (r.h.m.) lies in the floor of the stem-hemisphere fissure. Here four components of the stria medullaris converge, two of them—tr. olfacto-habenularis anterior and lateralis—passing up from the ventral surface externally of the lateral forebrain bundles.

Fig. 75.—In the region of the stem-hemisphere fissure the section passes medially of the lateral forebrain bundles, showing the connection of tr. olfacto-habenularis medialis and tr. septo-habenularis with the stria. Ventrally the plane of section is not far from that of figure 103; dorsally it is much more lateral.

Fig. 76.—At this level tr. cortico-habenularis medialis enters the stria and ascends as its most rostral member. All components of the stria are distributing fibers in the habenular neuropil, posteriorly of which several of the tracts retain their identity. Fibers of tr. habenulo-thalamicus which recurve spinalward over the gray of the dorsal thalamus have been seen to connect only with the ventral habenular nucleus. This tract includes also thalamo-habenular fibers. The boundary of the gray of the eminentia thalami is marked by a broken line.

Fig. 77.—A large proportion of the fibers of all components are here dispersed in the habenular neuropil. A residue of fibers is accumulated posteriorly in a commissural bundle, within which the following components can be identified: tr. cortico-habenularis lateralis and medialis, tr. olfacto-habenularis anterior, tr. amygdalo-habenularis.

Fig. 78.—In this section, 300 μ from the mid-plane, the limits of the gray of the dorsal habenular nucleus (corresponding with the medial nucleus of mammals) and of the pretectal nucleus are marked by broken lines. Most of the habenular neuropil has disappeared. Some fibers from this residue enter the habenular commissure, but most of the crossing fibers are derived from recognizable tracts. The thickest fibers cross at the anterior end of the commissure. These apparently are derived chiefly from tr. amygdalo-habenularis, with additions from other components. The thinner fibers of the commissural bundle include clearly some residue of tr. cortico-habenularis lateralis and medialis and tr. olfacto-habenularis anterior.

Figures 79 and 80.—Two adjoining obliquely sagittal Golgi sections of the adult brain, illustrating some components of the interpeduncular neuropil. In this specimen the finest axons are richly impregnated, especially the deep neuropil of the gray substance, which is continuous throughout the tectum, peduncle, and dorsal, isthmic, and bulbar tegmentum. In the area illustrated it receives fibers from the fields dorsally and anteriorly, and from it axons descend to the

ILLUSTRATIONS

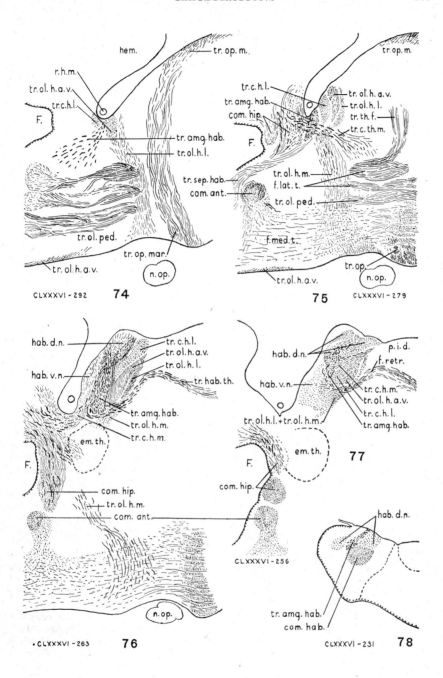

isthmic and bulbar tegmentum. All cell bodies are imbedded in this dense fibrillar reticulum. On these small-scale drawings it is impossible to portray the delicacy and complexity of this tissue, and the attempt is made to indicate the general trend of its coarser fibers.

Fig. 79.—This section is median in the floor of the ventricle from the fovea isthmi to the level of the IX nerve roots. × 40. The ventricular floor from the fovea to the tuberculum posterius is projected from two adjacent sections, as indicated by a dotted line. This section is very thick and is so inclined dorsoventrally as to include above the ventricular floor a thin slice of the nucleus of the tuberculum posterius, tegmentum isthmi, and rostral end of the trigeminal tegmentum of one side. Ventrally of the ventricular floor in the isthmus region the section cuts the interpeduncular nucleus and neuropil close to the mid-plane on the opposite side. The level of the VIII nerve roots is indicated by the decussation of Mauthner's fibers, from which one of the fibers is seen descending. The other fiber appears in the adjacent section 24. From the latter section the portion of tr. interpedunculo-bulbaris dorsalis ($tr.inp.b.d.$) spinalward of this decussation is added to the drawing, showing that some fibers of this tract descend as far as the level of the IX nerve roots. None of its fibers have been seen to extend farther spinalward. The ventral division of this tract ($tr.inp.b.v.$), on the contrary, goes much farther, perhaps as far as the spinal cord.

At the fovea isthmi several ependymal elements are impregnated, and a small artery here enters the foveal pit at the ventral surface. These ependymal elements are crossed by unimpregnated myelinated fibers of the ventral tegmental fascicles ($f.m.t.(1)$), as indicated by dashed lines. A single small neuron of the nucleus of the tuberculum posterius is impregnated. Its dendrites spread downward among slender tortuous axons of tr. mamillo-peduncularis, which form a dense neuropil in the gray of the peduncle. In this plane no fibers of tr. mamillo-interpeduncularis are stained, but they are abundant farther laterally.

Unimpregnated cell bodies of the interpeduncular nucleus are clearly visible, arranged as indicated by the dotted outlines. Among these and ventrally of them is dense neuropil, very inadequately shown in the drawing. These fibers are derived in part from tr. mamillo-interpeduncularis and from the overlying tegmentum, and in larger part they are axons of the cells of the interpeduncular nucleus. These axons take tortuous courses, mainly directed dorsalward into the isthmic and trigeminal tegmentum and spinalward into the dorsal and ventral interpedunculo-bulbar tracts. From them arise numberless collaterals which ramify in the interpeduncular neuropil and enter the glomeruli (fig. 83).

Fig. 80.—The adjoining section. × 75. The ependymal floor of the ventricle is slightly to one side of the mid-plane, ventrally of which the section is inclined laterally. Heavy broken lines mark the upper and lower limits of the zone of unimpregnated cell bodies of the interpeduncular nucleus. Scattered cells of this nucleus are distributed in the underlying neuropil except its ventral part. Above this zone of densely crowded cells are the less crowded cells of the isthmic and trigeminal tegmentum, none of which are impregnated. One neuron of the interpeduncular nucleus is impregnated with dendrite directed ventrally into the neuropil. Fragments of other dendrites, with tufted terminals in glomeruli, are spread throughout the neuropil, some of which are drawn (compare figs. 65, 66, 81-84). There is no other impregnation in the ventral interpeduncular neuropil containing the unimpregnated spiral endings of the f. retroflexus, but dorsally of this there are stained axons of various sorts among the dendrites. These are not drawn, except those lying between the cells of the interpeduncular nucleus and the myelinated fibers of the ventral commissure, the most dorsal bundles of which are outlined with dotted lines. Most of these unmyelinated fibers enter tr. interpedunculo-bulbaris dorsalis. The interpeduncular neuropil is broadly connected with the overlying tegmental neuropil by fibers passing in both directions. In this plane (not drawn) there are a few fibers of tr. mamillo-interpeduncularis and more of them farther laterally.

Figure 81.—Obliquely sagittal Golgi section of an adult brain taken close to the mid-plane. × 60. Some other sections of this specimen have been published (fig. 101; '25, figs. 14, 40). The plane of section of this figure is similar to that of figure 79, but somewhat more oblique. The ventral surface is nearly median in the isthmus, and dorsally and anteriorly of this the plane is inclined laterally. Since the sections are very thick, the dorsal ventricular border of the tuberculum posterius (indicated by the dotted line), and a slice of the lateral wall dorsally of it are included in the section outlined. The detail is drawn from this section and the adjacent section laterally. The four components of the commissure of the tuberculum posterius (p. 302)

ILLUSTRATIONS

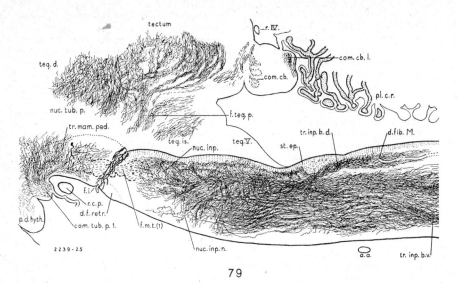

79

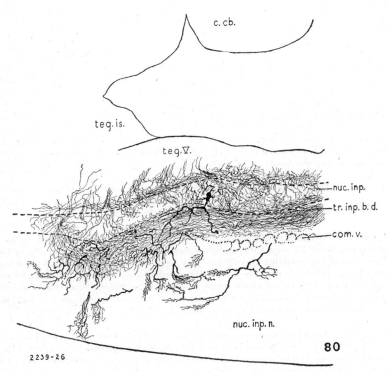

80

and the dorsal fascicles of the ventral commissure posteriorly of the fovea isthmi are outlined with dotted lines. The decussation of the anterior division of tr. tecto-bulbaris cruciatus in the ventral median tegmental fascicles ($f.m.t.(1)$) is indicated by dashed lines (compare '36, fig. 2).

A neuron of the dorsal (mamillary) part of the hypothalamus and the dendrite of another are impregnated. Their axons branch, and one branchlet enters tr. mamillo-interpeduncularis. This tract partially decussates in component 1 of the commissure of the tuberculum posterius (the retroinfundibular commissure of the literature), and its slender unmyelinated fibers arborize in the rostral part of the interpeduncular neuropil chiefly laterally of the plane here shown (compare '36, figs. 3, 8, 20). Two ependymal elements are drawn. These lie near the ventral median raphe under the cerebellum (compare figs. 63, 64, 70).

Unimpregnated cell bodies of the interpeduncular nucleus are scattered under the ependymal surface and among the myelinated median fascicles and ventral commissure bundles. A well-impregnated dendrite of one of these is drawn; it has a tufted terminal near the ventral surface (compare figs. 62, 65, 66, 80, 82). There is scanty and very clear impregnation of some of the coarser fibers of the interpeduncular neuropil. Those in its anterior part are derived chiefly from tr. mamillo-interpeduncularis and probably some from tr. olfacto-peduncularis, which is partially impregnated more laterally (fig. 101). More posteriorly the visible fibers all seem to be axons of cells of the interpeduncular nucleus which assemble to descend in the dorsal and ventral interpedunculo-bulbar tracts.

Figure 82.—A semidiagrammatic drawing of a thick Golgi section from an obliquely sagittal series of an adult brain. $\times$ 37. The section is nearly median posteriorly (at the right), and anteriorly it is much more lateral, including the f. retroflexus in the peduncle. Some fibers of this fasciculus descend uncrossed into the interpeduncular neuropil. Mingled with the uncrossed fibers of the f. retroflexus are thicker and smoother fibers of tr. olfacto-peduncularis, some of which enter the interpeduncular neuropil. Dorsally of these is the mamillo-peduncular tract, comprising ventral tegmental fascicle (3), and still farther dorsally are fibers of tegmental fascicle (6), which come from the postoptic commissure. These tegmental fascicles contain both myelinated and unmyelinated fibers, and both sorts are here darkened by the Golgi treatment. Between the two tegmental fascicles just mentioned are the darkened thick myelinated fibers of the anterior division of tr. tecto-bulbaris cruciatus ($f.m.t.(1)$), some of which decussate near the fovea isthmi dorsally of the olfacto-peduncular fibers and others descend in the ventral median fascicles. One neuron of the interpeduncular nucleus is well impregnated and dendrites of several others, all of which exhibit the characteristic dendritic tufts. No ependyma, spiral fibers, or axonic tufts are impregnated in the interpeduncular neuropil. The scanty axonic neuropil is probably composed exclusively of terminals of uncrossed fibers of the f. retroflexus and olfacto-peduncular tract.

Figure 83.—Diagram of a neuron of the interpeduncular nucleus, seen as projected upon the median sagittal section. $\times$ 40. This is a composite drawing from observations made on many sections cut in various planes. Compare figures 19 and 84.

Figure 84.—Diagram of the composition of the interpeduncular glomeruli. $\times$ 45. This drawing, like figure 79, shows the median section of the floor plate and dorsally of this an oblique slice of the overlying tegmentum. Two neurons of the interpeduncular nucleus are drawn. Tufted collaterals from the axon of the anterior element engage dendritic glomeruli of the posterior element. Tufted axonic terminals enter glomeruli from small cells of the isthmic and trigeminal tegmentum and also from collaterals of tr. tegmento-bulbaris, which arises from large cells of the same tegmental areas. Compare figures 19 and 83.

Figure 85.—The lateral aspect of the adult brain, drawn from the same wax model as figure 1A. $\times$ 15. This drawing shows the courses of the four most superficial components of the stria medullaris thalami seen as projected upon the lateral surface (fig. 20 and chap. xviii). Tractus olfacto-habenularis lateralis (2) goes directly dorsally from the preoptic nucleus. Tractus olfacto-habenularis anterior arises in ventral (3) and dorsal (4) divisions from the anterior olfactory nucleus. Tractus cortico-habenularis lateralis (5) arises from the primordial piriform lobe and joins no. 4 before entering the stria medullaris.

ILLUSTRATIONS

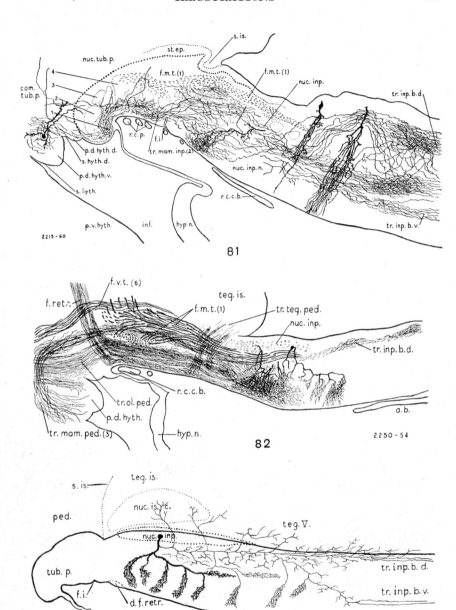

81

82

83

368 THE BRAIN OF THE TIGER SALAMANDER

Figure 86A, B, and C.—The brains of A. tigrinum and Necturus drawn to approximately the same scale.

A.—Lateral aspect of the brain of A. tigrinum and its arteries. After Roofe ('35). × 7.5. Drawn after fixation in 10 per cent formalin for six weeks.

B.—Lateral aspect of the cerebrum of Necturus. This and figure 86C are copied from 1933c, figures 3 and 4. × 8.

C.—Median section of the brain of Necturus (compare fig. 2). In the thalamus and midbrain the section is median, and the dotted lines indicate ventricular sulci which mark the boundaries of the chief cellular areas. In the cerebral hemisphere the section is slightly to one side of the mid-plane, and the dotted lines indicate the boundaries of cellular areas with limits marked internally by ventricular sulci.

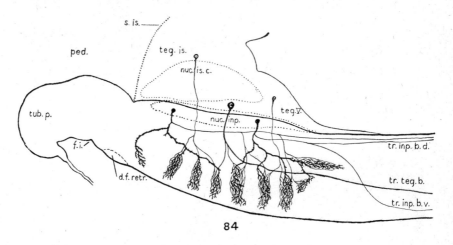

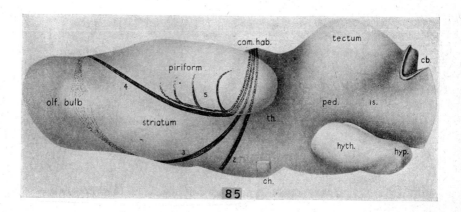

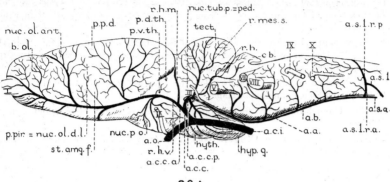

86 A

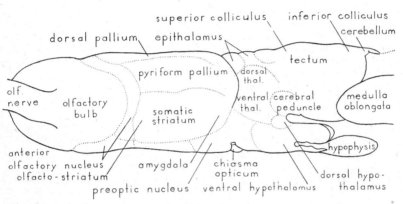

86 B

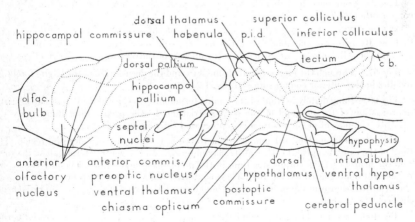

86 C

The section pictures which follow are here reproduced from previous publications (with minor alterations in some instances), and more complete descriptions will be found in the references cited.

Figures 87 to 91.—Five transverse sections of the adult medulla oblongata. Method of Wiegert. × 37. These are copies, respectively, of figures 5, 3, 2, and 1 of 1944b and figure 16 of 1936, drawn from the type specimen, no. IIC.

Fig. 87.—Section taken immediately below the calamus scriptorius through the commissural nucleus and the com. infima of Haller.

Fig. 88.—Through the lower vagus region and the most anterior ventral rootlet of the first spinal nerve.

Fig. 89.—At the level of the IX nerve roots. A rootlet of the VI nerve emerges in this section. The two lateral-line roots of the vagus enter immediately rostrally of this level.

Fig. 90.—Section taken immediately below the superficial origin of the V nerve roots, including the posterior part of the motor V nucleus.

Fig. 91.—Section through the cerebellum, auricle, and rostral end of the trigeminal tegmentum.

Figure 92.—Transverse section immediately spinalward of the nucleus of the IV nerve. × 37. Ventrally it passes through the posterior end of the infundibulum and dorsally through the junction of superior and inferior colliculi. The Arabic numbers in parentheses refer to the tegmental fascicles described in chapter xx. This is a copy of figure 14 of 1936, drawn from a reduced silver preparation, method of Rogers.

Figure 93.—Semidiagrammatic transverse section at the level of the III nerve roots. × 37. Copied from figure 14 of 1942. The layers of the tectum and some of its afferent tracts are shown on the left side and efferent tracts on the right.

Figure 94.—Section through the rostral border of the commissure of the tuberculum posterius. × 37. Copy of figure 10 of 1936 and drawn from the same specimen as figure 92.

Figures 95 to 100.—These are obliquely transverse sections (ventral side inclined spinalward), copied, with some alterations, from the paper of 1927, where full descriptions will be found. × 25. They are from the brain of a recently metamorphosed adult from Colorado, prepared by the Golgi method. The drawings are composite, the outlines and much of the detail being derived from the specimen mentioned, supplemented by additional details selected from 20 other specimens cut in the transverse plane and stained by the methods of Golgi, Weigert, and Cajal.

Fig. 95.—Through the habenular commissure and the posterior border of the postoptic commissure ('27, fig. 20).

Fig. 96.—Through the rostral border of the chiasma ridge and the middle of the eminentia thalami. By reason of the obliquity of the section, it passes rostrally of the habenulae ('27, fig. 15). The cellular area marked *p.v.th.* is the nucleus of the olfacto-habenular tract (p. 248).

Fig. 97.—Through the hippocampal commissure and the decussation of the lateral forebrain bundles in the anterior commissure ('27, fig. 12).

Fig. 98.—Immediately in front of the lamina terminalis. On the right side is a characteristic impregnation of the neuropil of the corpus striatum. In the preparation drawn, only the dendritic component of the neuropil is impregnated ('27, fig. 9).

Fig. 99.—Section about 0.1 mm. anterior to the last, through the mid-septal region. The head of the caudate nucleus is here at its maximum size. In the ventrolateral wall on the right side the neuropil of the corpus striatum, in the preparation here drawn, has only the axonic component impregnated (cf. fig. 108). These contorted and branched axons are interlaced with the dendrites shown in figure 98, and the boundary of this area of dense neuropil is sharply defined ('27, fig. 8).

Fig. 100.—Section through the middle of the olfactory bulb and the extreme anterior border of the primordium hippocampi. It includes also the posterior border of the f. postolfactorius ('27, fig. 4).

ILLUSTRATIONS

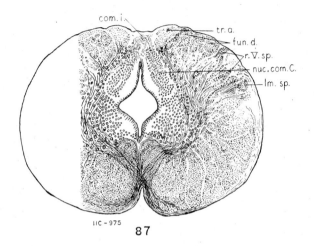

87

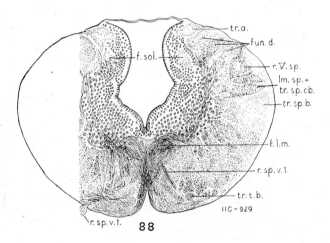

88

372   THE BRAIN OF THE TIGER SALAMANDER

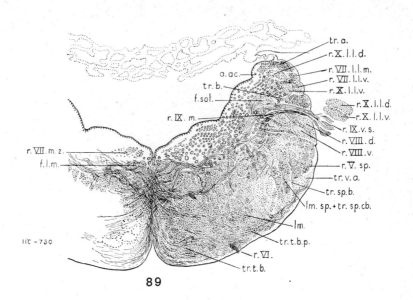

89

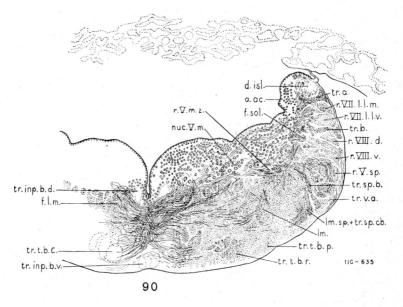

90

ILLUSTRATIONS

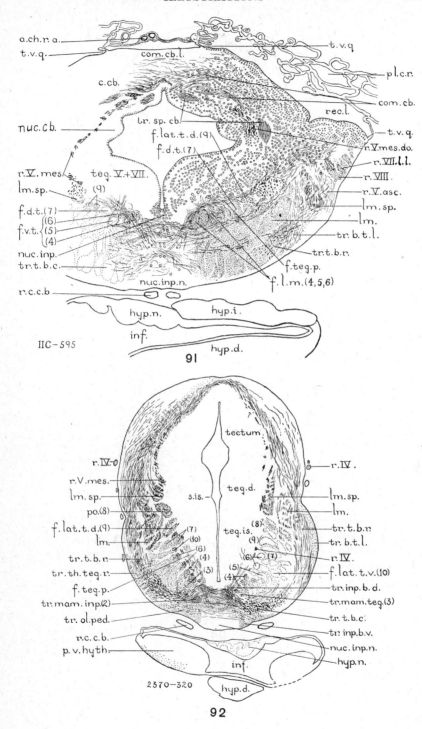

91

92

374   THE BRAIN OF THE TIGER SALAMANDER

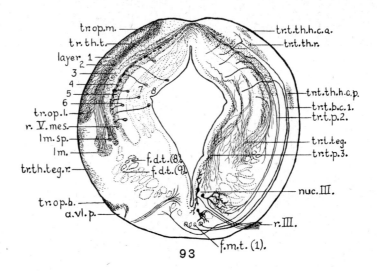

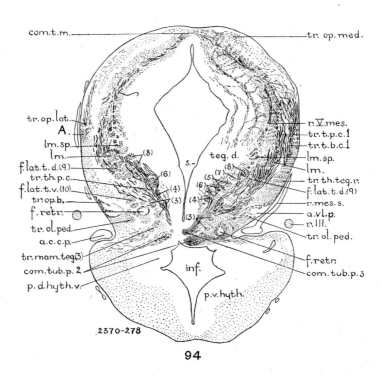

ILLUSTRATIONS

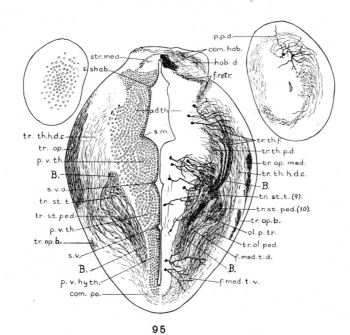

95

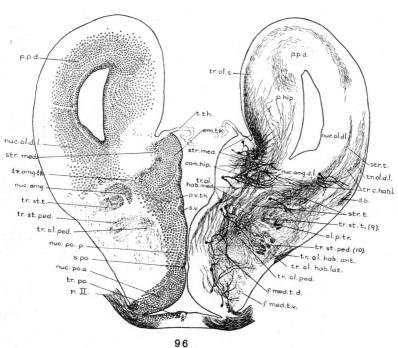

96

376   THE BRAIN OF THE TIGER SALAMANDER

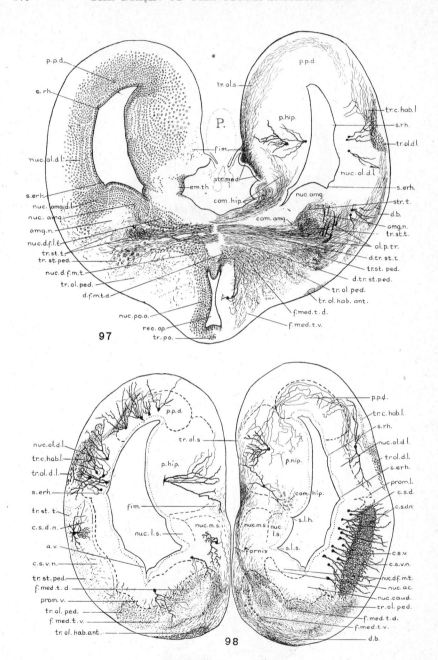

ILLUSTRATIONS

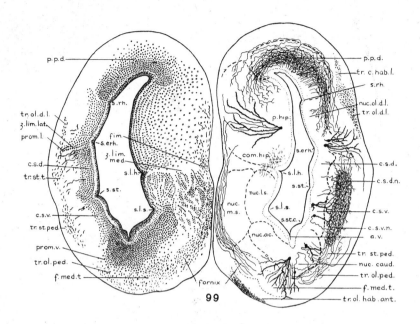

99

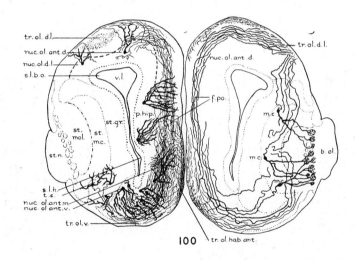

100

378 THE BRAIN OF THE TIGER SALAMANDER

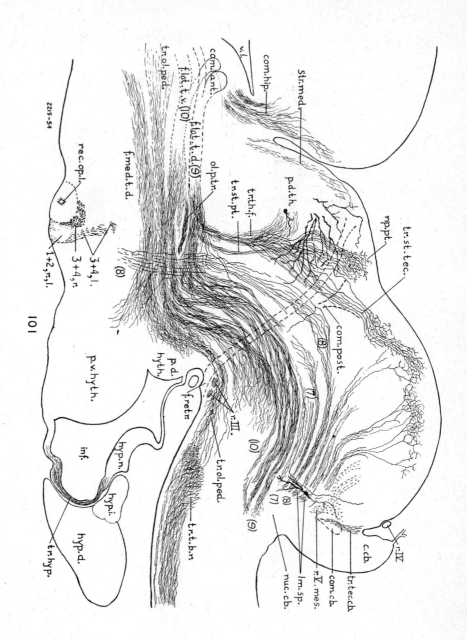

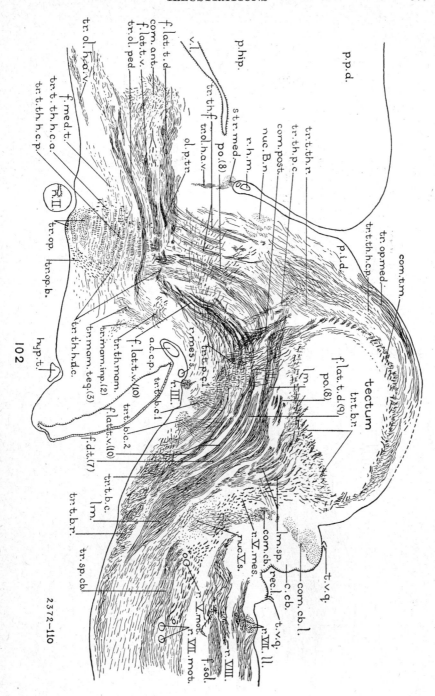

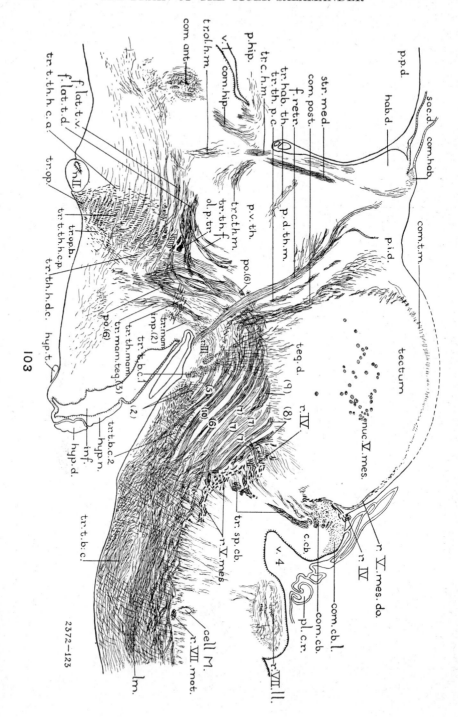

ILLUSTRATIONS

Figure 101.—Semidiagrammatic sagittal section of the adult brain, Golgi method. × 35. The section is oblique, with the dorsal side somewhat more lateral, cut in a plane which includes almost the whole length of the lateral forebrain bundles and tr. olfacto-peduncularis. In the tectal neuropil there are typical endings of the spinal lemniscus and tr. strio-tectalis (copied from '42, fig. 17).

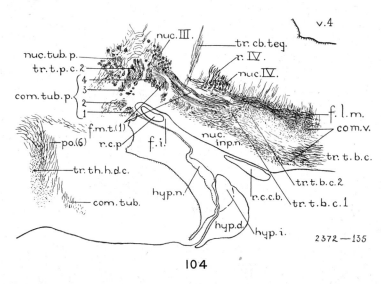

104

Figures 102, 103, 104.—Three sagittal sections from the brain of a specimen from Colorado prepared about two weeks after metamorphosis by the reduced silver method of Rogers. The sections are slightly oblique, with the dorsal surface and caudal end more medial. × 37. In this series of sections the courses of the tegmental fascicles can be clearly followed. These pictures are selected from a series published in 1936.

Fig. 102.—The section passes through the middle of the gray substance of the tectum opticum at its widest part and shows the courses of the lateral forebrain bundles and lemniscus systems ('36, fig. 17).

Fig. 103.—Section taken 0.17 mm. more medially and showing the arrangement of the more medial tegmental fascicles ('36, fig. 18).

Fig. 104.—Section taken 0.16 mm. more medially and about 0.1 mm. from the mid-plane at the tuberculum posterius. It includes the interstitial nucleus of the f. longitudinalis medialis (*nuc.tub.p.*) and the nuclei of the III and IV nerves ('36, fig. 22).

382   THE BRAIN OF THE TIGER SALAMANDER

Figures 105, 106, 107.—These figures, copied from the paper of 1934, show details of the structure of the olfactory bulb, anterior olfactory nucleus, and the related neuropil.

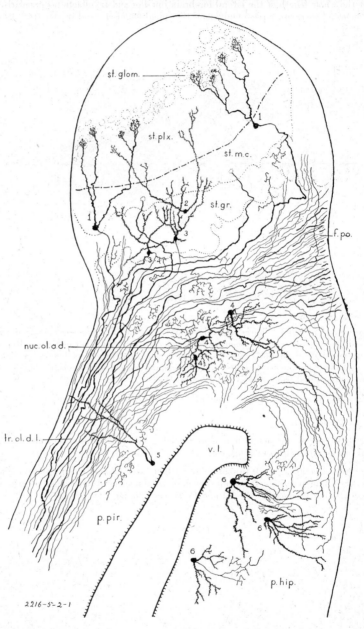

ILLUSTRATIONS

Fig. 105.—Semidiagrammatic horizontal section through the dorsal border of the olfactory bulb and dorsal sector of the anterior olfactory nucleus of the adult. × 50. The medial border (at the right) is slightly inclined ventrally. The diagram is based on sections of a single specimen. The details pictured have all been observed, but the assembling is a schematic composition ('34, fig. 1). The histological structure of the olfactory bulb as seen in horizontal sections is shown in figure 110. Figure 105 is a similar diagram at a more dorsal level. Two mitral cells (1) are drawn and a typical granule cell (2). Two elements of transitional type (3) in the stratum granulare send slender axons to the anterior olfactory nucleus, three neurons (4) of which are drawn. Farther back there are three typical neurons of the primordium hippocampi (6) and one of the primordium piriforme (5).

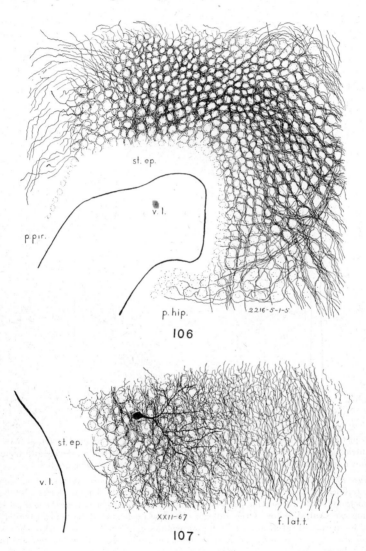

Fig. 106.—Detail of the neuropil of the gray substance of the dorsal sector of the anterior olfactory nucleus in the plane of figure 105 and from the same specimen ('34, fig. 2). Golgi method. × 100. Only the axonic component of the neuropil is here impregnated. The positions of the unimpregnated cell bodies are indicated by the dotted outlines.

Fig. 107.—This is a detail of the neuropil of the gray substance of the rostral part of the corpus striatum of the adult frog, Rana pipiens ('34, fig. 3). Golgi method. × 142. The

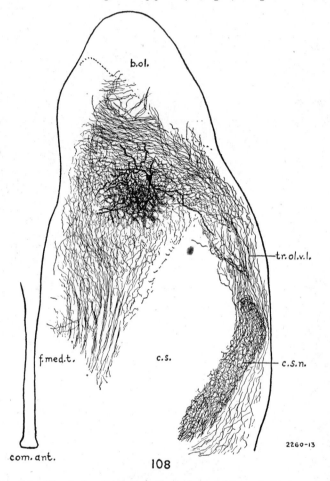

section pictured is obliquely longitudinal, inclined about 45° from the horizontal plane, with the ventricular border more ventral. The rostral end is above. The impregnation is similar to that shown in figure 106. One small neuron of the striatal gray is impregnated.

Figure 108.—Neuropil of the anterior olfactory nucleus and corpus striatum, as seen in an obliquely horizontal section of the adult brain. The right side is more dorsal ('42, fig. 15). Golgi method. × 50. The section passes through the right hemisphere near the ventral surface, including the ventral sector of the anterior olfactory nucleus and the ventral border of the gray of the corpus striatum (*c.s.*, here unimpregnated). Laterally of this gray is the striatal neuropil (*c.s.n.*), in which only the axonic component is impregnated (cf. fig. 99). Rostrally of the striatal gray is the extreme ventral border of the neuropil of the caudate nucleus, the structure of which is shown in figure 109.

Figure 109.—An obliquely longitudinal section through the ventrolateral border of the adult hemisphere ('42, fig. 16). Golgi method. × 80. The section is inclined to the sagittal plane, with the anterior and dorsal sides much more lateral. It cuts the accessory olfactory bulb and the anterior end of the caudate nucleus, whose gray (*nuc.caud.*) is unimpregnated. Above and externally of this is a dense impregnation of the caudate neuropil with sharply defined borders. Some unimpregnated cell bodies are enmeshed within this neuropil, only the axonic component of which is impregnated. This neuropil is continuous with that of the remainder of the striatal complex shown in figure 108. Anteriorly of it, six neurons of the anterior olfactory nucleus are impregnated.

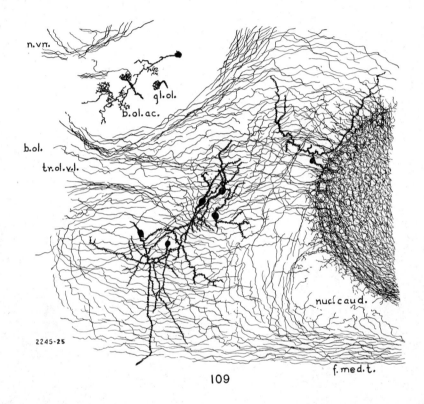

109

Figure 110A, B.—Diagrams of the structure of the olfactory bulb as seen in horizontal section of the adult brain.

A.—The outline is taken from a section at the widest part of the olfactory bulb and the elements are drawn to scale from several Golgi sections, most of them from the same specimen ('24b, fig. 1). × 50. Glomeruli are outlined with dotted lines. The approximate locus of figure 110B is indicated. The following types of neurons have been described: (1) periglomerular cells; (2) mitral cells; (3) subglomerular tufted cells; (4) granule cells; (5) cells intermediate between granules and neurons of nucleus olfactorius anterior; (6) transitional cells of nucleus olfactorius anterior.

B.—The layers of the olfactory bulb as seen in horizontal section after fixation in formalin-Zenker followed by Mallory's stain ('24b, fig. 4). × 150.

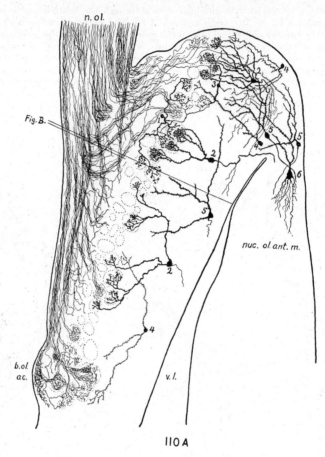
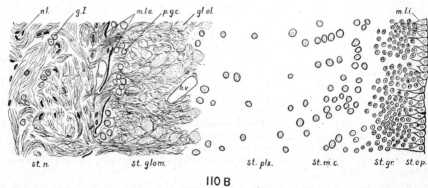

ILLUSTRATIONS

Figure 111.—Lateral aspect of the forebrain of Necturus (cf. fig. 86B), upon which some olfacto-somatic pathways are projected ('33e, fig. 1). × 8. Olfactory tracts are drawn in broken lines, descending tracts in solid lines, and the thalamic radiations and their connections are in red lines (p. 102). Connections with the epithalamus and hypothalamus are omitted.

Figure 112.—Median section of the brain of Necturus (cf. fig. 86C), upon which some of the descending pathways from the cerebral hemisphere to the hypothalamus and epithalamus are projected (p. 268) ('33e, fig. 2). × 8.

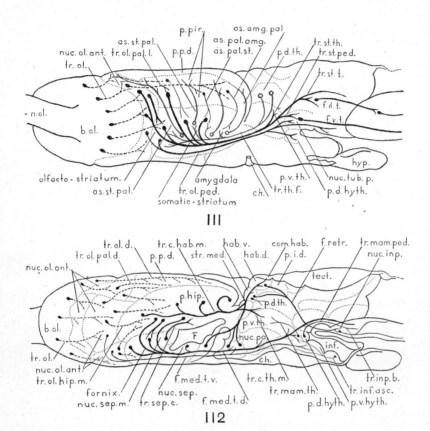

Figure 113.—Diagram illustrating the relations of projection and association fibers in the cerebral hemispheres of Necturus as seen in transverse section a short distance rostrally of the lamina terminalis (p. 102). The gray substance is outlined by a broken line. Descending fibers are drawn on the left, ascending and pallial association fibers on the right ('34, fig. 7). × 16.

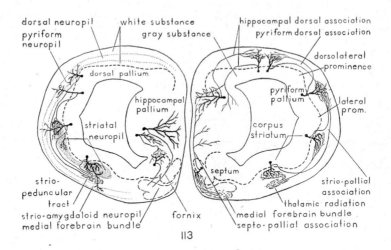

ABBREVIATIONS FOR ALL FIGURES

ABBREVIATIONS FOR ALL FIGURES

The following list includes all abbreviations on the figures and a few others used in previous publications. Reference may also be made to the list appended to the monograph on Necturus ('33b, pp. 280-88), where synonyms of many of these terms are given, with references to the literature. For reference to published lists of tracts see page 270, and for nomenclature of blood vessels see the paper of 1935 and Roofe ('35).

A., division A of *tr.th.teg.d.c.*
a.a., arteria auditiva
a.ac., area acusticolateralis
a.b., arteria basilaris
a.c.c., arteria carotis cerebralis
a.c.c.a., arteria carotis cerebralis, ramus anterior
a.c.c.p., arteria carotis cerebralis, ramus posterior
a.c.i., arteria carotis interna
a.ch.r.a., arteria chorioidea rhombencephali anterior
a.gen., area geniculata = *np.gen.*
a.l.t. = *a.vl.p.*
a.o., arteria ophthalmica
a.s.a., arteria spinalis anterior
a.s.1., arteria spinalis 1
a.s.1.r.a., arteria spinalis 1, ramus anterior
a.s.1.r.p., arteria spinalis 1, ramus posterior
a.v., angulus ventralis
a.vl.p., area ventrolateralis pedunculi = *a.l.t.*
amg., amygdala
amg.n., neuropil of amygdala
as.amg.pal., amygdalo-pallial association
as.pal.amg., pallio-amygdaloid association
as.pal.st., pallio-striatal association
as.st.pal., strio-pallial association
aur., auricle

B., division B of *tr.th.teg.d.c.*
b.ol., bulbus olfactorius
b.ol.ac., bulbus olfactorius accessorius
b.v., blood vessel
b.v.III., V., VII., IX., X., blood vessels associated with cranial nerve roots
br.col.; br.col.s.; br.col.inf., brachia of superior and inferior colliculus = *tr.th.r.*
br.conj., brachium conjunctivum

c.cb., corpus cerebelli
c.s., corpus striatum
c.s.d., corpus striatum, pars dorsalis
c.s.d.n., neuropil of *c.s.d.*
c.s.v., corpus striatum, pars ventralis

c.s.v.n., neuropil of *c.s.v.*
cb., cerebellum
cell M., cell of Mauthner
ch., chiasma opticum
col.inf., colliculus inferior = *nuc.p.t.*
col.sup., colliculus superior; tectum opticum
com.amg., commissure of amygdalae
com.ant., anterior commissure
com.cb., commissura cerebelli
com.cb.l., commissura vestibulo-lateralis cerebelli
com.cb.l.l., lateral-line component of *com.cb.l.*
com.cb.V., trigeminal component of *com.cb.*
com.cb.VIII., vestibular component of *com.cb.l.*
com.hab., habenular commissure
com.hip., hippocampal commissure
com.i., commissura infima Halleri
com.po., commissura postoptica
com.post., commissura posterior
com.t.d., commissura tecti diencephali
com.t.m., commissura tecti mesencephali
com.tub., commissura tuberis
com.tub.p., commissura tuberculi posterioris
com.v., commissura ventralis

d.b., diagonal band of Broca
d.f.l.t., decussation of *f.lat.t.*
d.f.m.t.; d.f.med.t., decussation of *f.med.t.*
d.f.retr., decussation of *f.retr.*
d.fib.M., decussation of Mauthner's fibers
d.isl., dorsal island of Kingsbury
d.r.IV., decussation of IV nerve roots
d.rinf., decussatio retroinfundibularis; components 1 and 2 of *com.tub.p.*
d.tr.st,ped., decussation of *tr.st.ped.*
d.tr.st.t., decussation of *tr.st.t.*

em.com.p., eminentia commissurae posterioris
em.s.t., eminentia subcerebellaris tegmenti = *nuc.cb.* + *teg.is.*
em.th., eminentia thalami
em.V., eminentia trigemini
ep., epiphysis

F., foramen interventriculare
f.d.t., fasciculi dorsales tegmenti
f.i., fovea isthmi
f.l.m., fasciculus longitudinalis medialis
f.lat.t., fasciculus lateralis telencephali; lateral forebrain bundle
f.lat.t.d., dorsal fascicles of *f.lat.t.* = *tr.st.t.*
f.lat.v., ventral fascicles of *f.lat.t.* = *tr.st.ped.*
f.m.t., fasciculus medianus tegmenti
f.med.t., fasciculus medialis telencephali; medial forebrain bundle
f.med.t.d., dorsal fascicles of *f.med.t.*
f.med.t.v., ventral fascicles of *f.med.t.*
f.po., fasciculus postolfactorius
f.retr., fasciculus retroflexus
f.sol., fasciculus solitarius
f.sol.pf., prefacial fasciculus solitarius
f.teg.p., fasciculus tegmentalis profundus
f.v.t., fasciculi ventrales tegmenti
fib.arc., arcuate fibers
fib.M., fiber of Mauthner
fim., fimbria
fun.d., funiculus dorsalis

g.I., ganglion cells of nervus olfactorius
gl.ol., glomeruli olfactorii

hab., habenula
hab.d., nucleus dorsalis habenulae
hab.d.n., neuropil of *hab.d.*
hab.v., nucleus ventralis habenulae
hab.v.n., neuropil of *hab.v.*
hem., cerebral hemisphere
hyp., hypophysis
hyp.d., hypophysis, pars distalis
hyp.g., hypophysis, pars glandularis
hyp.i., hypophysis, pars intermedia
hyp.n., hypophysis, pars nervosa
hyp.t., hypophysis, pars tuberalis
hyth., hypothalamus

I to *X*, cranial nerve roots
inf., infundibulum
inp.n., interpeduncular neuropil
is., isthmus rhombencephali

lam.t., lamina terminalis
lm., general bulbar lemniscus
lm.sp., lemniscus spinalis

M., mouth of paraphysis
m.c., mitral cells
m.l.e., membrana limitans externa
m.l.i., membrana limitans interna

n.II.; n.op., nervus opticus
n.ol., nervus olfactorius
n.vn., nervus vomeronasalis
nl., neurilemma nuclei
np.gen., geniculate neuropil

np.pt., pretectal neuropil
nuc.ac., nucleus accumbens septi
nuc.amg., nucleus amygdalae
nuc.amg.d.l., nucleus amygdalae dorsolateralis
nuc. B., nucleus of Bellonci
nuc.B.n., neuropil of Bellonci
nuc.caud., nucleus caudatus
nuc.cb., nucleus cerebelli
nuc.com.C., nucleus commissuralis of Cajal
nuc.d.f.l.t., bed-nucleus of decussation of *f.lat.t.*
nuc.d.f.m.t., bed-nucleus of decussation of *f.med.t.*
nuc.Dark., nucleus of Darkschewitsch
nuc.ec.mam. = *a.vl.p.*
nuc.f.sol., nucleus of fasciculus solitarius
nuc.III., nucleus of oculomotor nerve
nuc.inp., nucleus interpeduncularis
nuc.inp.n., neuropil of *nuc.inp.*
nuc.is.c., nucleus centralis isthmi
nuc.IV., nucleus of trochlear nerve
nuc.l.s., nucleus lateralis septi
nuc.m.s., nucleus medialis septi
nuc.ol.a.; *nuc.ol.ant.*, nucleus olfactorius anterior
nuc.ol.a.d.; *nuc.ol.ant.d.*, nucleus olfactorius anterior, pars dorsalis
nuc.ol.ant.m., the same, pars medialis
nuc.ol.ant.v., the same, pars ventralis
nuc.ol.d.l., nucleus olfactorius dorsolateralis = *p.pir.*
nuc.ol.p.tr., nucleus of olfactory projection tract
nuc.p.t., nucleus posterior tecti
nuc.po., nucleus preopticus
nuc.po.a., nucleus preopticus, pars anterior
nuc.po.p., nucleus preopticus, pars posterior
nuc.pt., nucleus pretectalis
nuc.sep., nucleus septi
nuc.sep.m. = *nuc.m.s.*
nuc.tr.ol.h., nucleus of tractus olfactohabenularis
nuc.tub.p., nucleus tuberculi posterioris; peduncle
nuc.v.l., nucleus ventrolateralis of cerebral hemisphere; amygdala
nuc.V.m., motor nucleus of trigeminus
nuc.V.mes., mesencephalic nucleus of trigeminus
nuc.V.s., superior nucleus of trigeminus
nuc.vis.n., neuropil of *nuc.vis.s.*
nuc.vis.s., nucleus visceralis superior; superior gustatory nucleus

o.s.c., organon subcommissurale
ol.p.tr., olfactory projection tract

ABBREVIATIONS

P., paraphysis
p.d.hyth., pars dorsalis hypothalami
p.d.hyth.d., dorsal lobe of p.d.hyth.
p.d.hyth.v., ventral lobe of p.d.hyth.
p.d.th., pars dorsalis thalami
p.d.th.m., pars dorsalis thalami, area medialis
p.d.th.p., pars dorsalis thalami, area posterior
p.g.c., periglomerular cells
p.hip., primordium hippocampi
p.i.d.; p.i.th., pars intercalaris diencephali
p.p.d., primordium pallii dorsalis
p.pir., primordium piriforme = nuc.ol.d.l.
p.v.hyth., pars ventralis hypothalami
p.v.hyth.a., anterior lobe of p.v.hyth.
p.v.hyth.p., posterior lobe of p.v.hyth.
p.v.th., pars ventralis thalami
p.v.th.a., pars ventralis thalami, area anterior
p.v.th.p., pars ventralis thalami, area posterior
par., paraphysis
pars.d.l., pars dorsolateralis of hemisphere = p.pir.
pars.v.l., pars ventrolateralis of hemisphere = c.s.
ped., pedunculus cerebri
pl.c.r., plexus chorioideus rhombencephali
po.(4),(6),(8), fibers from postoptic commissure entering tegmental fascicles
prom.l., prominentia lateralis
prom.v., prominentia ventralis

r.c.c.b., ramus communicans cum arteria basilaris
r.c.p., ramus communicans posterior
r.h., ramus hypothalamicus
r.h.m., ramus hemisphaerii medialis
r.h.v., ramus hemisphaerii ventralis
r.III., root of oculomotor nerve
r.IV., root of trochlear nerve
r.IX., root of glossopharyngeal nerve
r.IX.m., motor root of IX nerve
r.IX.v.s., visceral sensory root of IX nerve
r.l.; rec.l., recessus lateralis rhombencephali
r.mes.s., ramus mesencephali superior
r.p.m., recessus posterior mesencephali
r.po., recessus preopticus
r.sp., root of spinal nerve
r.sp.v.1., ventral root of first spinal nerve
r.V., root of trigeminal nerve
r.V.asc., ascending fibers of sensory V root
r.V.m.2., posterior motor V root
r.V.mes., mesencephalic root of V nerve
r.V.mes.do., dorsal division of r.V.mes.
r.V.mes.v., ventral division of r.V.mes.
r.V.mot., motor root of V nerve
r.V.sen., sensory root of V nerve
r.V.sp., spinal root of V nerce
r.VI., root of abducens nerve

r.VII., root of facial nerve
r.VII.l.l., lateral-line roots of VII nerve
r.VII.l.l.d., dorsal lateral-line VII root
r.VII.l.l.m., middle lateral-line VII root
r.VII.l.l.v., ventral lateral-line VII root
r.VII.m.; r.VII.mot., motor root of VII nerve
r.VII.m.2., posterior motor VII root
r.VII.v.s., visceral sensory root of VII nerve
r.VIII., root of VIII nerve
r.VIII.d., dorsal root of VIII nerve
r.VIII.v., ventral root of VIII nerve
r.X., root of vagus nerve
r.X.l.l., lateral-line roots of vagus nerve
r.X.l.l.d., dorsal lateral-line root of vagus
r.X.l.l.v., ventral lateral-line root of vagus
r.X.v.s., visceral sensory root of vagus
rec.l., recessus lateralis rhombencephali
rec.op., recessus preopticus
rec.op.l., recessus opticus lateralis
rec.p.m., recessus posterior mesencephali
rec.pcm., recessus precommissuralis
rec.pin., recessus pinealis

s., limiting sulcus of nuc.tub.p.
s.d., sulcus dorsalis thalami
s.erh., sulcus endorhinalis
s.hyth., sulcus hypothalamicus
s.hyth.d., sulcus hypothalamicus dorsalis
s.hyth.p., sulcus hypothalamicus posterior
s.ih., sulcus intrahabenularis
s.is., sulcus isthmi
s.l.b.o., sulcus limitans of olfactory bulb
s.l.h., sulcus limitans hippocampi
s.l.s., sulcus limitans septi
s.lat.mes., sulcus lateralis mesencephali
s.m., sulcus medius thalami
s.o., sinus obliquus
s.po., sulcus preopticus
s.rh., sulcus rhinalis
s.shab., sulcus subhabenularis
s.st., sulcus striaticus
s.st.c., sulcus strio-caudatus
s.v., sulcus ventralis thalami
s.v.a., sulcus ventralis accessorius thalami
sac.d., saccus dorsalis
sac.v., hypophysis, pars nervosa
sep., septum
sep.ep., septum ependymale
sep.m., septum mediale
sep.t.p., septum transversum paraphysis
st.amg.f., strio-amygdaloid field
st.ep., ependymal layer
st.glom., layer of glomeruli
st.gr., granular layer
st.m.c., layer of mitral cells
st.mol., molecular layer = st.plx.

st.n., layer of nerve fibers
st.plx., plexiform layer
str.med., stria medullaris thalami
str.t., stria terminalis

t.c., transitional cells
t.f., taenia fornicis
t.th., taenia thalami
t.v.q., taenia ventriculi quarti
tec.d., dorsal thickening of tectum
tect., tectum mesencephali
teg.d., tegmentum dorsale mesencephali
teg.is., tegmentum isthmi
teg.is.m., tegmentum isthmi, pars magnocellularis
teg.V., tegmentum trigemini
teg.VII., tegmentum facialis
th., thalamus
tr.a., dorsal correlation tract of Kingsbury
tr.amg.hab., tractus amygdalo-habenularis
tr.amg.th., tractus amygdalo-thalamicus
tr.b., ventral correlation tract of Kingsbury
tr.b.is., tractus bulbo-isthmialis
tr.b.sp., tractus bulbo-spinalis
tr.b.t.l., tractus bulbo-tectalis lateralis
tr.c.h.l.; *tr.c.hab.l.*, tractus cortico-habenularis lateralis
tr.c.h.m.; *tr.c.hab.m.*, tractus cortico-habenularis medialis
tr.c.th.m., tractus cortico-thalamicus medialis
tr.cb.teg., tractus cerebello-tegmentalis
tr.hab.t., tractus habenulo-tectalis
tr.hab.th., tractus habenulo-thalamicus
tr.hy.ped., tractus hypothalamo-peduncularis
tr.hy.teg., tractus hypothalamo-tegmentalis
tr.hyp., tractus hypophysius
tr.inf.asc., tractus infundibularis ascendens
tr.inp.b., tractus interpedunculo-bulbaris
tr.inp.b.d., tractus interpedunculo-bulbaris dorsalis
tr.inp.b.v., tractus interpedunculo-bulbaris ventralis
tr.mam.inp., tractus mamillo-interpeduncularis
tr.mam.ped., tractus mamillo-peduncularis
tr.mam.teg., tractus mamillo-tegmentalis
tr.mam.th., tractus mamillo-thalamicus
tr.ol., tractus olfactorius
tr.ol.d., tractus olfactorius dorsalis
tr.ol.d.l., tractus olfactorius dorsolateralis
tr.ol.h.a.; *tr.ol.hab.ant.*, tractus olfacto-habenularis anterior
tr.ol.h.a.v., ventral division of *tr.ol.h.a.*
tr.ol.h.l.; *tr.ol.hab.lat.*, tractus olfacto-habenularis lateralis
tr.ol.h.m.; *tr.ol.hab.med.*, tractus olfacto-habenularis medialis

tr.ol.hip.m., tractus olfacto-hippocampalis medialis
tr.ol.pal.d. = *tr.ol.d.*
tr.ol.pal.l. = *tr.ol.d.l.*
tr.ol.ped., tractus olfacto-peduncularis
tr.ol.s., fasciculus olfactorius septi
tr.ol.v., tractus olfactorius ventralis
tr.ol.v.l., tractus olfactorius ventrolateralis
tr.op., tractus opticus
tr.op.ac.p. = *tr.op.b.*
tr.op.ax., axial bundle of tractus opticus
tr.op.b., basal bundle of tractus opticus
tr.op.l; *tr.op.lat.*, tractus opticus lateralis, or ventralis
tr.op.m.; *tr.op. med.*, tractus opticus medialis, or dorsalis
tr.op.,mar., tractus opticus marginalis
tr.ped.mam., tractus pedunculo-mamillaris
tr.po., tractus preopticus
tr.pt. hy., tractus pretecto-hypothalamicus
tr.pt.tec., tractus pretecto-tectalis
tr.pt.th., tractus pretecto-thalamicus
tr.sep.c., tractus septo-corticalis
tr.sep.hab., tractus septo-habenularis
tr.sp.b., tractus spino-bulbaris
tr.sp.cb., tractus spino-cerebellaris
tr.sp.t., tractus spino-tectalis = *lm. sp.*
tr.st.ped., tractus strio-peduncularis = *f.lat.t.v.*
tr.st.pt., tractus strio-pretectalis
tr.st.t.; *tr.st.teg.*, tractus strio-tegmentalis = *f.lat.d.*
tr.st.tec., tractus strio-tectalis
tr.st.th., tractus strio-thalamicus
tr.t.b., tractus tecto-bulbaris
tr.t.b.c., tractus tecto-bulbaris cruciatus
tr.t.b.p., tractus tecto-bulbaris posterior
tr.t.b.r., tractus tecto-bulbaris rectus
tr.t.cb.; *tr.tec.cb.*, tractus tecto-cerebellaris
tr.t.hab., tractus tecto-habenularis
tr.t.hy.a., tractus tecto-hypothalamicus anterior
tr.t.p.; *tr.t.ped.*, tractus tecto-peduncularis
tr.t.p.c., tractus tecto-peduncularis cruciatus
tr.t.pt., tractus tecto-pretectalis
tr.t.sp.; *tr.tec.sp.*, tractus tecto-spinalis
tr.t.teg., tractus tecto-tegmentalis
tr.t.teg.c., tractus tecto-tegmentalis cruciatus
tr.t.th.h.c.a., tractus tecto-thalamicus et hypothalamicus cruciatus anterior
tr.t.th.h.c.p., tractus tecto-thalamicus et hypothalamicus cruciatus posterior
tr.th.r., tractus tecto-thalamicus rectus = *br.col.*
tr.teg.b., tractus tegmento-bulbaris
tr.teg.inp., tractus tegmento-interpeduncularis

ABBREVIATIONS

tr.teg.is., tractus tegmento-isthmialis
tr.teg.p.; *tr.teg.ped.*, tractus tegmento-peduncularis
tr.th.b., tractus thalamo-bulbaris
tr.th.f., tractus thalamo-frontalis
tr.th.h.d.c., tractus thalamo-hypothalamicus dorsalis cruciatus
tr.th.hab., tractus thalamo-habenularis
tr.th.mam., tractus thalamo-mamillaris
tr.th.p.; *tr.th.ped.*, tractus thalamo-peduncularis
tr.th.p.c.; *tr.th.ped.c.*, tractus thalamo-peduncularis cruciatus
tr.th.p.d., tractus thalamo-peduncularis dorsalis
tr.th.t., tractus thalamo-tectalis
tr.th.teg.d.c., tractus thalamo-tegmentalis dorsalis cruciatus, with divisions *A* and *B*
tr.th.teg.r., tractus thalamo-tegmentalis rectus
tr.th.teg.v.c., tractus thalamo-tegmentalis ventralis cruciatus
tr.th.teg.v.r., tractus thalamo-tegmentalis ventralis rectus
tr.v.a., tractus visceralis ascendens
tr.v.d., tractus visceralis descendens
tr.v.t., tertiary visceral tract
tub.p., tuberculum posterius

v.az.sep., vena azygos septi
v.l., ventriculus lateralis
v.m.a., velum medullare anterius
v.par., paraphysial veins
v.4., fourth ventricle

z.lim.lat., zona limitans lateralis
z.lim.med., zona limitans medialis

(*1*) to (*10*), tegmental fascicles

INDEX

INDEX

[Only the first page of extended references is cited. Some more important page references are printed in bold-face type.]

Activation
  general, 72, 75, 208, 221
  vs. inhibition, 205, 253
Adaptation, 76, 111
Addens, J. L., 249, 250
Adelmann, H. B., 231
Agar, W. E., 89, 111
Agassiz, L., 109
Alligator, cortex of, 103
Amblystoma, Ambystoma, 3
Ameiurus, 147, 167, 244
Ammocoetes; *see* Cyclostomes
Amphibia
  evolution of, 13, 93
  in general, 1
  regression of, 16, 81, 113
Amphioxi, 16, 41, 92
Amygdala, 20, **52**, 95, 238, 248, 266, 272, 273
Analyzers (*see also* System, analytic), in general, 67, 70, 149
Ansa lenticularis, 97, 272
Anura
  auditory apparatus, 165
  basal optic nucleus, 36
  chorioid plexuses, 27
  embryology of thalamus, 232, 233
  in general, 4, 11, 98, 155, 265
  mamillo-cerebellar tract, 170
  mating behavior, 207, 215
  meninges, 26
  nervus terminalis, 267
  nucleus of Bellonci, 250
  nucleus isthmi, 182
  olfactory projection tract, 242
  postoptic commissure, 295
  primary motor fibers, 129
  prominentia fascicularis, 247
  visceral-gustatory tract, 170
  visual apparatus, 35, 219, 222, 227, 228, 238, 239
Aqueductus cerebri, 25
Archetype, 109
Archipallium, 100
Arcuate fibers; *see* Nerve fibers, arcuate
Area
  acusticolateralis, 22, 43, 137, 153, 291
  dorsalis pallii, 21, 55, 259, 266
  geniculata; *see* Corpus geniculatum
  hippocampal; *see* Hippocampus
  lateralis tegmenti, 36
  olfactoria, 213
  perirhinalis, 266
  piriformis, 21, **55**, 96, **101,** 259, 266, 269, 272
  strio-amygdaloid, 266, 301
  subtectal; *see* Tegmentum dorsale
  ventrolateralis pedunculi, **35,** 50, 167, 170, 218, 221, 224, 234, 262, 279, 301
Ariëns Kappers, C. U., 4, 5, 62, 103, 139, 146, 162, 165, 188, 189
Aronson, L. R., 11, 98, 100, 190, 207, 215
Association, 65, 66, 102
Aula, 291
Auricle, 19, 20, 44, 139, 172
Autonomy, 112, 114

Bagley, C., 103
Bailey, P., 90, 208
Baker, R. C., 12, 117, 213
Barnard, J. W., 166, 167, 168
Barrera, S. E., 161
Basal plate, 60
Bellonci, J.; *see* Nucleus of Bellonci
Benedetti, E., 11
Benzon, A., 11
Beritoff, J. S., 80, 227
Bindewald, C. A. E., 11, 270
Birds
  cortex, 91, 105
  fasciculus solitarius, 149
  gustatory nucleus, 168
Bishop, S. C., 3
Blood vessels; *see* Brain, blood vessels
Bodian, D., 79
Body; *see* Corpus
Bonin, G. von, 35, 90, 98, 115
Brachia of colliculi; *see* Tractus tecto-thalamicus rectus and thalamo-tectalis rectus
Brachium conjunctivum, 45, 169, 175, **176,** 287
Brain
  architectonics, 70
  blood vessels, 19, 26
  development; *see* Embryology
  evolution of, 13, 33 (*see also* Phylogeny)
  form of, 18
  subdivisions of, 18, 20, 40
  variability of, 19
Brickner, R. M., 168

Broca, diagonal band of, 53, 255
Brodal, A., 173
Bulb; *see* Medulla oblongata
Bulbus olfactorius, 20, 29, 54, 267
Bulbus olfactorius accessorius, 54, 94
Bundle (*see also* Fasciculus)
  basal forebrain, 94, 247, 253, 271, 276 (*see also* Fasciculus lateralis et medialis telencephali)
  chiasmatic, 234, 358
  lateral forebrain; *see* Fasciculus lateralis telencephali
  medial forebrain; *see* Fasciculus medialis telencephali
  of Meynert; *see* Fasciculus retroflexus
Burr, H. S., 12

Cajal (*see also* Ramón y Cajal)
  commissural nucleus of, 42, 125, **129**, 133, 163, 166
  interstitial nucleus of, 217, 223, **281**
Calamus scriptorius, 42, 129, 130, 167, 192, 294
Calderon, L., 198
Campion, G. G., 76
Capsula interna, 256
Carp, 149, 167
Catfish; *see* Ameiurus
Caudata, 4
Ceratodus; *see* Lungfishes
Cerebellum
  cortex of, 174
  general, 20, 24, **44**, 119, **142**, **172**
  subdivisions of, 44, 172
Cerebrum (BNA), 40, 116, 119
Chemical senses; *see* System
Chezar, H. H., 135
Chiasma opticum, 219, 222, 228, 294
Child, C. M., 111
Circular conduction, 76, 98
Clark, G., 90
Clark, W. E. L., 87, 188
Clepsine, 84, 147
Cochlea; *see* Ear
Coghill, G. E., 6, 12, 15, 22, 24, 31, 38, 42, 63, 72, 73, 77, 84, 113, 114, 118, 120, 125, 127, 128, 131, 133, 141, 142, 150, 151, 153, 190, 191, 204, 205, 213, 219, 221, 265, 270
Colliculus inferior; *see* Nucleus posterior tecti
Colliculus superior; *see* Tectum opticum
Column
  dorsal, of spinal cord, 125, 129, 163
  primary motor, 78, 204
  ventral, of spinal cord, 126
Commissura
  aberrans, 253
  amygdalarum, 256, 294
  ansulata, 291, 303
  anterior, 24, 254, 267, 271, 291, **294**
  cerebelli, 173, 177, 293
  dorsalis of spinal cord, 291, 294
  of funicular nuclei, 294
  in general, 55, **289**
  habenularum, 234, 292
  hippocampi, 24, 254, 290, 291, **292**
  infima Halleri, 167, 294
  pallii posterior, 253, 257, 292
  posterior, 213, 217, 219, 223, 234, 281, **293**
  postoptica, 246, 283, 290, **295**
  of secondary visceral nuclei, 169
  superior, 234, 292
  superior telencephali, 253, 257, 292
  supra-optica, 246, 295
  tecti diencephali, 234, 292
  tecti mesencephali, 213, 222, 234, **293**
  tuberculi posterioris, 217, 218, 278, 279, 301, **302**
  tuberis, 280, 295
  of velum medullare anterius, 293
  ventralis, 287, 291, 302, 304
  vestibulolateralis cerebelli, 136, 178, 293
Conditioning; *see* Reflex, conditioned
Conel, J. L., 84
Co-ordination, 58, 65, 66
Corbin, K. B., 142
Corpus
  cerebelli, 44, 173
  geniculatum, 33, 49, 214, 221, 225, **238**, 273, 297
  mamillare, 241, **246**, 255, 278, 279
  pineale; *see* Epiphysis
  posticum, 215
  quadrigemina, 21
  striatum, 20, **52**, **96**, **103**, 105, 238, 239, 266, **272**, 273
Correlation, 58, 65
Cortex
  cerebelli, 174
  cerebri, 16, 35, **55**, 66, 88, **91**, **98**, 100, 228, 231
Craigie, E. H., 26, 105
Crest
  cerebellar, 136, 138
  neural, 132, 141, 142
Crosby, E. C., 5, 62, 85, 88, 103, 223, 228
Cryptobranchus, 11, 153, 211
Cushing, H., 148
Cyclostomes
  basal optic tract, 221
  cerebellum, 173, 174
  cerebral hemispheres, 99
  chorioid plexus of tectum, 290
  embryonic sulci, 118
  fasciculus retroflexus, 191
  general, 13, 84
  nucleus of n. trochlearis, 150

stria medullaris thalami, 256
tr. olfacto-habenularis lateralis, 259
visceral-gustatory system, 167

Darkschewitsch, nucleus of, 217, 223
Darwin, C., 109
Davis, E. W., 208
Decussatio (*see also* Commissura)
  in general, 289
  hypothalamicus posterior, 302
  retroinfundibularis, 302
Dempster, W. T., 11, 26, 322
Dendy, subcommissural organ of, 25
Detwiler, S. R., 62, 71, 118, 136, 209, 227
Dewey, J., 76, 98
Didelphis; *see* Opossum
Diemyctylus; *see* Triturus
Diencephalon
  boundaries, 21
  general, 23, 230
Dimension, definition of, 108
Dipnoi; *see* Lungfishes
Dominance, physiological, 41, 51, 86, 100
Dow, R. S., 44

Ear (*see also* System, acusticolateral)
  cochlear part, 22, 135, 137, 138, 165, 214, 239
  vestibular part, 118, 135
Economo, C., 34
Edinger, L., 4, 251, 252
Effectors; *see* System, motor
Elasmobranchs
  cerebral cortex of, 104
  nucleus of f. solitarius, 149
Ellison, F. S., 227, 228
Embryology
  of cerebellum, 22, 172
  of cranial nerves, 131
  of diencephalon, 230, 238, 239
  general statements, 6, 11, 17, 60, 71, 111, 120
  of hypothalamus, 232, 243
  of isthmus, 179
  of medulla oblongata, 153
  of midbrain, 212
  of optic system, 48, 51, 213, 219, 222
Emerson, A. E., 114
Eminentia
  cerebellaris ventralis, 175
  commissurae posterioris, 213, 222, 223, 261, 284
  subcerebellaris tegmenti, 175
  thalami, 51, 232, 239, 249, 257
  trigemini, 183
Endocrine organs; *see* System, neuro-endocrine

Endolymphatic organ; *see* Saccus endolymphaticus
Engram, 90
Enteropneusta, 16
Ependyma, 25, 150, 181, 193
Epigenesis, 112
Epiphysis, 20, 231, 235
Epithalamus, 21, 231, 234
Equilibration, 137
Equipotentiality, 82, 88, 159
Estable, C., 148
Evans, F. G., 16
Evolution (*see also* Brain, evolution of; Phylogeny), emergent, 108
Excitatory state, central, 65, 75, 78, 89, 171, 202, 231
Exteroceptors; *see* System, exteroceptive
Eye (*see also* System, optic), parietal, 59, 251

Facilitation, 81, 97, 99, 172, 209, 210
Fasciculus
  anterior tectal, 296
  communis, 145
  definition of, 9
  lateralis telencephali, 53, 94, 186, 247, **271**
  longitudinalis dorsalis (Schütz), 46, 49, 52, 185, 186, **202, 208,** 217, 281
  longitudinalis lateralis, 165
  longitudinalis medialis, **50,** 158, 161, 213, **216,** 223, 240, 272, 275, 277, 280, **281,** 283, 300, 301
  medialis telencephali, 53, 94, 242, 243, 247, 257, 269, 271, **273,** 301
  olfactorius septi, 53
  posterior tectal, 297
  postolfactorius, 55, 258, 269
  retroflexus, 46, 186, 197, 234, 235, 251, 253, **261**
  solitarius, 43, 133, **145, 148,** 149, 166, 294
  tegmentalis profundus, 169, 176, 186, 199, 200, 201, 278, 279, 280, 284, **286**
  tegmenti (nos. 1 to 10), 168, 186, 188, **275**
  uncinatus of Russell, 161
Feeding, apparatus of, **169, 190,** 204, 221, 226, 227, 252, 282, 284, 285
Ferraro, A., 161
Field (*see also* Area), dynamic, 89
Fimbria, 292
Fishes
  cerebral cortex, 104
  chemical senses, 146
  chorioid plexuses, 27, 93
  efferent fibers in optic nerve, 222
  evolution of, 13, 14, 27, 92, 155, 271
  postoptic commissures, 290
  thalamo-frontal tract, 238
  visceral-gustatory system, 166, 167, 168, 170, 189

Fissura
  definition, 9
  di-telencephalic, 116, 247, 257
  general arrangement, 19
  isthmic, 20, 21, 116, 179
  longitudinalis, 19
  stem-hemisphere, 116, 247, 257
Flechsig, P., 289
Flexure
  of midbrain, 19, 116, 179, 212
  of thalamus, 117, 212, 233
Flocculus; see Lobus flocculonodularis
Floor plate; see Plate, floor
Fluid, cerebrospinal, 27
Foramen
  interventricular, 24, 254, 291
  jugular, 26
Forel, A., 191, 195
Formatio reticularis, 44, **61, 64,** 79, 126, 155, 156, 162, 206
Fornix, 95, 242, 246, 248, **254,** 260, 274
Fovea isthmi, **21, 60,** 117, 118, 179, 191, 193, 302
Fox, C. A., 168
Francis, E. T. B., 23, 26, 126
Frey, E., 222
Frog; see Anura
Fulton, J. F., 79, 203, 208
Funiculus
  dorsalis spinalis, 125, 133, 137, 160, 161
  dorsolateralis, 133
  ventralis, 162

Ganglia
  of cranial nerves, 112, 132
  of spinal nerves, 23, 142
Ganser, S., 191, 198
Gaupp, E., 11, 36, 247
Gegenbaur, C., 109, 115
Gehuchten, A. van, 4, 125, 126, 191, 198, 253
Geiringer, M., 222
Gellhorn, E., 90
Gesell, A., 38
Gesell, R., 79
Gestalt, 89
Gillilan, L. A., 37, 200
Glasser, O., 107
Globus pallidus, 53, 96, 266, 272
Glomeruli
  of interpeduncular nucleus, 46, 191, 194, 195, 200, 202
  of olfactory bulb, 32, 54, 195
Goodrich, E. S., 126
Graves, G. O., 12, 117, 213
Gregory, W. K., 15

Griggs, L., 12
Gudden, B. von, 191, 198, 261
Gymnophiona, 11

Habenula, 20, 21, 51, 191, 231, 232, 234, **247, 250,** 261
Haeckel, E. H., 109
Hagfish; see Cyclostomes
Haller, commissure of; see Commissura infima
Hamburger, V., 129
Hansen, E. T., 79
Head, H., 90
Hearing; see System, acusticolateral
Hemisphere, cerebral, 20, 24, 265
Herrick, C. L., 4
Hippocampus, 21, 55, **101,** 242, 260, 266
His, W., 20, 117, 191
Histology of the brain, 9, 28
Hoagland, H., 136
Holmgren, N., 104, 174, 233, 264
Holtfreter, J., 111
Homologies, in general, 8, 18, 28, 110, 290
Hooker, D., 71, 73, 75, 84
Horst, C. J. van der, 264
Howell, A. B., 15
Huber, G. Carl, 5, 62, 88, 228
Humphrey, T., 84, 127, 128
Huxley, T. H., 109, 115
Hynobius, 11, 12, 25, 181
Hypopallium, 103
Hypophysis, 20, 25, 230, 231, 245
Hypothalamus, 21, 24, 25, **52,** 60, 95, 205, 207, 210, 220, 221, 224, 231, 232, **241,** 252, 254, 278, 301

Individuation, 77
Infundibulum, 25, 232
Inhibition, 38, 73, 75, **77,** 80, **99,** 190, **205,** 252
Integration, **33, 39,** 64, 65, **66,** 70, 74, 77, **85, 88,** 127, 149, **154,** 223
Interoceptors; see System, interoceptive
Island, dorsal, 136, 138, 139
Isthmus, 22, 23, 45, **118,** 119, **179**

Jacobsohn-Lask, 289
Jansen, J., 84, 173
Jeserich, M. W., 54
Johnston, J. B., 4, 103, 138, 167, 192
Jones, D. S., 141
Junction
  bulbo-spinal, 42, 118, 129, 167
  di-telencephalic, 116, 119, 247
  isthmic, 116, 118

INDEX

Kabat, H., 209
Kato, G., 207
Keefe, E. L., 129
Kingsbury, B. F., 11, 117, 159, 191
Kodama, S., 185
Kostir, W. J., 131
Kreht, H., 11, 160, 175, 255
Kuhlenbeck, H., 11
Kupffer, C. von, 117, 179, 212

Lamina terminalis, 24, 254, 291
Lamprey; see Cyclostomes
Landacre, F. L., 131, 132
Landmarks, morphological, 116
Langworthy, O. R., 103
Larsell, O., 11, 22, 44, 134, 135, 138, 139, 143, 150, 155, 160, 165, 170–78, 188, 189, 270
Lashley, K. S., 82, 90
Lemniscus
  bulbar, 156, **163**, 214, 222, 237
  in general, 65, **85**, 154, 156, **162**
  lateral, 139, **164**, 168, 182, 188, 214, 239
  medial, 129, 164
  spinal, 130, 143, **163**, 174, 188, 214, 222, 237
  trigeminal, 164
  visceral, 156, 166
Liggett, J. R., 99
Lillie, R. S., 111
Lobus
  flocculonodularis, 20, 45, 139, 172
  piriformis; see Area piriformis
  visceralis, 153
Localization of function
  in area pallialis, 92
  in cerebellum, 44, 172
  in corpus striatum, 52, 96
  in general, 28, 44, 70, 82, 85, 87
  in hypothalamus, 241, 243, 246
  in interpeduncular nucleus, 203
  in isthmus, 45
  of mating reactions, 207
  in medulla oblongata, 42, 154
  in olfaction, 268
  in peripheral nerves, 65
  in tectum, 48, 87, 220, 222, 223, 228
  in tegmentum, 44
  in thalamus, 49, 87, 231
Locomotor apparatus, 15, 40, 50, 63, 282
Locus coeruleus, 189
Loo, Y. T., 255
Lungfishes, 27, 53, 92, 104, 118, 233, 248, 264, 291

McKibben, P. S., 19, 199, 241, 267, 304, 321, 322, 326
Magoun, H. W., 79
Marburg, O., 260, 261

Mauthner's cells and fibers, 28, 79, 142, 156
Mayser's Uebergangsganglion, 189
Mechanics
  developmental, 112
  Newtonian and quantum, 106
Medulla oblongata, 22, 42, 118, 153
Medulla spinalis; see Spinal cord
Menidia, nerves of, 68
Meninges, 26, 181, 232
Mentation, 91, 107
Mesencephalon
  boundaries, 21,
  in general, 23, 63, 212
Metamerism 110, 220
Metamorphosis, 14, 40, 85, 153
Methods, 10
Mettler, F. A., 208
Meynert's fasciculus; see Fasciculus retroflexus
Midbrain; see Mesencephalon
Modulation, 62
Moncrieff, R. W., 146
Morphology, principles of, 109
Morphogenesis
  of cerebral hemispheres, 92
  in general, 29, 109, 271
Murphy, J. P., 90
Myotypic responses, 62
Myxine (*see also* Cyclostomes), cerebellum of, 173, 174

Naturphilosophie, 109
Necturus, 7, 11, 19, and *passim*
Neimanis, E., 19
Neopallium, 97, 100, 266
Neostriatum, 97
Neothalamus, 236, 240
Nerve fibers
  arcuate, 44, 153, 158, 159
  associational, 65, 96, 102, 105
  in general, **28**, 32, **34**, 37, **66, 220**
  internuncial, 65
  number of, in optic nerve, 220
  preganglionic, 69, 151
  primary and secondary, 127
  thick and thin, 220, 222, 228
Nerves
  components of, 65, 131
  cranial, 22, 40, 63, 65, 131
  development of, 131
  spinal, 23, 125, 126
Nervus
  abducens, 149
  accessorius, 22, 126
  acusticus, 22, 132, 165, 187
  cochlearis, 165
  facialis, 132, 145, 151, 187

Nervus—*continued*
  glossopharyngeus, 132, 145, 151
  hypoglossus, 23, 126
  hypophysius; *see* Tractus hypophysius
  liniae lateralis, 22, 40, 132, 187
  oculomotorius, 149, 151, 216
  olfactorius, 50, 267
  opticus, 25, **37**, 213, **219, 228**, 230, 236, 241, 294
  opticus, efferent fibers of, 222
  opticus, number of fibers, 220
  parietalis, 59, 151, 230, 235, 261
  terminalis, 36, 53, 59, 61, 94, 151, 199, 217, 230, 241, 247, **267**, 274, 294, 304
  trigeminus, 132, 140, 151, 187, 293
  trochlearis, 149, 151, 181, 293
  vagus, 126, 132, 145, 151

Neuroglia, 29
Neuropil
  of area pallialis, 101
  of Bellonci, 236, 250, 257, 260
  of chiasma ridge, 302
  of corpus striatum, 53, 96
  of epithalamus, 234
  in general, 24, 28, **29**, 59, 66, 80, 88, 227
  geniculate; *see* Corpus geniculatum
  of habenula, 251
  of hypothalamus, 243, 302
  of interpeduncular nucleus, 46, 194
  of isthmus, 187
  of medulla oblongata, 60
  of peduncle, 35
  posterior isthmic, 168, 182, 187
  of tectum, 48, 222
  of tegmentum, 79, 201, 208, 283, 285, 300
  of thalamus, 237

Nissl bodies, 28
Noble, G. K., 11, 16, 98, 100, 190, 207, 215
Nodus vasculosus, 26
Nomenclature
  of Amphibia, 4
  of the brain, 9
  of tracts, 9

Norris, H. W., 131
Nuclei
  in general, 28, 30, 32
  homologies of, 32
Nucleus
  accumbens septi, 266
  amygdalae; *see* Amygdala
  anterior superior corporis geniculati thalami, 250
  of Bellonci, 49, 232, **236**, 248, **249**, 257, 260
  caudatus, **53**, 79, 96, 206, 208, 242, **266**, 272
  centralis isthmi, 45, 182, **184**, 281, 284
  cerebelli, 44, 157, 172, 175
  cochlear, 138, 165, 174
  commissurae anterioris, 248, 260, 294
  commissurae hippocampi, 249
  commissuralis of Cajal, 42, 125, **129**, 133, 163, 166
  cuneatus externus, 129, 161

  of Darkschewitsch, 217, 219, 223, **281**, 293
  of dorsal funiculi, 42, 129, **160**, 161, 163, 175, 294
  dorsalis of lateral lemniscus, 188
  dorsalis of spinal cord, 42, 125, 175
  dorsalis tegmenti, 46, 201, 209
  ectomamillaris, 36
  entopeduncularis, 218
  of fasciculus solitarius, 153, 166
  habenulae; *see* Habenula
  interpeduncularis, 46, 63, 150, 170, **191**, 253, 263
  interstitialis of Cajal, 217, 223, **281**
  isthmi, 45, **182**, 188, 189, 224
  lentiformis, 53, 96, 266
  magnocellularis hypothalami, 221, 244, 274, 302
  magnocellularis isthmi, 184, 185
  magnocellularis trigemini, 183
  motorius tegmenti, 28, 62
  nervi abducentis, 149, 150
  nervi cochlearis, 165
  nervi facialis, 151
  nervi oculomotorii, 21, 36, 61, **150**, 217, 223, 224, **226**, 304
  nervi trigemini, mesencephalic, 45, 132, **140**, 150, 174, 181, 293
  nervi trigemini, motor, 151, 183
  nervi trigemini, spinal, 134, 163
  nervi trigemini, superior, 45, 134, 148, 173, 178, 188
  nervi trochlearis, 21, 45, 61, **150**, 182, 185, 226
  olfactorius anterior, 20, 53, **54**, 103, 258, 267, **268**, 272
  olfactorius dorsolateralis, 55
  of olfactory projection tract, 242, 301
  opticus basalis, 36, 51, 219
  pontilis, 173
  posterior tecti, **48**, 164, **174**, 176, 213, **214**, 297
  of postoptic commissure, 302
  preopticus, 20, 21, 221, 232, **241, 244**, 247, 248, 274
  pretectalis, 21, **47**, 220, 224, **226, 234**, 261, 273, 296
  raphis, 150, 191
  ruber, 45, 177
  sensitivus thalami, 49, 231, 236
  supraopticus, 244
  of tractus olfacto-habenularis, 232, 239, 241, 257
  tuberculi posterioris, 21, 215, 261, 279
  ventralis of spinal cord, 42
  ventralis tegmenti, 168
  vestibularis superior, 188
  visceralis superior, 45, 167, 182

Obersteiner, H., 4
Olfacto-striatum, 96
Olive, inferior, 161, 173
O'Neill, H. M., 26
Opossum, 41, 97, 208, 255, 268

Optocoele, 25
Organ
  endolymphatic; see Saccus endolymphaticus
  motor; see System, motor
  sensory; see Receptors
  subcommissural, 25
  vomeronasal, 54, 94
Orthogenesis, 113
Osborn, H. F., 145
Ostracoderms, 16
Owen, R., 109

Palay, S. L., 26, 147, 245
Paleopallium, 100
Paleostriatum, 96
Paleothalamus, 236, 240
Pallium, 21, **24, 55,** 57, **91,** 266, 291
Papez, J. W., 76, 86, 90, 98, 165
Paraphysis, 25, 27, 231, 250
Parker, G. H., 136, 146, 147
Pars
  dorsalis thalami; see Thalamus
  intercalaris diencephali, 21, 231, 232, 234
  libera hypothalami, 19
  pallialis; see Pallium
  ventralis thalami; see Thalamus
Pattern
  in general, **83,** 110
  partial, **15,** 29, 63, **72,** 220
  total, **15,** 29, 43, 63, 64, **72,** 154, 220, 282
Pearse, A. S., 4
Pearson, A. A., 127, 141
Peduncle
  of mamillary body, 278, 281
  of midbrain, 21, **35,** 37, **50,** 213, **216,** 220, 277
Pelobates, 118, 232
Phylogeny (see also Brain, evolution of), 13, 18, 75, 84, 109, 111, 115, 155
Piatt, J., 127, 132, 140, 141
Pineal body; see Epiphysis
Piriform lobe; see Area piriformis
Pituitary body; see Hypophysis
Placodes, ectodermal, 132
Plate
  alar, 117
  basal, 117, 216
  floor, 60, 78, 117, 150, 191, 290
  roof, 290
Plethodon, 131
Plexus chorioideus
  blood supply of, 27
  development of, 231
  in general, 25, 93, 290
  innervation of, 27, 181, 235

Plica
  encephali ventralis, 19
  rhombo-mesencephalica, 118, 180
Polyak, S. L., 79, 230, 268
Pons, 22, 44, 158, 172
Pool of physiological adjustment
  bulbar, 43, 60, 130, 159
  in general, 59, 81
  interpeduncular, 203
  tectal, 222, 223
  thalamic, 237
Preformation, 112
Primordium (see also under Area), in general, 8
Prionotus, 146
Prominentia fascicularis, 247
Proprioceptors; see System, proprioceptive
Proteus, 11, 153, 160
Protopterus; see Lungfishes
Psychology and physiology, 106
Pulvinar, 234
Purkinje, cells of, 174
Putamen, 53, 96, 266

Radix; see Nerves
Ramón y Cajal, P., 101
Ramón y Cajal, S., 4, 198, 289
Ramus
  auricularis vagi, 133
  palatinus, 148
Rasmussen, G. L., 76
Reaction
  anticipatory, 41, 78, 206
  circular, 39, 76, 169, 231
  consummatory, 41, 78, 206
  delayed, 41, 78, 99
  discriminative, 154
  regarding, 38, 78, 221
  stimulus-response; see Reflex
Recapitulation in evolution, 75, 84, 114
Receptors
  acoustic, 135
  chemical, 132, 146
  contact, 41
  distance, 41
  in general, 132, 146
  gustatory; see System, gustatory
  lateral-line, 135, 227
  proprioceptive; see System, proprioceptive
  somatic; see System, somatic sensory
  visceral; see System, visceral sensory
Recess
  lateral of fourth ventricle, 25, 153
  lateral optic, 25, 241
  pineal, 234, 235
  posterior mesencephalic, 213
  precommissural, 24, 232, 291
  preoptic, 25, 232, 294

Reflexes
  circular, 39, **76,** 169, 231
  conditioned, **31, 38,** 62, **66,** 74, 221, 230, 234, 273
  definition of, **76,** 205
  in general, 29, 39, **72, 76,** 155
Reinforcement, 81, 99, 202
Reiser, O. L., 108
Reissner's fiber, 25
Reptiles
  corpus striatum of, 97, 103
  cortex of, 91, 100, 102
Reserves, nervous, 66, 114
Reticular formation; *see* Formatio reticularis
Retina, 51, 79, 219, 222, 230, 231, 268
Rhombencephalon, 40, 116, 119
Ridge
  anterior commissure, 24, 232, 254, 291
  chiasma, 213, 232, 280
  di-telencephalic, 232, 233
  dorsal tectal, 222
  dorsal ventricular, 103
Röthig, P., 11, 255
Rohon-Beard cells, 127, 141, 142
Romer, A. S., 14
Roofe, P. G., 11, 18, 25, 26, 322
Rudebeck, B., 12, 53, 92, 104, 117, 118, 232, 233, 248, 250, 291
Russell's fasciculus, 161

Saccus
  dorsalis, 25, 231, 250
  endolymphaticus, 25, 26, 27
  vasculosus; *see* Hypophysis, pars nervosa
Salamanders, in general, 3, 6
Salamandra, 11, 23, 26, 117, 125, 126, 153, 160, 175, 253, 255
Salientia, 4
Scharrer, E., 52, 136, 147, 227, 245
Schema, 90
Schriever, H., 136
Schütz's fasciculus; *see* Fasciculus longitudinalis dorsalis
Selachians
  cerebral cortex of, 104
  nucleus of f. solitarius of, 149
Senescence of tissue, 112
Sense organs; *see* System
Septum, 20, 53, 242, 258, 260, 266, 274
Shanklin, W. M., 37
Sheldon, R. E., 146
Sherrington, C. S., 41, 78, 143, 178, 203, 206
Siredon, 4
Siren, 11, 153
Smell, organs of; *see* System, olfactory

Smith, G. E., 76, 103
Smith, S. W., 147
Söderberg, G., 12, 104
Somatic organs; *see* System, somatic
Sosa, J. M., 125
Speidel, C. C., 135
Spelerpes, 11
Sperry, R. W., 113, 137, 228, 229
Spinal cord, 40, 41, 125
Spontaneity, 75, 81, 114, 209
Stensiö, E. A., 115
Stimulus-response; *see* Reflexes
Stone, L. S., 132, 227, 228
Stria
  cornea; *see* Stria terminalis
  medullaris thalami, 94, 242, 244, 247, **256**
  olfactoria lateralis, 269
  semicircularis; *see* Stria terminalis
  terminalis, 242, 248, **255,** 273, 274
Ströer, W. F. H., 228
Strong, O. S., 68, 131
Subiculum, 104, 266
Substantia
  alba, 28, 34
  grisea, 28, 34
  nigra, 185, 189, 218
Subthalamus, 21, 51, 240
Sulci
  arrangement of, 19, 233, 322
  definition of, 9
Sulcus
  diencephalicus ventralis, 232, 250
  dorsalis thalami, 237
  hypothalamicus, 232, 241
  hypothalamicus dorsalis, 241
  intraencephalicus anterior, 117, 212, 232
  intraencephalicus posterior, 117, 179, 212
  intrahabenularis, 250
  isthmi, 20, 21, 116, **118, 179,** 193, 213, 215
  lateralis mesencephali, 25
  limitans, 117
  limitans pedunculi (*s.*), 21, 215
  medius thalami, 237, 239
  paramedianus, 183
  posthabenularis, 232, 250
  preopticus, 118, 241
  striaticus, 96
  subhabenularis, 250
  subpallialis, 232
  ventralis thalami, 232, 239
Sumi, R., 12, 25, 181
Summation, 80, 202
Suppressor bands, 79, 206
Suprasegmental apparatus, 119
Synapse, 29, 31, 79, 138, 203
Synthesis; *see* Integration

System
  acusticolateral, 40, 69, 135, 172, 178, 227
  analytic, 67, 70, 85, 88, 149
  of chemical senses, 146
  cochlear; *see* Ear
  communis, 145
  exteroceptive, 23, 48, 63, 69, 144, 210, 251, 254
  extrapyramidal, 53, 94, 205, 271
  of fibers, 270
  functional, 57, 65, 67, 132
  general cutaneous, 69, 132, 133, 147
  gustatory; *see* System, visceral sensory
  integrative; *see* Integration
  interoceptive, 52, 144, 210, 251, 254
  neuro-endocrine, 52, 221, 245
  olfactory, 41, 69, **92**, **93**, **99**, 105, 210, 243, **266**
  olfacto-somatic, 211
  olfacto-visceral, 211, 217
  optic, 23, **41**, **48**, 69, **87**, 173, **219**, 251
  proprioceptive, 44, 63, 69, 134, 137, 142, 174, 178, 214
  proprius, 145
  somatic, 14, 67, 83, 149
  somatic motor, 63, 69, 149
  somatic sensory, 43, 69, 132
  synthetic (*see also* Integration), 66, 85
  thalamo-striatal, 271
  visceral, 14, 63, 67, 83, 149, 242, 302
  visceral motor, 61, 69, 151
  visceral sensory, 43, 68, 69, 145, 242

Taenia
  semicircularis, 256
  thalami, 250
  ventriculi quarti, 153, 159

Taste, organs of; *see* System, gustatory

Taste buds, 68, 146

Taylor, A. C., 128

Tectum
  mesencephali, 20, 23, 24, **47**, 60
  opticum, **48**, 87, 213, 214, 219, **222**, 273, 296

Tegmentum
  bulbar, 61, 157, 211
  dorsale, 21, 64, 215, 284
  isthmi, 36, 45, 62, 63, 157, 177, 179, **182**, **184**, 189, 196, 273, 278, 279, 282, 284
  rhombencephali, 44
  spinal, 61
  trigeminal, 183, 273, 282, 285

Teleology, 77

Terminal buds, 146

Thalamus
  in general, 21, 49, 220, 231
  pars dorsalis, 21, 49, 60, 231, **236**, 261, 284, 299
  pars ventralis, 21, 51, 231, **239**, 277, 281, 297, 300

Thompson, d'A. W., 107

Thompson, E. L., 185, 202, 208, 217

Tonus, 81

Torus
  longitudinalis, 222
  semicircularis, 215

Tracts
  of amygdala, 52, 248
  associational, 33
  of cerebellum, 172
  of corpus striatum, 53
  definition of, 9
  of dorsal funicular nuclei, 42
  in general, **9**, 28, 30, **32**, 33, 71, 87, **270**
  of hypothalamus, 52, 241
  of interpeduncular nucleus, 46, 197
  of isthmus, 186
  of medulla oblongata, 153, 158
  of midbrain, 214, 220
  of olfactory system, 266
  of peduncle, 50, 217
  of septum, 53
  of tectum, 48, 220, 223
  of tegmentum isthmi, 179
  of thalamo-cortical projection, 95, 227
  of thalamus, 49, 221, 236

Tractus
  a, of Kingsbury, 125, 153, 159, 188
  amygdalo-habenularis, 259
  amygdalo-thalamicus, 248, 256
  b, of Kingsbury, 125, 153, 160, 188
  bulbo-isthmialis, 164, 168, 187
  bulbo-spinalis, 156, 162
  bulbo-tectalis lateralis; *see* Lemniscus, lateral
  bulbo-tegmentalis, 156
  cerebello-spinalis, 175
  cerebello-tegmentalis, 45, 176
  cortico-habenularis lateralis, 258, 259
  cortico-habenularis medialis, 254, 260
  cortico-mamillaris, 255
  cortico-pontilis, 173
  cortico-thalamicus, 255
  cortico-thalamicus medialis, 250, 254, 255, 260
  dorsoventralis thalami, 237, 240
  gustato-tectalis, 168
  habenulo-interpeduncularis, 191, 197, 261
  habenulo-tectalis, 222, 261, 292
  habenulo-thalamicus, 237, 260, 261
  hypophysius, 52, 151, 221, 230, **244**, 274, 302
  hypothalamo-cerebellaris, 172
  hypothalamo-peduncularis, 217, 226, 243, 244, 279, **280**, **301**, 302, 303
  hypothalamo-tegmentalis, 243, 244, 280, **301**, 303
  infundibularis ascendens, 243
  interpedunculo-bulbaris, 46, 201
  interpedunculo-tegmentalis, 47, **201**, 287
  mamillo-cerebellaris, 170, 176, 244
  mamillo-interpeduncularis, 46, **198**, 200, 244, 253, **278**, 303
  mamillo-peduncularis, 217, **244**, **279**, 280, 303

Tractus—*continued*
  mamillo-tegmentalis, 244, **278**, 279, 288, 303
  mamillo-thalamicus, 239, 244, 278, 279
  olfacto-habenularis anterior, 53, 258, 259
  olfacto-habenularis lateralis, 248, 257, 259, 274
  olfacto-habenularis medialis, 248, 257, 274
  olfacto-peduncularis, 36, 46, 53, 94, 186, **199**, 217, 230, **242**, 269, 272, **274**, 286
  olfacto-thalamic, 260
  olfactorius, 269
  olfactorius dorsolateralis, 258, 269
  olfactorius ventralis, 269
  olfactory projection, 242, 256, 273, **301**
  opticus, 87, **219**, 235, 236, **237**, 238, 245
  opticus basalis, 36, 200, 219, 221
  pallido-habenularis, 261
  pedunculo-bulbaris ventralis cruciatus, 277
  pedunculo-hypothalamicus, 301, 303
  pedunculo-mamillaris, 279
  pedunculo-tegmentalis, 278, 279, 281, 288
  preoptico-hypothalamicus, 242, 243
  preopticus, 242, 244, 274, 294
  pretecto-hypothalamicus, 226, 234, 240, 296
  pretecto-peduncularis cruciatus, 224, 298
  pretecto-tectalis, 234
  pretecto-thalamicus, 234, 237, 240, 296, 297
  primary motor, 213
  septo-habenularis, 53, 260
  septo-hypothalamicus, 274
  spino-bulbaris, 162, 163
  spino-cerebellaris, 130, 161, 163, 172, 175, 177
  spino-tectalis; *see* Lemniscus, spinal
  strio-habenularis, 259, 260
  strio-peduncularis, 272, 286
  strio-pretectalis, 234, 237, 273
  strio-tectalis, 222, 234, 273
  strio-tegmentalis, 272, 285
  strio-thalamicus, 240
  tecto-bulbaris et spinalis, 186, 219, **224**, 226, 277, 303
  tecto-bulbaris posterior, 165, 188, 214
  tecto-bulbaris rectus, 168, 188, 189, **225**, 284
  tecto-cerebellaris, 172, 174, 176, 214, 293
  tecto-habenularis, 261, 292
  tecto-hypothalamicus anterior, 224, 234, 243, **296**
  tecto-hypothalamicus posterior, 297
  tecto-peduncularis profundus, 226, 304
  tecto-peduncularis rectus et cruciatus, 186, 223, 226, 237, 298, 303
  tecto-pretectalis, 234
  tecto-spinalis; *see* Tractus tecto-bulbaris
  tecto-tegmentalis, 186, 283, 288
  tecto-tegmentalis cruciatus, 224, 283, 297
  tecto-thalamicus et hypothalamicus cruciatus
    anterior, **224**, 234, 242, **296**
    posterior, 214, **224**, 240, 243, 282, 283, 284, **297**
  tecto-thalamicus rectus (brachium of colliculi), 214, 221, **225**, 237, 238, 240
  tegmento-bulbaris, 46, 157, 185, 199, 283
  tegmento-interpeduncularis, 46, **199**, 287
  tegmento-isthmialis, 185, 215, 283, 288
  tegmento-peduncularis, 215, 287
  thalamo-cortical, 231, 238
  thalamo-frontalis, 49, 53, 95, 221, 238, 271, 273
  thalamo-habenularis, 237, 260, 261
  thalamo-hypothalamicus, 242
  thalamo-hypothalamicus dorsalis cruciatus, 299
  thalamo-hypothalamicus et peduncularis cruciatus, 299
  thalamo-mamillaris, 243, 279
  thalamo-peduncularis, 237, 281, 303
  thalamo-peduncularis cruciatus, 224, 234
  thalamo-peduncularis dorsalis superficialis, 224, 226, 234, 237
  thalamo-tectalis rectus (brachium of colliculi), 222, 225, 237
  thalamo-tegmentalis dorsalis cruciatus, 186, 221, **299, 300**
  thalamo-tegmentalis rectus, 186, 221, 237, 283, **286**, 300
  thalamo-tegmentalis ventralis cruciatus, 282, 283, **284**, 286, **300**
  rectus, 283, **285**, 286
  trigemino-cerebellaris, 172, 177
  vestibulo-cerebellaris, 161, 178
  vestibulo-spinalis, 161
  visceralis ascendens, 36, 130, 166, 182, 187
  visceralis tertius, 36, **169**, 176, 200, 287

Trigla, 146
Triton; *see* Triturus
Triturus, 11, 12, 25, 118, 160, 170, 172, 219, 227, 228, 232
Tube, neural, 92, 110, 117
Tuber cinereum, 20
Tuberculum
  olfactorium, 53, 266, 268
  posterius, 21, 212
Tunicata, 16

Uebergangsganglion, 189
Uncus, 100
Urodela, 4

Van Heusen, A. P., 136
Variability, 19, 43, 90, 112, 212
Velum medullare anterius, 45, 140, 150, 176, 181, 293
Ventricles, 24, 93, 269
Vermis cerebelli, 172
Vestibular apparatus; *see* Ear

Vicq d'Azyr's tract, 244
Visceral system; *see* System, visceral
Visual system; *see* System, optic
Vitalism, 106

Wallenberg, A., 34
Weiss, P., 62, 84, 89, 113
Whitman, C. O., 78, 84, 147
Woodburne, R. T., 85, 134, 164, 223
Wundt, W., 289

Yntema, C. L., 128, 129, 132
Youngstrom, K. A., 127, 128

Zauer, L. S., 228
Zones, functionally defined
  of cerebellum, 172
  of cerebral hemispheres, 59, 61
  of diencephalon, 230
  in general, 57
  intermediate, 64
  of isthmus, 181
  of medulla oblongata, 153
  of midbrain, 214
  motor (ventral), 60
  sensory (dorsal), 58
  of spinal cord, 125